ADVANCED MATERIALS INNOVATION

ADVANCED MATERIALS INNOVATION

Managing Global Technology in the 21st century

SANFORD L. MOSKOWITZ

WILEY

Library of Congress Cataloging-in-Publication has been applied for.

ISBN: 9780470508923

Printed in the United States of America

10 9 8 7 6 5 4 3 2 1

In memory of my parents,

Dr. Fred Moskowitz (1919–1964) & Rose Surval Moskowitz (1925–2004)

CONTENTS

PREFACE

Few areas of human endeavor play such a large role in the technological and economic activity of society today as the field of advanced materials. More scientists
and engineers devote their professional lives in pursuit of them, more businessmen
and government agencies have taken a serious interest in nurturing them, and more
venture capitalist and financial firms have poured money into startups formed around
them than ever before. And never before has the advanced materials arena extended
so globally, with Europe and even more Asia playing an increasingly important part
in this century's advanced material saga.

Certainly, advanced materials have been with us since man's earliest experiments
in creating civilization. Whole eras in fact have been named for them, as our classifica-
tions of the "iron" age, "copper" age and "bronze" age attest. Historians of technology
tell us of the great importance of advanced materials—especially the new steels coming out of the giant blast furnaces of England and the US—in the industrial revolution
of the ninetieth and early 20th centuries. But today, in the first quarter of the 21st
century, never has the potential of advanced materials seem so important, and indeed
crucial to human existence. Mobile electronic technology, molecular and quantum
computing, optical communications, alternative energy sources, the biotechnology
revolution, robotics and automation, virtual reality and three dimensional printing,
not to mention fundamental transformations in transportation, manufacturing and
infrastructure all receive significant attention by the business, trade and general press
because they are poised to fundamentally alter the nature of our lives in this century
and beyond. All of these technologies and the changes they make in our world depend
vitally on progress made within the advanced material landscape. At their core are
increasingly powerful microchips, tiny lasers, energy-efficient batteries, smart textiles and metals, super strong and lightweight composites and alloys, superconducting

materials, electronic displays, conducting polymers, all of which depend directly or indirectly on our ability to come up with new and more sophisticated materials and the processes to make them. In 2006, a study by the Rand Corporation called *The Global Technology Revolution 2020* identified 16 major areas of innovation that would by 2020 " … enjoy a significant market demand … affecting multiple sectors (e.g., water, food, land population, governance, social structure, energy health, economic development, education, defense and conflict, and environment and pollution)."[1] A full 75% of the innovations on the list depend directly on progress in advanced material technology. A recent White House report on innovation and industrial competitiveness reminds us of the critical importance of advanced materials technology for the present century:

> … the development of advanced materials will fuel many of the emerging industries … Advanced materials are essential to economic security and human well-being, with applications in multiple industries, including those aimed at addressing challenges in clean energy, national security, and human welfare … Since the 1980s, technological change and economic progress have grown ever more dependent on new materials development.[2]

It should come as no surprise then that companies, governments, and investors should be scouting around for the next big thing in advanced materials. Many research initiatives are emerging from universities, others from corporate research and development departments and a few from government laboratories. The list is long and varied and includes such intriguing prospects as shape memory polymers, special ceramics for absorbing oxygen, architectural "smart" materials, power generating fibers, new films and coatings, fire-resistant paper, biorefined adhesives, and structural nanocomposites. One of the most fascinating possibilities is something which Caltech researchers call "3-Dimensional Architected Structural Meta-Materials." These potential materials of the future will, if realized, allow planes, cars, bridges and other structures to be hundreds of times lighter than they currently are but with the same strength and superior operating characteristics.[3]

P.1   SOME QUESTIONS AND WHY THIS BOOK?

So what are we to make of this abundance of possibilities? Certainly, the sort of the aforementioned research holds great scientific interest. But so many of these projects never successfully leave the university, the company research department or government laboratory. In other words, they never are able to cross that "dead man's zone" separating promising science from real-world technology. This doesn't mean attempts aren't made. In many cases, such research leads to the creation of a startup, usually close by a major university. Patents are licensed out, investors pursued, personnel rounded up, and facilities and equipment secured. In some cases, a larger, established firm buys the technology and creates or makes use of an internal venture group with the aim of nurturing, developing and commercializing the product or process.

Recently, for example, Bayer Chemical created a special group to commercialize nanotube technology. But despite these efforts, few of these well-intentioned ventures ever see the light of day in the marketplace. Bayer understands this all too well, for in 2014 it abandoned its nanotube venture. In other cases, the company that invents the new material often is not the one to market it, thus missing out on the fruits of its creation. While RCA, for instance, invented the first liquid crystal display, it never commercialized the most important liquid crystal technology, flat panel displays for TVs, nor, as it turns out, did any American firm. This honor went to the Japanese and South Koreans.

What then determines whether or not a promising laboratory invention actually becomes a robust, fully formed market technology "with legs" and whether this transition from idea to working system can be accomplished in a timely manner? Certainly venture capitalists and corporate investors need to know where and when to place their bets; established firms looking to "buy into" new technology need to know what sort of startup it should purchase or what sort of partner it should do business with; For its part, the startup and small firm focused on that one technology and looking to share the risks with—and obtain more resources from—a corporate sponsor, need to know which company offers the best opportunity for its "baby" to grow into a fully mature and profitable adult; Academics and business scholars specializing in the innovation process need—or at least want—to know what factors are responsible for the rate, direction and timing of new technology; and government very much needs to know how to keep advanced materials innovation, so important for a nation's global competitiveness, flowing from the research laboratory to the production plant and market arena. In that same White House report quoted above, we also can read the importance the US government places in finding ways to keep new materials technologies pouring into the American economy:

> ... the time it takes to move a newly discovered advanced material from the laboratory to the commercial market place remains far too long. Accelerating this process could significantly improve U.S. global competitiveness and ensure that the Nation remains at the forefront of the advanced materials marketplace ... [it is important that the US finds a way] to discover, develop, manufacture and deploy advanced materials in a more expeditions and economical way.[4]

And so I come to the reason for this book, which, simply put, hopes to find some of the answers to the questions posed above. The possible causes for success or failure, speed or sluggishness of innovation are well known and focus on such things as knowledge, capital, resources, organization, demand, skills, foresight, and risk-taking. It is not my intention to pit one or two of these against the others and declare these favorites to be *the* answer to the innovation question. Nor is it my aim to construct an all-encompassing "model of innovation" so much favored by academics and business scholars—and many business writers outside the Academy as well—or to attack as insufficient any such models that others have carefully and scrupulously developed from their specific case studies and data bases. My goal is much less ambitious. I simply want to narrate in as much detail as I can stories of important 20th and 21st century advanced material innovations and from these tales of technology creation—and failure—glean some common themes and conclusions that can help venture capitalists, established corporations, startups, government and even business scholars find some of the answers to why and how early and exciting research efforts manage—or mismanage—their way across the dreaded "valley of death". Success in surviving the voyage often prove a boon to the companies that made it across and, in some cases, transform the world; those that fumble just as often drag the innovators down with them and are soon forgotten as just once great ideas that failed to launch. Whether then a promising material such as a graphene shakes up the world or is relegated to the dust bin of history will likely depend on whether its champions attend to the lessons that emerge from the case studies presented in the following pages.

P.2   SELECTING THE CASE STUDY INNOVATIONS

But, if I am going to tell the stories of advanced materials innovation, which ones are to be selected for the spot light? So many of the most interesting ones—and the ones that receive so much attention in the press—are still in process, their outcomes uncertain. A case in point is graphene mentioned above. It has received so much publicity of late that it would seem to be a good candidate for a case study. But whether or not it actually becomes a major commercial product is still in question. Thus there is no way to know whether the history of this material—or any of the many other excellent advanced material prospects—as of this writing (2016) will lead to success, failure or just plain limbo (neither a clear success or failure) and so no way to make reasonable conclusions over what makes a company—or country—rise or fall in its quest for a particular new material technology. But what we do have is a record of technologies in the 20th and 21st centuries that have gone through all or most of their innovation cycles. These provide enough material to decide which companies make good on their early research and which falter, and why some flourish as others stumble. These then allow conclusions to be drawn as to the likely commercial success of current and future research and development efforts, where they are most likely to succeed and where to fail.

This book is interested in telling the stories of, quite simply, the most important advanced material innovations of the last few decades. Recent discussions on the distinctions between radical, breakthrough and disruptive technology, important as they are from a theoretical viewpoint, do not much concern us here. Suffice it to say, the innovations highlighted in this book are, arguably, the chief advanced material technologies in recent times. I have relied on the trade and business press and on historical opinion to tell me that, when it comes to advanced materials, these are "the essentials." They are generally characterized as a "revolution" or "high water mark"—or some such expression—in the history of modern innovation. They mark a big leap in technological knowhow and generally have made a major impact on the market and even society. Of course, in any such list, a bone or two could be

TABLE P.1   Classification of the Major Advanced Materials Innovations

Categories of Advanced Materials	Advanced Materials and Related Devices
Structural: metals	Thin slab, thin strip, and ultrathin steel and microalloys
Structural: polymers	Nylon, Orlon, Dacron, block copolymers, polysilicones, aramids, advanced polyolefins, metallocenes
Functional: semiconductors	Transistor, "dram" memory chip, "EPROM" memory chip, microprocessor, semiconductor laser, silicon–germanium chip
Hybrids and new forms of matter	Thin-film transistors, liquid crystals and conducting polymers, "buckyballs," nanotubes, nanocomposites, graphene

picked over a particular entry included, another left out. However, I believe that, on the whole, this list of innovation will stand the test of scrutiny.

Table P.1 presents the classification of the advanced materials (left column) and the specific major advanced material innovations that we will explore in this book.

The narratives of these important advanced material technologies, while each interesting in and of itself, together paint a picture and expose a common pattern of how and why some companies succeed while others fail in creating and commercializing crucial advanced materials. But which ones are the likely winners in the innovation game is still in debate; each type of firm has its strengths and its shortcomings. It is a common belief that large established corporations are too bureaucratic and tied to their current product line to innovate while smaller firms are more nimble and more likely to take the risks to introduce radically new, truly groundbreaking technologies. However, these startups and small and medium-sized enterprises (SMEs) often lack the necessary resources available to the big organization and so often fail to achieve their ambitious goals. The idea of combining the large corporation and high-technology start up—to form what one book calls the "Start Up Corporation"—is very intriguing.[5] A resource-rich organization possessing the flexibility and far-sightedness of a small firm is clearly an enticing prospect and certainly a very promising approach. But even in this hybrid situation, doubts arise, especially when the company faces financial pressures. In this case, with immediate problems with which to deal, it is likely not to care very much about its future technological position. Just as a drowning man is more likely to grab a life jacket rather than a volume of Shakespeare, so a company in trouble—needing all the resources it has to maintain a current market position—will abandon the incubator, innovation hub and other experiments in internal innovation and divert the men and money being devoted to them to more pressing needs of the moment, that is, existing customers willing to spend their money now on proven products and services. Moreover, what constitutes hardship varies from one organization to another depending on external conditions, such as the general economy and a company's customer base, and its internals, which is its management, culture and organization. In other words, many situations will exist that will provoke even the most technologically savvy start up corporation to slip into a deep conservatism in response to what it perceives to be a major and immanent threat to its very existence.

The creation and use of incubators, innovation hubs and other approaches to heighten the technological dynamism of an established company tell managers what they *should* do to remain competitive in the 21st century. But what these putative solutions do not address is the question of how to keep the innovation process moving forward when a company—or a venture group within that company—faces the reality of desperate times and senior management feels compelled—even against its better judgement—to dismantle the very structures that many believe keep innovation moving forward.

But, in fact, a close look at the course of innovation within our advanced material technology cases underscore a very important point about real-world innovation: it often takes place under quite difficult conditions. Thus, the company faces declining revenue, the initial reason for conducting an R&D project evaporates, no business unit can be found to support a project, valued customers demand a project terminated because the new product will directly compete in their markets, these are just some of the roadblocks put in the way of new material creation. Yet, in many of these cases the established company, even without the benefit of incubators, innovation hubs, and internal venturing groups ended up producing—and profiting from—a break-through advanced material technology, while many companies that did have these structures in place did not. What accounts for these strange, unexpected results and how can investors, companies and governments best direct their resources to opti-mize the chance of successfully creating important 21st century advanced materials and rapidly introducing them into the market? These are the questions that concern us throughout the remainder of this book.

P.3 BOOK ORGANIZATION AND SOURCES

Organizing a book such as this poses some unique difficulties. One obvious organiza-tional approach is to proceed chronologically. But the types of material technologies that would be included in any reasonable time period seemed to be somewhat chaotic as they encompass such wildly different products as, say polymers and steel. It would also be difficult to find conceptual ideas that might link different sorts of materials over time. It would be convenient, for example, to show how the earlier innovations came out of large corporate R&D while, as time went on, the university-based startup took over advanced material innovation. While there is some truth to this in that many startups and smaller firms attempted new technology over the course of the last cen-tury, we still find very important instances of large corporate R&D coming out with breakthroughs, such as Union Carbide's revolution in the polyolefins and metallocene polymers fairly late, that is in the 1980s and 1990s. An alternative, and more prof-itable, organizational approach is by type of advanced material technology. While there are different ways this can be done, it seemed most economical to divide the major sections or parts of the book into the three main product groups—structural, functional and hybrid—and the chapters themselves into the key processes used to create these products.

P.3.1 The Main Sections (or Parts) of the Book: The Product Categories

This book arranges the vast universe of advanced materials into three major categories: structural, functional, and hybrids. The first refers to metals, plastics and composites that are generally used in making parts and components for cars, planes, buildings, bridges, roads as well as for such biomedical applications as synthetic bones, muscles, tissues and organs. Functional materials are those semiconductor technologies that power solid state electronic devices such as transistors, microchips and solid state lasers. The third category—the hybrids—highlight the new forms of matter that have multiple uses. These materials enjoy applications that span both the structural and functional spaces. Nanotubes, for example, can be used to make the so-called nanocomposites (structural) as well as advanced microchips (functional).

P.3.2 The Chapters: The Major Processes

The advanced materials sector subsumes a wildly varied group of products to the extent that it is at first difficult to see how they can be studied in any coherent way at all. Yet, on closer study, we see that they can be grouped and categorized according to a relatively few common processes, or ways in which they are made. The chapters within each of the three sections of the book then are focused on such important process technologies as high-pressure catalysis, fluidization, low-temperature polymerization and so forth. It is important to stress that the advanced materials produced by a single process can be very different with respect to their applications and markets but can all be linked together by that common mode of production. For example, high pressure technology has produced structural materials as different as synthetic ammonia, nylon and nanotubes. Understanding that our material of interest, as important as it may be, is actually just one in a series of products turned out by a particular underlying process over time affords us a much deeper perspective on where our material came from, what went into its making, and how and why it itself spawns future breakthroughs. Take nylon for example. Contrary to popular belief, its story did not begin when Wallace Carothers began his experiments on a new type of polymer fiber. Because, as we shall see later on, the fundamental process behind nylon is high-pressure technology, we can only begin to understand DuPont's success in fibers by considering its earlier work in synthetic ammonia and organic chemicals, which made the company the world leader in high-pressure production and provided it with the essential knowledge, skills and technology to move into the first miracle fiber. In the same way, DuPont's ability to quickly build its famous Hanford plutonium plant for the Manhattan Project during World War II rested in no small measure on the high-pressure work it had perfected for nylon a few years earlier. Similarly, we cannot fully appreciate how Intel came up with its most important invention, the microprocessor, without grasping the company's earlier success in working out and applying the silicon gate process; this is the core method that first gave Intel the edge in semiconductor memories and only after learning how to accomplish this, allowed it to take on, a few years later, even bigger game, the microprocessor itself.

This more profound understanding of breakthrough innovation provided by a consideration of fundamental processes also, and very crucially, allows us to have a much better sense of the patterns of success and failure at the firm level. For example, both DuPont's triumphs (e.g., nylon) and its disappointments (e.g., polyethylene) depended a great deal on its attachment to its high pressure process. In a like manner, the legendary accomplishments of Intel and its total annihilation of its nearest and very capable competitor, Mostek, comes down to one simple fact: Intel held the secret to the silicon gate process (and thus to the microprocessor) while Mostek did not. Accordingly, each chapter shines the spotlight on a major process technology, its origins and its evolution as it creates different but technically related materials within a given product grouping (structural vs. functional), as displayed in Table P.1 above. The story then of each breakthrough material (e.g., polysilicones) will be told within the context of its associated process (e.g., fluidization).

P.3.3 Narratives of the Innovation Cycle

But how should the narratives of these materials be structured? Because we are talking about an evolution from idea to product, chronologically makes the most sense. However, rather than the units being days, months and years, we will follow each innovation project in units comprised of the "phases" of the innovation lifecycle. And so the case history narrative for each new material innovation discussed within its respective chapter follows the same general pattern. This "innovation cycle" extends from project initiation to production and first entrance into the market. The concept of "Research and Development" is a fairly wooly one. It is often used to mean conducting scientific investigations and turning the results into marketable products. But this is far from the case. In fact, conducting "R&D" gets us only half the way to that goal. The full innovation cycle consists of "Research," "Development," "Scale-Up," and "Commercialization." Accordingly, the story of innovation for each advanced material encompasses all four of these phases (or "RDS&C"). First exploring why and how a company began researching that particular advanced material technology, the story moves through the "research phase", ending with creation of a working laboratory model, such as a reactor or device, demonstrating nothing less (or more) than feasibility of concept. A discussion of the "development phase" follows, which takes that laboratory model and works it up into a functioning prototype potentially capable of being expanded into a commercial unit. This prototype generally looks and behaves very differently from the crude laboratory model. The development phase can be viewed as the transition stage from laboratory science to real-world technology and, as such, often involves scientists and engineers working closely together. The "scaling up" phase follows and closes in on the all-critical transformation of the promising exemplar into a fully realized commercial technology, generally consisting of a working production unit making an advanced material (e.g., nylon, ultrathin Steel) or advanced material device (e.g., microprocessor, liquid crystal display). The "commercialization phase" completes the innovation lifecycle. It is concerned with the final formulation of a business model for the new technology, which is closely linked with a company's cultural make up and strategic outlook. Commercialization is generally an activity that is solidified and carried out just prior to entering the market. Certainly, while management—and even researchers—may begin thinking about and even developing portions of a business model as early as the research phase, they often have to alter their strategic vision for the technology over the course of the innovation sequence. During development and even scale up, for example, as they learn more about the nature of the technology and its place within the company's product portfolio, project champions and their superiors may change the type of markets to be approached or the position the technology ought to have on the relevant value chain. They may even decide against making the material or device after all, possibly sitting on its patents and preventing competitors from pursuing it. Whatever the case, only after (or possibly during) the building of a commercial plant will the corporate powers pin down the precise commercial strategy to be pursued; it is this latest iteration as the company first enters the market that defines the final phase of our innovation cycle.

Tracking these four phases of the innovation cycle for each of our major advanced material case studies is the best way to appreciate the various and changing difficulties that are placed in the way of success as a company struggles to bring a new and promising material to commercial existence. The difficulties and thus risks involved are unique across the different phases; the barriers encountered in conducting research are not at all the same as those facing the development team nor those challenging the engineers in charge of scaling a manufacturing facility. The different set of problems encountered along the road to market requires different talents and even personalities. Appreciating this reality as we study the rise (and often fall) of the innovative effort, allows us to see how the more agile firms manage to successfully negotiate the various difficulties they are forced to encounter throughout the cycle and, for those less nimble companies, to identify where such efforts tend to falter and why such failures happen.

P.3.4   Sources

This book ranges widely over the advanced materials landscape and so relies on many types of sources. Industry sources from my network of contacts in the field supplied important insights into the technical and economic aspects of both current and emerging materials. It has also benefitted immensely from the published work of others, especially those that are deeply researched and touch directly on areas crucial to our case studies. These include full-scale books, company histories and personal memoirs. Articles too contribute greatly to this work, including those that appear in the trade and business press. Primary material also comes into play throughout the book as do unpublished interviews of the major participants. For obvious reasons, these latter have supplied some of the most valuable details and insights, especially for those technologies that have not been given very much attention by scholars in the past. They are invaluable in coming to grips with the motivations and thinking of scientists, engineers, managers and businessmen who witnessed—and actively took part in—the innovation process first hand. These are vital in forcefully drawing attention to how the forces of risk, process and championship creatively interacted to introduce into the world new and groundbreaking advanced materials technology.

REFERENCES

1. Silberglitt, R. (2006), *The Global Technology Revolution 2020: Bio/Nano/Materials/ Information Trends, Drivers, Barriers, and Social Implications—Executive Summary*, Santa Monica, California: The RAND Corporation, p. 2.

2. National Science and Technology Council. (2011), *Materials Genome Initiative for Global Competitiveness*, Executive Office of the President's National Science and Technology Council: Washington, DC, pp. 3, 5.

3. Bourzac, K. (2015), "Nano-Architecture: A Caltech Scientist Creates Tiny Lattices with Enormous Potential," *MIT Technology Review*. http://www.technologyreview.com/featuredstory/534976/nano-architecture/. Accessed November 25, 2015.

4. National Science and Technology Council. (2011), p. 3.

5. See particularly Leifer, R. et al. (2000), Radical Innovation: How Mature Companies Can Outsmart Upstarts. Boston, Massachusetts: Harvard Business School Press and Davila, T. and Epstein, M. (2014), *The Innovation Paradox: Why Good Businesses Kill Breakthroughs and How They Can Change*: San Francisco, California: Barrett-Koehler Publishers.

ACKNOWLEDGMENTS

Many of the unpublished interviews of scientists, engineers, managers, and corporate executives—as well as the papers of the nanomaterials pioneer and Nobel Laureate Richard Smalley—that have played an important part of this book belong to the Othmer Library and the Oral History Program of the Chemical Heritage Foundation (CHF, Philadelphia, Pennsylvania). The Foundation, as always, has been very helpful to me in accessing this material. I want to especially thank David J. Caruso, Director, The Center for Oral History, CHF, for his help in obtaining interview transcripts. DuPont's Hagley Library has also been of great help to me over the years in examining archival documents related to the chemical industry and technology, a number of which I have incorporated into this book. I want to also thank Melanie J. Mueller, Acting Director, Niels Bohr Library and Archives, American Institute of Physics (College Park, Maryland), for permission to use excerpts from their oral history of Eugene Gordon and Robert Colburn, Research Coordinator, IEEE History Center (Hoboken, New Jersey), for permission to quote from their oral history of Richard Petritz.

This book has made extensive use of published research when such materials have proven useful to the overall narrative of advanced materials. I need to particularly acknowledge the contributions of four very helpful works: David A. Hounshell and John Kenly Smith whose exhaustive study of DuPont R&D in their book *Science and Corporate Strategy: DuPont R&D, 1902–1980* (Cambridge University Press, 1988) proved so useful to me in the chapters devoted to polymers within the section on "Structural Materials"; Ross Knox Bassett whose book *To the Digital Age: Research Labs, Start-up Companies, and the Rise of MOS Technology* (Johns Hopkins University Press, 2002) added so much to my discussion of semiconductor technology in the chapters related to "Functional Materials"; Jeff Hecht and his story of fiber optics as told in *The City of Light: The Story of Fiber Optics* (Oxford University Press, 1999) made an important contribution to my narrative of the semiconductor laser and its role in fiber optics technology; and Bob Johnstone and his excellent account of how Asia became important competitors in advanced electronics in his book *We Were Burning: Japanese Entrepreneurs and the Forging of the Electronic Age* (Basic Books, 1999), a work that guided me in analyzing the rise and global development of liquid crystal device technology in the section of "Hybrids and New Forms of Matter."

I also want to especially acknowledge the help of Robert Burgelman, Edmund W. Littlefield Professor of Management and Director of the Stanford Executive Program at Stanford University, during the writing of this book. Professor Burgelman was kind enough to review parts of my manuscript and to offer important insight that led especially to the structure and content of the final chapter. He has also been very helpful in directing me to extremely useful articles that have added considerably to the quality of the book.

The school where I teach, St. John's University and the College of St. Benedict (Collegeville, Minnesota), has helped me in many ways during the writing of this book. Through university grants, it provided me funds for research travel. Its inter-library loan program under the coordination of Janine Lortz has proven invaluable in securing for me crucial articles and books I needed in a most timely and efficient manner. Its teaching assistants—and particularly Mattie Lueck—helped me design a number of tables and figures that I have used in this book. And its faculty and my colleagues have, as always, been totally supportive. I need to particularly single out Economics Professor Louis Johnston and Global Business Professor Lisa Lind-gren for leading me to important documents and ideas and concepts that have greatly enriched my discussions and been of enormous help to me in forming my conclusions.

My editor for the book at John Wiley, Dr. Arza Seidel, has, as always, been supportive and very patient. While it has taken me longer than expected to finish this book, she has stood by it from the very beginning. She has, as with my first book, offered wise counsel, support, and encouragement throughput this project. I am most appreciative. Finally, I would like to thank my wife Becky for her sage advice and patient understanding during my writing of the book. As her mentor Pearl S. Buck wrote of her in *For Spacious Skies: Journey in Dialogue*, "Give us a hundred Beckys."

PART I

INTRODUCTION AND BACKGROUND

1

ADVANCED MATERIALS INNOVATION

An Overview

The world is in the midst of a global technology revolution ... with the potential to bring about radical changes in all dimensions of life.

The RAND Corporation, 2006

1.1 THE ADVANCED MATERIALS REVOLUTION

The aforementioned quote comes from a 2006 study that focused most of all on advanced materials and their economic and social impact worldwide. The statement thus gives us a fair sense of the importance of advanced materials to man's future economic progress. Actually, advanced materials technology has been an integral part of society and its evolution for centuries. It is embodied in the extracting of coal or iron ore from the earth or creating new materials from combinations of the old, such as iron and carbon to produce steel. Less well known but extremely important were the German coal tar-based synthetics—dyes, drugs, industrial gases, and explosives—that dominated the world's demand for chemicals in the last quarter of the 19th and the first part of the 20th century. Germany's chemical supremacy culminated in the industry's greatest achievement up to that point: the Haber's synthetic ammonia process (1913).[1]

But even before the First World War, the United States had begun its ascendance in advanced materials. It had of course by then a large and technically sophisticated iron and steel industry in Western Pennsylvania. But by the 1890s, another region had opened up a whole new world of materials. Niagara Falls, because of the cheap energy it provided, developed into the first major US industrial cluster of the 20th

Advanced Materials Innovation: Managing Global Technology in the 21st century, First Edition.
Sanford L. Moskowitz.
© 2016 John Wiley & Sons, Inc. Published 2016 by John Wiley & Sons, Inc.

century, companies like Alcoa, Union Carbide, and Carborundum first turned out advanced nonferrous metals, particularly aluminum, the first nickel "superalloys," and the carbide family of metals for a growing number of industrial applications. Soon chemical companies moved in to produce organic synthetics using Niagara's cheap electrical power.[2]

The Niagara Falls area, in industrial decline for decades, would not be the last important advanced material center. The following table displays many of the major advanced material innovations according to year of introduction, category, company, and country (and in the case of the United States, region) from the start of the First World War to the present (2016). A number of trends can be identified. As expected, structural materials continued to control advanced materials innovation until the late 1940s. Polymers (and the intermediates that went into making them) soon began to dominate. This was the age of macromolecular technology, and the two ruling powers in this field were DuPont and (surprisingly given the nature of its core business) General Electric (GE). DuPont particularly—along with Union Carbide—created a very important advanced material region in West Virginia's Kanawha Valley. Raw materials in the form of coal and natural gas furnished the raw materials. Carbide depended on the ethane-rich gas to make its ethylene-based chemicals and plastics, while DuPont, much like the Germans, opted for the coal as its basic starting point. This region would prove remarkably fertile over the years as research conducted there turned out some of the most important new materials of the age, including the most prominent of them all, nylon. But Kanawha, like Niagara, turned out not to be the last word in American advanced materials. General Electric certainly proved this: it came out with revolutionary new polymers through the 1940s and 1950s without having to dip in the Kanawha well to do so. The table also shows the growing importance of the southwest and its oil fields in Texas, Louisiana, and Oklahoma, which is where research on and early production of high octane fuel using fluid catalytic cracking—one of the most powerful advanced material processes—took place.[3]

We see then that during the first half of the 20th century, American advanced materials shifted geographically from the northeast (Pittsburgh and Niagara Falls) to the south (Kanawha Valley) and southwest (Gulf States) and did so in pursuit of abundant and cheap resources, whether energy or raw materials. Beginning in the late 1950s, the center of advanced materials—never content to stay in one place for too long—was on the move again, this time headed due west. The reason this time was to take advantage of another type of resource involving neither cheap power nor abundant fossil fuels but the free movement of ideas and knowledge and a growing source of capital—venture money—specifically tailored for high-technology enterprises. A whole new type of advanced material now entered the scene. Whereas the metals and polymers were made by the advanced materials producers and sold to fabricators of components and structures that in turn went to the construction, transportation, textile, machine tool, and a host of other industries large and small, the semiconductor company synthesized advanced semiconductor composites and from them created actual working devices—transistors, memory chips, microprocessors—that then were sold to original equipment manufacturers (OEMs), notably personal computer manufacturers. Semiconductor firms thus are active further up that value chain than are the steel and chemical companies. These functional materials began their upward ascent into history in the late 1940s with the invention of the transistor, which, along with nylon, is ranked as one of the most important inventions of the century. A whole new era now came to the fore. Today, we think of Silicon Valley as the place where money is made in the software field dominated by video games, social media, and internet services of all kinds. But Silicon Valley was created around companies like Shockley Semiconductor (defunct for decades), Fairchild Semiconductor (still with us as of 2017), and the mighty Intel (still the king of the chip). The people who powered these companies were not software designers but chemists, chemical engineers, applied physicists, and electrical engineers, and their business was materials and creating new and more sophisticated semiconductor devices. The transistor, integrated circuit and microprocessor, solid-state laser, and silicon–germanium chip are all advanced material composites made by very intricate processes with such names as metal on oxide, silicon gate, and epitaxy. This advanced material technology which creates ever more powerful microchips is absolutely necessary before there can be software, cloud computing, or any of the ever-growing number of social media venues that populate the 21st century IT landscape. We can say then that from the 1960s Silicon Valley evolved into another major American advanced materials region.

By the late 1970s, the advanced material situation becomes a great deal more complicated. New materials innovation was now more dispersed geographically, originating not only in the United States but internationally. Within the United States itself, former technology centers revived, at least temporarily, to issue extremely significant—indeed breakthrough—new technologies. In the late 1970s, many years after the great achievements of DuPont and GE, that other veteran chemical company Union Carbide introduced its pioneering polyolefin technology that breathed new life into American polymers. In the 1980s, long after the United States had been written off as a serious competitor in global steel, a relatively small company made a big breakthrough in advanced steelmaking and does so in a plant in Indiana, far from Pittsburgh and Big Steel. Thin slab (and later thin strip) casting technology that emerged from here completely revitalized steelmaking and put the United States back on the map as a major metals producer.

And this was not all. By the 1980s, we begin seeing the ascendance of an advanced material group including long overlooked materials and totally new forms of matter with commercial potential. These materials combined elements of both the structural and functional groups. Their increasing commercial importance underscores two significant trends that remain essential to this day: the growing ability of other countries—particularly Asian countries—to take what the United States and Europe invent but discard and make something economically important out of it (e.g., liquid crystal displays) and the rise of the "university research–technology transfer office–high-tech start-up" network as an engine for advanced materials innovation, as in the case of nanomaterials. While the small start-up commands growing attention in recent years, we cannot by any means discount the role of the big corporation in 21st century innovation.

A glance at Table 1.1 shows that large, established companies continued to turn out new and breakthrough materials into the last quarter of the 20th century. Today, companies like IBM and 3M have not shied away from pushing the envelope of new and sometimes path breaking technology. At the same time, many start-ups with loads of technical and business talent, impressive patent positions, and ties to deep-pocketed investors never did make it past the development stage. Understanding and accounting for these successes and these failures from such unexpected quarters is an important part of this book and is something we will take up after a brief look at the present and future impacts of advanced materials on major sectors of the economy.

1.2   THE ECONOMIC IMPACT OF ADVANCED MATERIALS

New materials have had a major economic impact on society for millennia. This influence has been especially strong over the course of the 20th century. To be sure, the very symbols of modern life—such as automobile, airplane, skyscraper, man-made textiles, computer, and atomic energy itself—all are deeply dependent on the regular availability of new and pioneering materials. There is little question that the improved performance that has occurred in most areas of economic life has been increasingly governed and even determined by breakthrough metals, polymers, semiconductors, and other advanced material technologies. In their book *Abundance: The Future Is Better Than You Think*, Peter Diamandis and Steven Kotler, quoting from a National Science Foundation report, emphasize the economic and social importance of just one class of advanced material: "Nanotechnology has the potential to enhance human performance, to bring sustainable development for materials, water, energy and food, to protect against unknown bacteria and viruses, and even to diminish the reasons for breaking the peace [by creating universal abundance]."[5]

But what has been and is likely to be the impact of advanced materials on the economy and society? There are of course many ways in which a sector or industry or individual company for that matter expands that may not involve the use of new material at all including innovations in design, organizing work, and revamping and applying old technology. But there is little question that advanced materials plays a growing role in the economic performance of the major industrial sectors operating within the developed world within the 20th and early 21st centuries and that this trend will likely continue—and indeed some believe accelerate—as the present century unfolds.

Table 1.2 shows the growing influence of advanced materials on the economic performance within six major US industrial sectors from 1980 to 2050. The table has evolved from discussions with experts in the field, conference presentations, as well as from data, information, and analysis from secondary literature including government studies and articles within the technical and business press. The numbers shown are the percentages of total progress in performance accounted for by the introduction of advanced materials alone within each industrial sectors over those seven decades. Naturally, the measure of performance differs across industries and technologies—improvement in the performance of automobiles is measured very

TABLE 1.1 Major Advanced Material Innovations: Materials and Related Devices[a] 1913–2015

Advanced Material and Devices	Year	Advanced Material Category	Company	Country (Region)
Synthetic ammonia	1913	Structural	Badische (BASF)	Germany
Synthetic methanol and the advanced alcohols	1926	Structural	DuPont	United States (NE)
Nylon	1938	Structural	DuPont	United States (NE)
Polysilicones	1942	Structural	General Electric	United States (NE)
High octane ("fluid catalytic") gasoline	1943	Structural	Jersey Standard (EXXON)	United States (SW)
Transistor	1947	Functional	AT&T/Bell Labs	United States (NE)
Rayon and Dacron	1953	Functional	DuPont	United States (NE)
Aramid fibers	1996	Functional	DuPont	United States
Polycarbonate plastics	1996	Functional	General Electric	United States (NE)
Integrated circuit: "DRAM" memory chip	1970	Functional	Intel	United States (SV)
Integrated circuit: microprocessor	1973	Functional	Intel	United States (SV)
Integrated circuit: "EPROM" memory chip	1974	Functional	Intel	United States (SV)
Advanced polyolefin plastics	1977	Structural	Union Carbide	United States (NE)
Liquid crystals: small panel displays	1975	Hybrid and new form of matter	Seiko	Japan
Nanomaterials: multiwalled nanotubes (MWNTs) and nanotube composites	1983	Hybrid and new form of matter	Hyperion Catalysis	United States (NE)
The metallocene polymers	1998	Structural	Union Carbide	United States (NE)
Germanium arsenide semiconductor complex: solid-state laser	1998	Functional	AT&T/Bell Labs	United States (NE)
Thin film transistor + liquid crystals: large panel display	1999	Hybrid and new form of matter	Sharp	Japan
Thin slab steel	1987	Structural	Nucor Steel	United States (MW)
Ultrathin steel and the microalloys	1997	Structural	Nucor Steel	United States (MW)
Silicon–germanium chip	1999	Functional	IBM	United States (NE)
Nanomaterials: "buckyballs" and single-walled nanotubes (SWNTs)	2000	Hybrid and new form of matter	Carbon Nanotechnologies	United States (SW)
Graphene	2016	Hybrid and new form of matter	Cambridge Nanosystems	United Kingdom

MW, Midwest United States; NE, Northeast United States; SE, Southeast United States; SV, Silicon Valley; SW, Southwest United States.

TABLE 1.2 Impact of Advanced Materials in Major Sectors of the US Economy of the Developed[6] (% Contribution to Performance Growth)

	1980	1990	2000	2010	2020	2030	2040	2050
Information and computer technology	25	40	55	60	65	68	73	75
Energy*	15	30	45	50	55	60	65	70
Biotechnology and health care	10	13	20	25	33	45	55	65
Transportation	8	10	15	20	30	40	45	60
Construction and infrastructure	5	8	10	15	25	30	40	55
Manufacturing	3	6	8	12	20	25	35	50

*Not including energy savings realized by the transportation, construction, and infrastructure sectors.

differently than advances made in the efficiencies of solar panels. It is also important to understand that whether these projections actually come to fruition depends greatly on whether the promising advanced materials that are still in the research phase actually enter the market in a timely manner. It should not be a surprise that advanced materials impact industries with different levels of intensity. While information and communications technology (ICT) depends mightily on new materials, manufacturing and construction have more performance-improving options (such as new ways of organizing work, new building and machine design, and so forth). Nevertheless, throughout all these industries, the profound effect of new materials technology and the intensification of their influence over the next few decades are clearly evident.

The sections that follow discuss in more detail the current and projected advance material trends in these various industries. These promising materials are the ones that need to escape the confines of the research lab and enter the real world of the market if the trends depicted in the table are to come about.

1.2.1 Information and Computer Technology

Much will be said of the role of advanced materials in the information and computer technology sector in Part II of this book. For now, we can say that no area of modern technology has been so profoundly influenced by new materials as ICT. By all accounts, the key measure of progress in ICT is the ability of the central processing unit—the microprocessor—to continue to follow the path dictated by Moore's Law, that is, the statement that the microchip contains ever greater processing speed and power on a smaller and smaller piece of silicon.

Certainly, other factors must be considered as important for the growth of this sector, notably advances in circuit design and software engineering. However, there is little doubt that these essential technical activities must take a back seat to new materials for they would not be able to progress without the creation of more powerful microprocessors, and these, in turn, depend mightily on the development of new and more sophisticated advanced material processes capable of making breakthrough semiconductor composites that maintain the integrity of Moore's Law. The personal computer, mobile device, Internet, and even social media would not be possible without this. It is for this reason that many in the ICT industry assign such a large and rapidly increasing proportion of the sector's growth to new materials past, present, and yet to come. Regarding this last category, research is being conducted at universities in the United States and globally on new materials that will keep the ICT sector progressing toward ever smaller and more powerful devices. One example is the work being conducted at the University of Chicago on nanocrystals capable of being designed for extremely powerful computing technology known as quantum computing. By midcentury, it is fair to say that up to 80% of all ICT growth will depend on new materials, such as new types of silicon (e.g., strained silicon), nanomaterials (nanotubes and graphene), conducting polymers, and even DNA-based circuits. Without these and other advanced material innovations, Moore's Law will falter and no viable replacement will be forthcoming, a situation that will cause stagnation in every other activity associated with information and communications.

Nor does the influence of advanced materials in the ICT industry end with the wonders of the microchip. The personal computer and mobile device revolution could not exist without the flat screen. In this realm of ICT, performance is measured by a display that consumes less power and allows greater visual power and flexibility. Liquid crystals and, in a smaller way, light-emitting diodes (LEDs), have been the driving force in this technology. A newcomer in the field is the organic light-emitting diode (OLED). By 2050, these and other new materials will account for up to 75% of performance improvements made by the ICT industry.

1.2.2 Energy

Next to ICT, the energy field will be the main beneficiary of progress made in advanced materials. The economic and social consequences of climate change is a major factor in pushing efforts to find new and advanced energy technologies. Probably the area of energy technology receiving the most attention is solar. Efficiency, storage capacity, and cost-effectiveness are the critical areas of research, and this generally involves working in the advanced material arena. Over the past three decades, the costs of solar cells have been coming down because of progress made in materials engineering. Most importantly are advances made in making the silicon wafer that goes into the solar cell thinner and thinner. The crystalline silicon from which the wafer is made is the most expensive part of the device, and the less of it needed, the cheaper the cell. MIT engineers, for example, are exploring ways to increase the efficiency of solar panels so that, instead of constructing an entire roof as a solar panel, homeowners could have small solar cells attached in discrete locations on existing roof structures.[7] Another research approach consists of converting ordinary windows into solar cell panels using new types of conducting polymers and advanced thin film technology that can generate electricity (to operate the cell) from both artificial and natural light and provide power for buildings and mobile devices.[8] The University of Chicago is currently exploring the use of nanocrystals to significantly improve the efficiency of solar cells.[9]

Energy-savings technology also involves efforts to produce carbon-neutral fuels (biofuels) that can be stored. A possible path is being pursued by Caltech, Berkeley, and Lawrence Livermore National Laboratory who are conducting joint research to develop an artificial photosynthesis process that can turn sunlight, water, and carbon dioxide into "solar" fuels that can be readily transported and stored and are capable of powering cars and heating buildings to the point of making fossil fuels obsolete.[10] Advanced materials research is also exploring improved batteries that are cheap and have very high storage capacities capable of stowing electrical energy for use by utilities and their customers.[11] New structural materials in the form of nanocomposites, light- and high-strength polymers, ultrathin steel alloys, and so forth will also be necessary to continue to improve the performance and efficiency of structures such as wind turbines as well as making automobiles and airplanes more fuel efficient.[12] Another area of energy research, currently being undertaken by Northwestern and Michigan State University, explores novel thermoelectric materials that can efficiently convert waste heat into electricity. These materials could theoretically be placed in car exhaust systems and smokestacks of factories, refineries, and power generating plants to collect and convert up to 20% of the heat contained in the waste streams into useful electrical energy.[13]

1.2.3   Biotechnology and Health Care

Growth in biotechnology and health care depends on many factors. Structural materials have already made important inroads in the design and manufacture of artificial limbs and organ components, such as heart valves as well as new and more precise surgical tools. Three-dimensional printing holds the promise of actually creating complete organs and being able to do that where the patient is located, thus eliminating the need for organ waiting lists and the uncertain and time-consuming transportation of organs over long distances. Functional materials are playing a role as well in new types of electronic equipment in use in hospitals and clinics and in robotics in surgical work.

Those who work in the field believe a veritable explosion of new materials technology will completely transform health care over the next few decades. The field of diagnostics will be particularly affected. Researchers are testing certain nanomaterials that, when ingested by—or injected into—a patient, can more accurately track even the very early stage cancers. New types of polymers and advanced sensors are being designed to more accurately, efficiently, and cheaply test blood, urine, and skin cells for certain diseases than is currently possible. Future sensing systems that depend on advanced biomaterials and mobile electronics and that are small and cheap will allow individuals to monitor important health-related variables—calories burned, blood sugar, cholesterol levels—extremely precisely and in real time. This technology will greatly aid in disease prevention—and thus lowering society's medical costs—by allowing people to "make important health decisions day by day, moment by moment."[14] Two promising routes toward portable, "zero-cost" diagnostics available to all are "radiological suite in a suitcase" and "lab on a chip." Experiments on the former involve making use of the so-called "triboluminescent" materials. These generate X-rays when crushed together potentially leading to very small, portable, and cheap radiological equipment (such as X-ray machines).[15] Research is also underway at the Polymer Electronics Research Center at University of Auckland to develop a relatively quick and cheap way to test DNA using a new type of sensing technology based on conducting polymers and nanocrystals (also known as "quantum dots").[16] Not only would such systems reduce the costs of medical treatment, but also they could revolutionize medical care in the home and in poor and currently underserved regions of the world. The concept of "lab on a chip" comprises the testing of DNA and treating such diseases as HIV, cancer, and even those arising out of bioterrorism using a small optical chip linked to a central supercomputer.

Advanced materials are likely to infiltrate other areas of human health as well. "Smart" drug delivery systems, which release their pharmaceutical agents in response to the changing chemistry of the body, will depend on advanced polymers and nanoparticle technology. Robotics also is likely to reimagine medical care by taking human error out of the surgical theater and by facilitating surgery at a distance.[17] New materials for lasers, sensors, and computing power—not to mention the advanced metals, polymers, and composites that will be needed for construction of the robotic device itself—will all come into play in this new surgical world. Overall, by midcentury, up to 60% of the potential benefits of modern medicine and health care—lower costs and effectiveness in serving a greater number of people globally—are likely to come in one way or another from new materials technology.

1.2.4 Transportation

Progress in the transportation sector can be measured in terms of safety (number of deaths per mile) and efficiency (number of miles per gallon of fuel). The industry makes headway in these metrics for a number of reason, including superior designs. Substitution of the new, stronger, more durable, and lightweight materials for the heavier metals has been an important trend in both the automotive and aircraft industries. In 2010, the total consumption of lightweight materials used in all transportation equipment was 46.7 million tons ($95.5 billion). This figure is expected to grow to over 100 million tons by 2030 given current technology and more than this with more rapid development of new and high-performimg structural materials.[18] In addition to the use of lightweight metals, polymers, and carbon fiber-reinforced composites, advanced coatings will improve engine durability and efficiency and significantly reduce the rate of corrosion of automotive parts and components. Some industry observers believe that between 2010 and 2030, the use of these new materials can result in a 50% reduction in fuel consumption and carbon dioxide emissions. Others maintain that this is a conservative estimate, assuming even more radical new material technologies now on the boards can be placed on a commercial basis and can convince the transportation industry to use them in place of more familiar and tested materials. One stumbling block is the high cost of carbon fiber, a critical input into advanced automotive structural composites. *Scientific American* recently reported the formation of a consortium of national research labs, industry, and academia to find a way to make cheaper carbon fiber materials.[19]

Beyond, carbon fiber composites, new types of steels are in the offing, particularly the so-called "TWIP" plastic steels, as are next generation metal matrix and polymer matrix composites, nanocomposites, and advanced nickel and titanium alloys.[20] There are other and very promising structural material technologies that will, if realized commercially, make a major impact throughout the transportation, construction, and infrastructural industries. Caltech's "3D Architectured NanoStructural Metamaterials," already mentioned and potentially capable of making extremely light vehicles, planes, and structures, is certainly a possible game changer. New materials also power the search for improved battery systems for the all-electric car. One possibility of creating an environmentally friendly vehicle is the substitution of the hydrogen fuel cell for the traditional alkaline or lithium battery. Since fuel cells need platinum as the central catalyst, new ways to easily and inexpensively apply this metal need to be found. New methods are then being explored of quickly and cheaply making and depositing very thin layers of catalytic platinum to the point of making fuel cell technology sufficiently cost-effective in automobiles.[21]

1.2.5   Construction, Infrastructure, and Manufacturing

The construction, infrastructural, and manufacturing industries have their own measures of technical progress. In manufacturing, efficiency and safety are key. In construction, stability, strength, and energy efficiency take precedence. In the realm of infrastructure, stability and strength are also important as are durability under increasing traffic and variable climate conditions. All three industries have become more and more dependent on advanced materials to improve performance across all these metrics. Advanced materials in civil engineering projects include new types of composites. Nanoconcrete, made by adding carbon nanofibers to concrete, is much stronger and has higher tensile strength than the traditional material. Since buildings account for up to 30% of all energy consumed by society, energy efficiency in the construction industry has taken on intense importance in the 21st century.[22] Advanced materials—such as new type of glass windows—play a very important role in reducing interior energy use. Smart materials will become more prominent in the design and construction of buildings, homes, factories, bridges, and other structures in the 21st century. These materials can "sense" and respond to changes in the environment in such a way to mitigate damage to structures. In case of earthquakes or high winds, for example, the material that goes into the critical internal structures of a building becomes less rigid and more pliable and capable of "bending" with the shocks. In another instance, smart coatings alter their internal molecular structures to become more resistant to changes in the chemical composition of the air and to fluctuations in outside temperatures thus minimizing the effects of corrosion and thermal degradation to structures. Advanced composites, new pliable steels, shape memory alloys, conducting polymers, piezoelectric ceramics, multifunctional sensors, and other material technologies now being developed all will likely contribute to progress made in this field.

Manufacturing relies on advanced materials in a number of ways. Automation and robotics continue to transform the modern factory. New materials here run the gamut from the microchip and sophisticated sensors to metal and plastic composites that are being developed for the robotic parts and components. High-performance polymers have many other uses as well, such as in creating cost-effective "smart" packaging that responds optimally to the mechanical and chemical environment to maintain the integrity of its contents over long periods of time and types of transport conditions. Smaller and more powerful cutting lasers rely on progress in light-emitting substances. Stronger and more durable equipment also depends on new materials. Scientists and engineers from the national laboratories and the Colorado School of Mines are developing a new generation of extremely tough, cheap, and durable metal alloy coatings for industrial bits, bores, and cutters. These advanced coatings can handle the toughest loads without breaking and, if the economics comes through, may replace such traditional materials as tungsten carbide cobalt. These so-called "Nano Super Hard Inexpensive Laser Deposited (NSHILD)" coatings use a laser to fuse the alloy material onto the surface of the cutting or boring tool.[23] Overall, by midcentury, between 50 and 60% of performance improvements in construction, infrastructure, and manufacturing will come from the direct infusion of the very latest materials into these sectors.

1.3 ADVANCED MATERIAL INNOVATION: THE MAIN PLAYERS

The optimists of the world trust that the many interesting advanced materials research projects now underway in universities, government labs, and company R&D departments will somehow find their way into the economy and that the projections shown in the earlier table will no doubt come about and even be surpassed. But how do we know this is so? In fact, many, if not most, of these never do reach the commercial stage and so do not amount to anything approaching economic importance. Patents may flourish in the realm of advanced materials, but, after all, an invention is a long way from a workable technology. Even when the chasm between invention and real-world technology is ultimately bridged, delays by companies in bringing important research to market often occur, and these have serious repercussions, most importantly, loss of competitive position when rivals get to the finish line first. The issue then is how firms and nations can optimize the chances for successfully commercializing a promising research initiative and do so in a timely manner.

Before delving into our tales of advanced materials innovation to find why certain research efforts succeed while others fall by the wayside, I should point out that, like players in a repertory company, three "actors" take on important parts from one story to the next, interacting with one another in different ways as they move across the different innovations. Each one of these characters, and the way it relates to the other two, needs to be observed closely as the members of this compact little troupe alter their roles and the way they kick the drama forward over the course of these narrative histories. One of these actors is the *process*, which I have already introduced as a general method for making different products. The process not only helps to organize

TABLE 1.3   Advanced Materials and Their Processes

Advanced Material Process	Associated Advanced Materials and Related Devices
Thin slab casting	Continuous thin cast steel
Thin strip casting	Ultrathin steel, microalloys
Vacuum melting	Superalloys
Fixed-bed , high-pressure catalysis	Synthetic ammonia, nylon, orlon*, dacron*
Low-temperature polymerization	Polycarbonates, spandex, kevlar, aramids
Low-pressure fluidization	Fuels, polysilicones, polyolefins, metallocenes
Metal-on-oxide (MOS) process	Integrated circuits
Silicon gate process	"DRAM" memory chips, "EPROM" memory chips; microprocessors
Liquid-phase epitaxy	Heterojunction composites: Semiconductor lasers
Ultrahigh vacuum (UHV) vapor-phase epitaxy	Heterojunction composites: Silicon–germanium chips
Product orientation[†]	Liquid crystal displays
High-pressure fluidization	Nanomaterials (nanotubes)

*While high-pressure technology did not directly create these fibers, it indirectly influenced the rate and path of their innovation.

[†]Refers to the fact that no general process guided the creation of liquid crystal displays. Rather, they were fashioned via laborious "hunt and peck" methods that changed from one display to another. With product orientation, the "wheel" really did have to be reinvented over and over again.

these accounts but also injects itself centrally into the action of the storyline. The plot cannot proceed without an appropriate process in place. Table 1.3 displays the major advanced material processes and chief products that each generates. Except in the case of vacuum melt technology and the superalloys, the innovation history all of the processes and materials shown will be covered over the following chapters.

The second major force in our triumvirate, not too surprisingly, is "risk," which in our context, is determined by the likelihood a given project that is deemed strategically relevant will succeed at an acceptable cost and in a reasonable time frame. The concepts of "perceived" and "relative" risk are pertinent here. The first tells us that, whatever the real risks, it is what senior management *perceives* to be the risks—whether their perception is accurate or not—that counts in the final decision of a company pursuing, or not, a new technology. Taking into consideration the idea of relative risk is often critically important in understanding an organization's "go-no-go" decision-making process. Relative risk must also be taken into account, for even if senior management believes moving forward on an innovation is very risky, it still may do so if it also is firmly convinced that not proceeding constitutes an even greater risk. Both perceived and relative risk come into play as our stories unfold. The third character in our small, nimble ensemble of players is an actual human being, the "champion." I will have a lot more to say about the champion's role in the innovation process throughout the book and so will refrain from talking much now about this often mercurial, chameleon-like character. Suffice it to point out here that the champion fundamentally drives breakthrough innovation and we will witness how he does so by focusing in on the continually shifting ways he manages risk and process within very different organizational settings.

REFERENCES

1. Aftalion, F. (2001), *A History of the International Chemical Industry: From the "Early Days" to 2000*, Philadelphia, Pennsylvania: Chemical Heritage Press, pp. 82–91.

2. Trescott, M. (1981), *The Rise of the American Electrochemicals Industry, 1880–1910: Studies in the American Technological Environment*, Westport, Connecticut: Greenwood Press.

3. Spitz, P. (1988), *Petrochemicals: The Rise of an Industry*, New York, New York: John Wiley & Sons, pp. 449–456.

4. This table derives from the discussions to follow in succeeding chapters.

5. Diamandis, P. and Kotler, S. (2014), *Abundance: The Future Is Better Than You Think*, New York, New York: Free Press, p. 72.

6. This table is an extension and update of data originally presented in Moskowitz, S. (2009), *The Advanced Materials Revolution: Technology and Economic Growth in the Age of Globalization*, Hoboken, New Jersey: John Wiley & Sons, Inc., pp. 23–59. It also incorporates discussions from a variety of recent reports on advanced materials such as *New and Advanced Materials: Future of Manufacturing Project: Evidence Paper 10* (October 2013), London, UK: Foresight, Government Office for Science.

7. Schmalensee, R. (2015), *The Future of Solar Energy*, Cambridge, Massachusetts: MIT Energy Initiative (MITEI), MIT.

8. Lavelle, M. (August 5, 2015), "See-Through Solar Could Turn Windows, Phones into Power Sources," *National Geographic* (Online). http://news.nationalgeographic .com/energy/2015/08/150805-transparent-solar-could-turn-window-phones-into-power-generators/. Accessed November 26, 2015.

9. Koppes, S. (December 4, 2012), "Next Scientific Fashion Could Be Designer Nanocrystals," *UChicago News*, Chicago, Illinois: University of Chicago. http://news.uchicago.edu/article/2012/12/04/next-scientific-fashion-could-be-designer-nanocrystals. Accessed November 25, 2015.

10. Yarris, L. (June 10, 2013), "Testing Artificial Photosynthesis," Berkeley, California: Berkeley Lab News Center, US Department of Energy National Laboratory. http://newscenter.lbl.gov/2013/06/10/testing-artificial-photosynthesis/. Accessed November 25, 2015.

11. Chen, J. (2013), "Recent Progress in Advanced Materials for Lithium Ion Batteries" *Materials*, 6: 156–183.

12. Katz, J. (May 14, 2012), "Lockheed Martin Gives Innovation New Wings," *Industry-Week*. http://www.industryweek.com/leadership/lockheed-martin-gives-innovation-new-wings. Accessed November 25, 2015; Camargo, P. (2009), "Nanocomposites: Synthesis, Structure, Properties and New Application Opportunities" *Materials Research*, 12(1): 1439–1516; and Potts, J. (2011), "Graphene-Based Polymer Nanocomposites" *Polymer*, 30: 5–25.

13. Michigan State University. (2012), "New Thermoelectric Material Could Pave the Way for Low-Cost Energy Solutions," *MSU News*, East Lansing, Michigan: MSU College of Engineering. https://www.egr.msu.edu/news/2012/11/14/new-thermoelectric-material-could-pave-way-low-cost-energy-solutions. Accessed November 29, 2015.

14. Diamandis and Kotler (2014), p. 202.

15. Park, H., et al. (2011), "A New Triboluminescent Material" *Bulletin of the Korean Chemical Society, 32*(2): 401–402.

16. Kjällman, T. (December 14, 2006), "DNA-Sensors Based on Functionalized Conducting Polymers and Quantum Dots" in Proceedings of the International Society for Optics and Photonics (SPIE), Vol. 6416, Biomedical Applications of Micro- and Nanoengineering III, Nicolau, D. (ed), SPIE—The International Society for Optical Engineering: Washington (December, 2006) pp. 1–12.

17. Robotics Business Review. (2014), *Healthcare Robotics: 2014 – Leading Robotics Companies, Their Technology, Markets and Future*, Framingham, Massachusetts: EH Publishing, Inc.

18. BCC Report. (2011), "Lightweight Materials in Transportation." http://www.bccresearch.com/market-research/advanced-materials/lightweight-materials-ransportation-avm056b.html. Accessed November 25, 2015.

19. Scientific American. (2015), "9 Materials That Will Change the Future of Manufacturing." http://www.scientificamerican.com/slideshow/9-materials-that-will-change-manufacturing/. Accessed November 27, 2015.

20. Busse, M., et al. (2012), "Advanced Materials for Transportation – Euromat 2011" *Steel Research International, 83*(10): 937. See also Neu, R. (2013), "Performance and Characterization of TWIP Steels for Automotive Applications" *Materials Performance and Characterization, 2*(1): 244–284.

21. Bourzac, K. (2015), "Nano-Architecture: A Caltech Scientist Creates Tiny Lattices with Enormous Potential," *MIT Technology Review*. http://www.technologyreview.com/featuredstory/534976/nano-architecture/. Accessed November 25, 2015.

22. UNEP. (2009), *Buildings and Climate Change: Summary for Decision Makers*, Paris, France: United Nations Environment Programme, p. 3.

23. Walsh, J. (May 21, 2015) "NanoSHIELD Coating: The Future of Manufacturing," *OmniOne Information Technology* [Retrieved at http://www.omnione.com/nanoshield-coating-the-future-of-manufacturing/ on July 3, 2016].

PART II

STRUCTURAL MATERIALS

Metals and Polymers

2

ADVANCED CASTING TECHNOLOGY

Ultrathin Steel and the Microalloys

Much has been written about [my] ... decision [to build] ... our revolutionary new steel mill in Crawfordsville, Indiana ... Many accounts imply that we threw all caution to the wind, or that we were heroically courageous. We didn't, and we weren't ... We looked before we leaped

Ken Iverson, 1998

2.1 INTRODUCTION

It is appropriate that we begin with the metals, for this group was certainly the first advanced materials known to man. While a number of important new fields of advanced metals are ripe for examination—shape memory alloys, smart metals, new types of superalloys, metal-based composites—as of 2017, no area of advanced metals has played as important a role in the late 20th and 21st centuries than the introduction of thin slab and thin strip casting in the making of steel. Recently, *New Steel* magazine called thin slab casting one of the two most important developments of 20th century steel (the basic oxygen furnace was the other).[1]

These processes—both developed and commercialized by one company—were pivotal in revitalizing America's steel industry and in transforming steel production globally. In 2014, approximately 75% of all flat steel products made within the highly industrialized countries come from mini- or micromills that use either thin slab or thin strip casting technology, respectively. Emerging countries, such as China, have been slowly adopting the processes and today make about 30% of their flat steel products from minimill operations.[2]

Advanced Materials Innovation: Managing Global Technology in the 21st century, First Edition.
Sanford L. Moskowitz.

The importance of these processes extends beyond traditional flat steel products. They are also the essential manufacturing technology in the making of such advanced metals as ultrathin steel and the microalloys, materials with exceptional mechanical and thermal properties and which play an increasingly important role in 21st century structural applications. The story of thin slab and thin strip casting is a tale of David versus Goliath where a relatively small, outsider company challenges and—by innovating a breakthrough technology—beats the big, established, and hidebound corporation.

2.2   BACKGROUND

At the end of World War II, America's steel industry was the largest and most technologically advanced in the world.[3] But over the next few decades, the tide turned dramatically against US steel. Europe and Japan aggressively built up their own steelmaking capabilities. The fact that the war left these countries in rubble actually worked in their favor in the long run, for not having an existing and older infrastructure to deal with, they could begin anew, constructing their plants from scratch using the latest processes, equipment, and building designs. They were the first to adopt large-scale oxygen furnaces and successfully operate efficient electric steel minimills. American steel was slow to embrace such innovations. It had become complacent and eschewed any major effort to revamp in the face of a growing global competition. It continued to work an aged technology and rigid production system. The large steel companies held on tight to the old World War II vintage blast furnaces and spent little money on research and development. Improvements in the integrated steel mills of the Midwest came slowly and incrementally with money spent on older equipment and machinery that provided little competitive advantage.

By the early 1980s, to many observers in and out of the industry, American Big Steel had become a hulking, self-satisfied, and turgid entity especially in comparison to the new, gleaming, and nimble European and Japanese minimills. The result was predictable: a series of American plant closings and bankruptcies the likes of which had not been seen since the Great Depression. By the late 1970s, 320,000 US steelworkers lost their jobs to foreign competition. Between 1977 and 1987, 50 million tons of Big Steel's annual capacity had shut down. American Big Steel was heading rapidly toward irrelevancy in the face of a rising and technologically advancing global steel industry. The traditional way of making steel that had proven such a powerful force in the 19th and 20th centuries had lost its edge and could now barely meet the demands of America's aging industrial infrastructure.

2.2.1   Thick Slab Casting and "Big Steel"

A major steel mill is a sprawling industrial complex.[4] It occupies anywhere from 2 to 7 miles of real estate generally along a river. The plant is an integrated, self-contained production center capable of performing all the major operations of steel manufacture within one facility. It makes the metal itself and then casts it into desired shapes that are shipped to its customers in the construction, automotive, appliance, and other industries that require large volumes of high-quality, traditional steel.

The integrated steel mills have continued to employ the same process they have been using since the 19th century. They put coke and iron ore into a blast furnace and heat the mixture until a chemical reaction takes place to produce liquid pig iron, which is then poured into a second furnace to convert it into liquid steel. The casting process turns the steel into a variety of complex shapes. The hot, liquid metal is usually emptied into a mold, which contains a hollow cavity of the desired shape, and then allowed to solidify. The hardened part, the casting, is ejected or broken out of the mold. These ingots require further working. They are reheated and sent through a rolling mill half a mile in length to make sheet steel.

Traditionally, steelmaking was done in batches. Then, in the second half of the 20th century came a revolution in materials processing as the large steel mills adopted a technology to continuously make thick (200 mm wide) slabs. The heart of the process is a vertical box-shaped funnel, or mold, that oscillates up and down. Liquid steel is first poured into the funnel. Its up and down motion keeps the metal from sticking to its sides and coaxes a rectangular piece of steel "as thick and wide as a queen-sized mattress" to slide down to the bottom of the apparatus where it is withdrawn and sent through a massive, half mile-long hot rolling mill that pounds the slab into sheet steel between 3 and 20 mm thick. These sheets eventually become the steel used in such products as automotive bodies, appliances, and industrial machinery. The size and cost—which runs into the billions of dollars—of such a plant mean that only Big Steel companies could afford to build and operate such facilities. In addition to being capital intensive, such plants are also big consumers of energy and labor.

This is the process that Big Steel—its managers and production workers—have known very well. It is deeply rooted in the technology: its size, its organization, and its flow. It is proud of its history and past accomplishments and uncomfortable in new ways of making steel. Cannibalizing the capital and skills it has built up over the years has been anathema to its culture and to its strategic vision.

2.2.2   The Mini- and Micromill Revolution: Thin Slab and Thin Strip Casting

Minimills, a more recent development, are relatively compact and highly efficient steel mills.[5] Instead of huge blast furnaces heating massive amounts of coal and iron ore, the typical minimill uses smaller electric arc furnaces (EAFs) to melt scrap metal and continuously cast and roll the molten steel. A typical minimill has an annual capacity of between one-half and two million tons of steel annually. Unlike the large integrated steel companies, minimills serve customers who do not require large tonnages of steel possessing high surface quality.

By the late 1980s, innovation in minimill technology lent producers the ability to more easily and cheaply process thinner, high-quality metal sheet. Starting in the 1990s, these new types of minimills pioneered an important advanced metal process called "near-net-shape" casting, which produces the form and dimensions of sheet steel without the use of large and expensive rolling and machining operations thus eliminating up to two-thirds of the production costs in some industries. The two major types of near-net-shape casting are thin slab casting and thin strip casting. Both types of casting technologies integrate casting and rolling into a single process thus eliminating reheating and repeated hot rolling steps typical of the conventional large steel mill. This means lower capital and operating costs. Moreover, these thin slab and thin strip minimills produce much thinner, higher-valued sheet products with a greater range of application than the old steel plants and even than the traditional minimills. In thin slab casting, the liquid steel is cast directly into slabs with thicknesses between 30 and 60 mm or about five times thinner than those coming out of Big Steel. If the traditional casting technology of Big Steel produced slabs as thick as a "five high stack of Cleveland telephone books, the thin slab casting process made slabs as thin as a standard door."[6]

Thin strip casting, developed in the 1990s and early 2000s, could produce even thinner, more valuable steel than possible with the thin slab casting technology. Using a totally new type of mold design, it turns out what is called ultrathin cast strip (UCS) steel, one of the most advanced of the 21st century's mass produced materials. As the name suggests, UCS is far thinner—only a few pages thick—than any other casting technology known.

The greater economy and flexibility of steel production based on thin slab and thin strip casting is evident in the smaller sizes of the plants using these processes. The typical length of the conventional slab casting facility—from furnace to rolling mill—contained within a Big Steel plant is 700 m or close to 2300 feet. A thin slab casting plant—the typical minimill—is much smaller, approximately 375 m (1230 feet). In sharp contrast to both of these, strip casting mills stretches about 100 m (328 feet) or only one-seventh the length of a Big Steel plant. Because of their small size, thin strip casting mills are often referred to as *micro*mills.

Because of their relatively small dimensions, mini- and micromills are far more flexible in a number of ways compared to the conventional integrated steel plant. Whereas Big Steel can only locate their massive plants in certain areas—such as near water and natural resources—mini- and micromills have a much greater range of choice of location. This means they can be situated at optimal distances between raw materials (steel scrap) and customers for maximum economic and market advantages. Such localized operations not only reduce transportation costs but also facilitate just-in-time (JIT) possibilities and cut inventory costs. Mini- and (even more) micromills require less land and building space thus lowering site and development costs. Because their minimum economic capacities are significantly smaller than Big Steel plants, mini- and micromills can produce steel sheet profitably both for traditional steel customers—the ones formerly monopolized by large steel companies—as well as new markets too small to be economically viable for established integrated plants.

2.2.3 Ultrathin Steel and Microalloys

The ultrathin steel made by the mini- and micromills is very different material from the more traditional, thicker slabs that come out of the integrated steel mills. The key to the making of these newer materials is the quick solidification of the steel sheet as the molten metal travels through the caster. While it takes up to 18 min for solidification to occur in the conventional "Big Steel" slab casting process, it requires just 45 seconds in thin slab casting and only between 1/10 and 1/20 of a second in thin strip casting.[7]

These more rapid solidification rates means that there is not enough time for imperfections and other constituents in the metal to "clump together," which tends to weaken the entire structure. The final product possesses a much finer and homogenous internal microstructure than possible with traditional casting processes. These so-called "microalloys" have superior mechanical properties compared to conventionally produced cold roll steel. Alloying such ultrathin and ultrauniform steel with such elements as manganese, niobium, and vanadium during the thin strip casting process produces a very strong metal that is free of cracks and other imperfections. Rapid solidification and a uniform microstructure also mean that ultrathin steel is extremely tolerant of the presence of unwanted contaminants such as copper and tin. Because rather high levels of these "tramp" elements in ultrathin steel can be tolerated without sacrificing material quality, producers do not have to use high-cost premium scrap for their raw material. Moreover, because of the very fine grain structure of the ultrathin steel made from thin strip casting, a number of unique steel alloys have been made with exceptional engineering properties. In addition to their great strength—they are two to three times stronger than traditional steel—they are very tough but flexible, lightweight, and readily fabricated into different shapes. These advanced materials reduce the need for steel for any particular application by an average 30% and save 15% on costs. One of the most important applications for these materials is in automotive bodies for safer and more fuel-efficient cars.[8] As the 21st century advances, new applications for these metals are expected in aerospace, electronics, communications, energy, and biotechnology.

The center of this revolution in steel technology has been Nucor Steel, headquartered in Charlotte, NC. How this relatively small, rather unassuming company came to become the dominant player in global steel is one of the great success stories of the late 20th century advanced materials industry.

2.3   NUCOR STEEL: GROUND ZERO FOR THE MINI (AND MICRO-)-MILL REVOLUTION

The Nucor Corporation innovated both the thin slab and thin strip processes.[9] Both technologies operate at Nucor's Crawfordsville, Indiana plant. Today, Nucor controls most of the thin slab operations and the only two commercially operating thin strip plants in the United States (one facility at Crawfordsville and the other at its Blytheville, Arkansas site). Nucor's steelmaking technology is not only much more efficient than its competitors—by the 1990s, while the Japanese needed 5 h to make a ton of steel, Nucor could do this in only an hour—but it also can create important new thin and ultrathin steel alloys. Nucor is one of the great success stories of American business in the late 20th century.

2.3.1 Nucor's Flexible Structure

Ken Iverson, who became president of the company in 1966, was its guiding influence and responsible for shaping its innovative culture. Under his leadership, Nucor
became a highly decentralized company. The general manager of each plant ran that
site as a more or less independent business. He made the day-to-day decisions including who to hire, where to obtain needed supplies, how to keep production on schedule,
what technologies to purchase, and so forth. In order to limit the influence headquarters could have over operating sites, Nucor's management made a point of never
building operating plants in or around its Charlotte headquarters.

While such decentralized autonomy sometimes caused duplication of functions
and facilities from plant to plant, it also meant that decisions could be made quickly
and by those closest to the situation. Also important in Nucor's ability to act swiftly
to exploit opportunities and to deal with problems as they arose, the company had a
simple and streamlined bureaucracy: it adopted a relatively flat organizational structure with only three layers of management. Iverson designed the company this way
because he felt that a large bureaucracy hindered growth and created an authoritarian
organization that got in the way of people on the ground doing their jobs; they, he
believed, knew what needed to be done better than any executive sitting at headquarters. For him, multileveled hierarchy retards corporate growth:

> My longest-running battle against hierarchy has been holding the line against more lay-
> ers of management … I think adding more layers of management would wreck one of
> the great strengths of our company – very short lines of communication.[10]

In the 1980s, the top management of Nucor consisted of only three executives—
Iverson as chairman and CEO, David Aycock as president and chief operating officer,
and Sam Spiegal as chief financial officer—who also made up the board of directors.
Iverson kept the total numbers of corporate staffers to a bare minimum and at the factory level demanded fluid and informal lines of communication between employees
and supervisors.

Nucor's simple and elastic organizational structure helped the company become
extremely nimble and quick to adopt new approaches to steelmaking that challenged
the traditional Big Steel companies. Most importantly, Nucor was the first US company to build successful minimills. Its first minimill went on line in 1969 in Darlington, South Carolina, with Nucor purchasing its scrap from third party agents at open
market prices.

2.3.2 Ken Iverson and Nucor

Iverson began leading Nucor at a particularly critical point in the company's
history. Nucor is the successor to the Reo Motor Car Company, founded in 1904 by
Ransom Eli Olds, inventor of the Oldsmobile. By the early 1950s, Reo abandoned
automobiles to pursue other opportunities. The revamped company, called Nuclear
Corporation of America, exploited the postwar popularity of atomic energy; it built
and sold products and instruments that ran on or measured radioactive phenomena,

including a line of radiation sensors. It also purchased other companies with growth potential in new technology, such as an outfit called US Semiconductor Products. The company continued to lose money on all these endeavors. The one bright spot was the unexpected success of one of its acquisitions, Vulcraft, which specialized in the making of steel joists for roofing. Ken Iverson first came to the company as manager of this growing division.

Iverson is a perfect example of a person who easily moved between the worlds of science, engineering, business, and management. As such, he believed in the importance for the success of a firm in the close working relationship between these disciplines, especially as they come together on what counts in the end: what happens on the factory floor. Iverson grew up in the small farming town of Downers Grove, Illinois, 30 miles west of Chicago. His father, an electrical engineer, worked for Western Electric. Iverson studied engineering at college and went on to obtain his masters in mechanical engineering. He had a particular fascination with metallurgy and the properties of molten metal. His career began in the late 1940s as a research physicist at International Harvester in Chicago where he took pictures of the internal structure of metals with an electron microscope. In the early 1950s, he took a position as chief engineer at the Illium Corporation, a foundry specializing in casting technology. Here, he invented a casting machine to make nozzles used in the meatpacking industry (it made sausages). Iverson soon moved on to become sales manager and chief metallurgist at Cannon Muskegon Corporation in Michigan, a producer of specialized nickel, iron, and cobalt alloys. Among other responsibilities, he personally took part in the casting of liquid metal. He worked first hand with the pouring and casting of molten iron alloys and saw up close the potential as well as the dangers of the process when carried out on a large scale.

But Iverson's long-term goal was to run a company—or at least a critical piece of a company. In this capacity, he hoped one day to build and manage innovative greenfield steel plants. His ambition led him to Vulcraft. By the late 1960s, Nuclear Corporation had sold off its other divisions and focused on making steel joists. Ken Iverson became President and CEO of the new company, now called Nucor. He moved the company's headquarters to Charlotte to be near Vulcraft's joist plant, the first and last time any Nucor plant would be so near the company's head office. Over the years, as the company expanded, Iverson made it a point to visit annually each one of Nucor's plants to keep himself informed on the particulars of each site's structure and operations. He looked closely at each plant's manufacturing line. He knew many of the production men by name. He personally examined up close the equipment and processes.

2.3.3 Nucor Builds a Steel Minimill

One of the first critical decisions Iverson made for the newly organized company was to enter into steelmaking. Up to this point, the bar steel needed as raw material in the making of Nucor's roofing joists came from Big Steel companies and specifically US Steel. But US Steel continued to push the price of its steel up. While Nucor succeeded in finding alternate, and cheaper, steel overseas, the future uncertainty of this source pushed Nucor to secure and control its own steel supply. It integrated backward into steel production both for its own use and for external sales. In this way, Nucor would become both independent and a competitor of Big Steel. Of course, Nucor couldn't afford the $200 million required to purchase a blast furnace. Since it was cheaper to melt steel scrap in a minimill than to smelt iron ore, Iverson's route into steel production was through the EAF. He decided to locate his minimill in Darlington, South Carolina, an agricultural community close to the joist plant. He borrowed $6 million from Wachovia Bank to purchase the most modern, top-of-the-line equipment, including a Whiting EAF, the newest continuous casting machines then available, and an advanced Swedish rolling mill. The furnace produced liquid steel, which was poured into the continuous caster turning the molten metal into a steel billet. The rolling mill converted the hot, liquid metal into steel bars, which were transported to the Vulcraft plant to make roofing joists.

Iverson faced many difficulties in getting this new plant to operate properly. If placing this plant in an agricultural community helped keep the unions out, it also meant having to train farm workers—the only available labor—how to run an electric furnace and rolling mill, something most of them had never even seen before. Added to this was the difficulty in breaking in new machinery—always a problem even with the most experienced of personnel. Not unexpectedly, the project was plagued with numerous accidents. The most serious of these—the so-called "breakout"—involved molten steel taking wrong turns within the casting machine, pouring through drenching and melting the internal mechanism—the "guts" of the apparatus. However, these accidents did not deter Iverson from moving forward with his plans to make steel. He had been through similar start-ups before in his career; the problems encountered were not foreign to him. His experience as a hot metal engineer and as a businessman helped him find the way out of, what was to Iverson, just another thorny wood to be navigated.[11]

The Darlington minimill was up and running by the summer of 1969. Iverson well understood the central importance of this facility, and he said so to a group of financial analysts in Phoenix, Arizona in 1972:

> ... in recent years, the single most important event in the shaping of the future of the company has been the construction of the ... steel mill at Darlington, South Carolina ... This is no ordinary mill. It has been described as a forerunner of a new generation of minimills.[12]

The importance of the Darlington Mill is that it gave the United States—as well as Nucor—a strong foothold in the minimill industry. For Nucor, the success of this plant emboldened the company to move into even more cutting-edge minimill technology. Nucor established another site in the small Midwest town of Crawfordsville, Indiana dedicated to scaling up and operating these new processes. Iverson looked beyond the United States for a technology. Nucor did away with the risks associated with R&D by purchasing the rights to technology that others had already researched and partly developed. We now consider the R&D history of the thin slab and thin strip processes, which takes place prior to when Nucor acquired these technologies and mostly outside of the United States.

2.4 THIN SLAB AND THIN STRIP CASTING: RESEARCH AND DEVELOPMENT

Iverson—and his champions—often credited Nucor's success to risk taking. He is said to have told his plant managers never to be afraid of taking risks and failing, since a pioneering spirit needs to be embraced if the company was going to move forward.[13] While Iverson and Nucor did indeed take their share of risks in embracing new technology, we shouldn't overstress this factor as a key to the company's success. The fact is, Iverson did not dare delve into areas in which he was unfamiliar. A case in point is Iverson's position on research and development. In fact, Nucor did not have a separate, stand-alone R&D department or corporate engineering group or even a dedicated technology officer. Iverson had no intention of having Nucor get into the R&D game. Rather, it relied on outside companies to undertake this type of work on their own and create a potentially workable prototype. Nucor would then buy (or license) it and devote its time and money to scaling it up.

2.4.1 Thin Slab Casting

By the 1980s, what Nucor needed had become clear.[14] While the company pioneered the minimill in the United States, Iverson understood that his steel operations in Darlington were severely limited in scope, especially in its ability to go after the major steel markets. His minimill could only make steel for the nonflat, commodity portion of the market, such as reinforcing bar for construction and rods for pipe and rail. While Nucor's operations held decided cost advantage over the older integrated Big Steel plants, the output consisted of low-priced, low-end products. Nucor did not see significant profit from them. The "big game" in steel was the flat steel products required by the automotive and appliance customers and still monopolized by the big, integrated plants. But this was not Nucor's only problem. Even in the low-end markets, Nucor confronted increasing competition from a growing number of efficient minimills within the United States, Europe, and Asia. By the 1980s, there was little to distinguish Nucor's Darlington plant from others around the country: they all purchased their scrap at the same prices, consumed the same amount of energy at the same cost, and employed the same technology.

Then, in early 1987, Nucor had the opportunity to radically change the competitive landscape and move well ahead of the pack. Nucor executives, scouting around for promising steel processes, had learned of a new type of casting technology coming out of West Germany and a firm called Schloemann Siemag A.G. (SMS) located in Dusseldorf. Its inventor, Manfred Kolakowski, had come out with a wholly novel mold design for continuous casting. Kolakowski's idea for this radically new mold appears to have been a "eureka" moment that occurred in March 1983. He had seen the mold "in his mind's eye" and sketched it on a piece of paper. Then he built test models and tried them out in a small pilot plant. SMS had already spent $17 million on research and development and on building a working pilot plant. But the company did not have the resources to move into commercialization. Iverson was willing to look at the mold system as a possible technology for Nucor.

Iverson sent a team to Germany to see the technology first hand. The Nucor crew were sufficiently impressed with the pilot plant and they agreed to take it on. At the center of the process was Kolakowski's mold into which molten metal poured and from which a thin steel slab emerged. This mold was what made the process so valuable. Traditionally, the continuous steel casting machine had a mold shaped like a box-like funnel with flat sides. The problem was that this sort of device produced fairly thick slabs of steel requiring further hot and cold rolling down the line, thus increasing the costs of production. Kolakowski, in contrast, thought of a very different design: a swollen, curved funnel that was shaped like a lens or "squeezed envelope." This mold collected the liquid metal and sent the steel downward in a sheet through a roller system that squeezed the hot metal into a thin solid steel slab that would then be bent and continuously withdrawn horizontally from the bottom of the machine. Experts in the field warned Kolakowski that, on a large scale, his mold would produce nothing but cracked and warped—that is, useless—steel. But in fact, it worked and proved capable of turning out smooth, flat, and thin rectangular steel sheet desirable for many important applications. Essentially, the research and development—at least early development—of the new process had already been done by SMS.

The Nucor team in Germany had a pretty good idea of how to design a full-scale commercial plant around this novel mold, for they had planned, built, and managed other plants that incorporated other types of molds. Iverson understood that scaling the plant would not be an easy matter. It would take time and money but, based on past experience, was definitely a doable project. A risk here was that, once Nucor had built its steel plant, the planned market could shift thus forcing the company to completely revamp the process, which would involve additional and unbudgeted time and costs, or abandon the technology altogether. But in fact the target market—applications for low-carbon flat-rolled steel sheet—was clearly identified and quite substantial. It was also stable since low-carbon flat-rolled steel is one of the most common types of steel used for general purposes. Even though the thin slab process, like Henry Ford's assembly line, is a fairly inflexible process, it nevertheless could be used to make the steel for different shapes and applications. This is so since only small modifications in carbon content are required to prepare the metal in the making of structural plates or forgings, stampings, or pipes. Iverson knew this new technology could handle these incremental adjustments without much trouble.

The process itself seemed fairly clear. Two electric furnaces melt the scrap. The molten steel is then be drained from the furnace and poured into a large container called a ladle positioned just above the mold. From the ladle, operators pour the liquid steel into the mold's copper funnel where it flows down through the mold and into the casting machine, inside which it is hot rolled into steel sheet and withdrawn from the bottom of the caster. A further cold rolled operation thins the sheet further into a ribbon of "rich, glistening" steel to be coiled up "like a roll of paper towels," conveyed from the plant into trucks and shipped out to be sold to Nucor's customers.

2.4.2 Thin Strip Casting

The eventual success of thin slab casting emboldened Nucor a few years later to take on an even more advanced metal-making process called thin strip casting.[15] Once again the company sidestepped the risks of research and development by looking outside for a promising—but not yet scaled—process. As with thin slab technology, the idea of thin strip casting was known by steel companies large and small, none of whom succeeded in commercializing the technology. Nucor's interest in this process was that it would allow it to make much thinner steel than has ever been made by any company and, because of this, to enter into the higher-end steel products offering the greatest profit margins. Unlike the thin slab process, thin strip technology is a very elastic process, capable of making a wide range of very different ultrathin steels (steel sheet less than 1 mm thick) for many applications. The process operates at optimal efficiency when making thin steel at very high casting speeds. This property is of great advantage since it can turn out different types and grades of steel for a variety of markets that had been off-limits to Nucor (and many other steel makers). This range of application is made possible by the high cooling rates that are achieved during solidification, which, as described earlier, impart unique properties to the alloys. Because of this advantage, the cast strip process can realistically move into markets not within reach of either the traditional hot rolled methods or even the thin slab process.

For these reasons, in the 1980s, the large steel companies felt that thin strip casting had great commercial potential as a major new type of steel process. Through the decade, Big Steel companies such as Allegheny Ludlum, Armco, Inland Steel, Weirton, and Bethlehem Steel all tried but failed to find a process that could cast reproducible grades of thin steel sheet. By 1994, all R&D efforts by the American integrated steel outfits on thin strip casting ceased. Global R&D on such a process continued, however, especially in the United Kingdom, Japan, Scandinavia, Germany, Italy, and Canada but without much in the way of tangible results. One of the major problems encountered with existing designs was that both productivity and product quality declined as engineers scaled up the equipment.

Nucor however continued looking for a possible commercial process. In the late 1990s, it came upon a new technology being jointly developed by the Australian company BlueScope Steel, one of Australia's largest flat-rolled steel maker, and IHI Corporation (Ishikawajima-Harima Heavy Industries Co., Ltd) a Japanese producer of engines, industrial machines, power plants, and transportation-related machinery. Their strip caster pilot could produce steel sheet to ultrathin dimensions on the order of three-hundredth of an inch or less thick. What came to be called the CASTRIP process was a twin-roll casting technology with the two large copper rolls next to each other, their surfaces almost touching as one point and made to spin opposite directions. Liquid steel was poured between them from above, and a continuous thin strip came out of the rollers from below.

In addition to the economy with which the technology could make ultrathin steel, the process also proved to be quite flexible in that it could readily shift from making one type of steel to another. The "thinness" of the steel as well as its strength and bendability depended on the rate at which the liquid steel was poured onto the two copper rollers: the faster the spin, the thinner the steel. So Nucor could make many types of grades of steel for different applications by simply adjusting the casting speed. Because of the rapid cooling of the material, it could also add alloying elements to make many types of high-quality microalloyed metals. This meant the ability to quickly and cheaply create diverse types of steel for different customers and their requirements. This flexibility was an important reason the technology continued to be viewed as a viable commercial process. It was first designed to make a special type (called "304 grade") of stainless steel. But when the competition within this market from traditional producers intensified, champions of the process could rapidly shift its output to a far more promising product, ultrathin gauge low-carbon steel, for the many demands of the construction industry.

For a decade, BlueScope Steel and IHI worked on the process, proceeding from laboratory work to pilot plant. By 1993, the pilot facility turned out five ton coils of low-carbon steel two millimeters thick. Over the next 5 years, the Australians and Japanese further moved into the semicommercial phase producing enough steel with their enlarged plant to sell to the construction industry. To the Nucor team, this "gave the first glimpse of the technical potential" of the process. The team was particularly impressed with the elasticity and compactness of the process, two important points if it was to achieve the necessary economies of scale. It would also be much less expensive to build than any existing minimill because of lower capital and operating costs. In late 1999, Nucor joined the project. A few months later, the three companies formed Castrip LLC to commercialize the technology, with Nucor and BlueScope each having a 47.5% share and IHI the remaining 5% and the joint venture owning all of the intellectual property related to the technology. Nucor's Crawfordsville site was selected to be the location of the first CASTRIP plant. For Nucor, this endeavor would be another challenge and another potential game changer. Now that we have explored the R&D phase of both the thin slab and thin strip technologies, we will next examine the scaling up processes that took place for each project. In doing so, we shift our attention back to the US and Nucor's Crawfordsville plant.

2.5 THIN SLAB AND THIN STRIP CASTING: SCALE-UP

For both technologies, expanding the pilot plants into full-blown commercial machines proved to be no easy matter.[16] It took considerable skill, patience, and money. Nevertheless, Nucor executives had prior experience in similar ventures. They were really not entering into unknown territory.

2.5.1 The Challenges of Scaling

For both technologies, pouring and casting operations were extremely difficult to expand to commercial size given the materials being worked. Liquid steel does not behave well at all. The casting machine and its parts are themselves made of steel, and when trying to pour molten metal into the caster, it is "apt to weld itself into a lump [much like] trying to make ice by pouring water into a machine made of ice."[17] Liquid steel is also hard to contain because of its vicious reactivity. For this reason, it can easily break out of any container and spill over a plant floor leaving the dead and injured in its wake. Even when the pouring operation succeeds, the attempt to convert this to thinner and thinner steel is fraught with its own difficulties. These strips tend to move rapidly through the mechanism creating all sorts of internal havoc including fires, meltdowns, and explosions. There was always the potential for disasters due to excess oxygen within the molten steel, which made the metal bubble up and burst backwards onto equipment and people. In addition, the creation of solid "skulls" within the liquid metal blocked exit holes of the ladle and tundish thus stopping the entire operation. Leaving all this aside, the major issue that had to be faced was the extremely rapid rate at which heat was transferred from the molten metal to the giant rollers as the steel turned from liquid to solid, an especially difficult problem in thin strip casting. While this quick, almost instantaneous conversion helped keep the impurities from clumping, thus forming a homogeneous internal structure and a steel product with superior properties, it also rapidly released massive amounts of heat that could warp the copper rollers, which would then create a steel sheet that was an uneven, unusable "mess."

Despite these challenges to both the thin slab and thin strip processes, Iverson and his colleagues believed they could tame both technologies. The first order of business for them before they proceeded was to minimize the risks—at least as they perceived them.

2.5.2   Nucor and Reducing the Risks of Scaling

Even though Nucor did not have to grapple with the research and development phase of either technology, it had to face the challenges of scaling these difficult processes, something that soundly defeated metallurgists and engineers since the early 19th century.[18] There were indeed risks, mostly of time and money. The Big Steel companies and industry insiders knew all this and believed that there was a good chance Nucor's "folly" would fail and that its failure would destroy the company. Bethlehem Steel and other integrated mills thought so. Industry observers outside Big Steel thought so as well. Iverson knew this very well as he entered into the first of his two great projects, thin slab casting. The big questions for Iverson were the costs required and length of time it would take. Iverson was not prepared to embark on this new project if certain guarantees were not in place to mitigate these risks. Nucor's ability to face and neutralize them is an important part of the company's willingness to pioneer these radically new processes.

2.5.2.1   Structural Risks   One of the first risks in scaling up these processes was transferring the capabilities of the scientist and inventor to the manufacturing arena. This issue involves a structural risk, that is, whether the company, either formally through its organizational structure and procedures, or informally through unofficial networks, link up R&D and manufacturing. The organization must also have the ability to block knowledge and biases held by scientists from entering into and hindering—even derailing—the scaling effort. Such was the case with thin slab casting. While the German inventors had little idea of what it took in the real world to build a full-fledged plant from a laboratory experiment, they thought they did; this made them dangerous to have around the Crawfordsville plant during the critical months of building and testing. They thus posed a tremendous risk for Nucor, and the company made sure through contractual agreement that the Germans had limited access to the Crawfordsville site.[19]

Within the company itself, executives consciously blurred the line between research, development, and scale-up. Nucor did not have a separate R&D department to fall back on. All work and process improvements took place on site; there was no back looping to do more time-consuming research. Whatever it took to handle the bottleneck metallurgists and the production crew worked closely together—and faced the same physical dangers—as the plant went up. Problems that emerged could be quickly communicated down the line and promptly solved. This extremely fluid structure kept the project moving at a rapid clip and without undue delays.

2.5.2.2 Resource Risks: Capital, Raw Materials, and Labor

Scaling also puts pressure on resources. Purchasing the rights to the process itself cost money. Iverson and his people minimized the cost damage from this by carefully—and forcefully—negotiating the deal. With thin slab casting, for example, the German company caved into Nucor's demands and agreed to license the process for what Nucor executives called "chump change" ($70 million).[20]

Then there was the question of how much the scale-up process itself would cost. Despite the fact that the thin slab and thin strip projects represented breakthrough innovations, Iverson and his team—from their prior experience in building minimills—had a pretty good idea of what the final costs to build the plants and put them into operation would be, that is, somewhere in the neighborhood of between $250 million and $275 million, at the outside. While undoubtedly a large amount of money, Nucor was earning between $1 billion and $2 billion annually in the late 1980s and early 90s. That is to say, Iverson was not by any means risking the company in either of the Crawfordsville projects. In both cases, Nucor's job was to build—or contract out—plant infrastructure, reassemble the parts and components for the caster, and do whatever it took to convert the full plants into working, commercial operations. All the necessary parts for both plants were shipped to Nucor by the Germans and Australians for the thin slab and thin strip plants, respectively. Iverson stuck to the same game plan he had used in the successful scaling and start-up of other plants, including the first minimill at Darlington. In order to further reduce the risks of cost overruns and time delays, Nucor relied on outside contractors they knew and trusted to keep on schedule.

Capital was one thing, labor another. The demands and costs of labor can be a risky proposition. Labor unions can add to the problems borne by a company. Whether production workers were up to the tasks of building a totally new process was another concern. Iverson mitigated these risks in large measure in the location selected for the plants and the type of people hired on to the job. Crawfordsville in the 1980s—that is, just prior to scaling up the thin slab plant—resembled many Midwest farming towns with "Farmland stretch[ing] for miles and silos dott[ing] the landscape." In 1987, it had a population of only 13,000 souls—many of whom were farmers—and had only two industrial plants: a chemical facility and a printing operation. Situated in the middle of America's rust belt, it was not very far from a plentiful source of cheap scrap in the form of shredded automobiles, factory punchings, and demolished steel mills. The surrounding farmland had plenty of workers anxious to earn the wages of a steelworker. They were generally mechanically knowledgeable and so highly trainable in the ways of steelmaking. Iverson's famed incentive system also played its part in getting them to work both harder and smarter. A large incentive was monetary: 60% of the salary of each member of a team or crew came from bonuses Nucor gave them if they met production and product quality goals. Beyond the cash, Iverson's decentralized, egalitarian system instilled professional pride and team spirit in the Crawfordsville crews. Nucor employees felt they played a major part in the success of the company and so took personal satisfaction in the performance of their plant. The fact that orders did not come from the top and that managers, supervisors, and mill workers were all expected to be creative and entrepreneurial pressed them to be proactive when confronted with plant-related problems. The supervisor of the melt shop, the manager of the casting operation, and the chief of the rolling mill all understood that the success or failure of the whole plant depended on them and their workers. They had to make very sure they did not screw up and that they did everything they could to get their facility up and running. Iverson understood that, in fact, it was the guy on the ground at the plant who was the real "engine of progress" for the organization, for he or she was the one who performed the many incremental changes that made the new processes work.[21]

As in the past, Iverson optimized the conditions for creative thinking on the plant floor. He demanded that his people on the ground in Crawfordsville think on their feet and solve problems that arise right on site. If a difficulty arose, it must be addressed then and there—and quickly—by whatever means that offered a workable solution. Iverson demanded that both the thin slab and thin strip projects stick as closely as possible to the original schedule so that both plants were operational within a year and a half of the groundbreaking. While Iverson set the general tone and parameters of these projects, he made sure the project remained flexible by leaving the detailed decisions—and day-to-day problem solving—to the Crawfordsville team. While decision-making in the Big Steel companies generally came from the top with little creative leeway given to engineers and middle managers, Nucor granted its minimills (and other plants) practically carte blanche on how they dealt with issues and problems impacting their facility. This freedom is expressed well by the supervisor of the melt shop in one Nucor site:

> They [Nucor] put it all on my shoulders—the planning, the engineering, the contracting, the budgets ... I mean, we are talking about an investment of millions of dollars and I was accountable for all of it. It worked out fine ... because my team and I knew what not to do from our experience running the melt shop on the first line.[22]

This distribution of authority allowed plants to be built much quicker and cheaper than a comparable Big Steel facility. It spurred competition between plants—such as that between Crawfordsville thin slab crew and a team at Blytheville, Arkansas building a new plant to produce steel I-beams for skyscrapers—as well as between the teams working on the melting, casting, and rolling facilities within a plant. Such a pressured, "gung-ho" environment galvanized teams to action and stimulated creative energies to be the first to get their mill or plant on line.

2.5.2.3 Experiential Risks   Nucor also faced the risk that its personnel would not have the requisite experience and skills to make a new and untested technology work on the large scale. This uncertainty could be the biggest gamble of all, especially in an industry where accidents are not uncommon and injury, death, and lawsuits are a constant threat. The big question was whether Nucor's leadership believed these risks were not only manageable but (more or less) minimal. This in turn hinged on their own past experiences in handling molten metal and in assessing the technical issues involved. But, in the case of thin slab casting, Iverson could call upon his past experience—personally as a metal man and as someone who knew very well Nucor's proven capability. Iverson had extensive experience in the casting of metal and had already been through many plant start-ups including the Nucor minimills. He knew there would be problems as scaling proceeded—there always were—but that they could be solved by continuing at it and in due time. Iverson's confidence came in large part from knowing precisely the core capability of his company: building plants quickly and operating them efficiently.[23]

In the end, Iverson's belief in the company and its capability turned out to be more than justified. Both processes took about 14 months to scale-up, more or less according to plan. Both processes hit snags and bottlenecks along the way. In both cases, Iverson's system of "on-the-ground" innovation solved these problems fairly rapidly resulting in advances in productivity and product quality. In the case of the thin slab project, Nucor faced numerous breakouts and product quality issues. Redesign of some of the equipment, adjustment of procedures, and modification of the chemical composition of the steel all helped to put these problems to rest. Similarly for the thin strip process, a series of issues hampered progress. For example, the Crawfordsville crew found a way to quickly replace spent casting rolls with new ones to increase the capacity of the electric melt furnaces so that they kept up supplying enough molten metal to the rapidly growing CASTRIP mill and to extend the number of heats possible per cycle (by using improved refractory materials).[24]

The thin slab plant first went on line in August 1989; the thin strip plant did so in May 2002. Between 1990 and 1995, the former increased it shipments of finished product from about one million tons to 3.8 million tons per year. From 2002 to 2008 Nucor's CASTRIP process expanded its production from only 54,000 tons of steel annually to 540,000 tons per year of various grades of steel for different applications in the construction and other industries.[25] The great success of both of these plants prompted Nucor to build additional facilities in Arkansas: a thin slab plant in the town of Hickman (1992) and an additional CASTRIP facility in Blytheville (2009).

2.6   THIN SLAB AND THIN STRIP CASTING: COMMERCIALIZATION

Whether a company will actually continue to support a technology once scale-up is in process and even well on its way to production is an interesting question. The assumption that of course it will—otherwise why go to all the trouble and expense involved—needs to be carefully considered. A number of factors can come into play during and even after the scale-up phase that can hinder—even shut down—a project in motion. Management may change hands with the new regime advancing a strategy within which the new technology plays no role. Then too, top executives come to realize that the new process is so revolutionary that it will cannibalize all of the company's existing facilities around the world. Or important customers, suddenly realizing that the plant will turn out products that compete in their markets, threaten to take away their business. Internally, a powerful department—such as marketing—feels deeply threatened by the company changing strategic directions. Why these voices did not rise up earlier during the research and development phases has to do with the nature of the scaling process. Prior to this, scientists, engineers, and selected managers have been busy building and studying experimental pilot plants. But once the focus is on the design and building of a large-scale, industrial facility, the project "goes public," and the outside world—including customers—becomes more intimately acquainted with the technology and its broader implications. Whether a firm chooses to embrace, marginalize, or even snuff out its own growing technology depends a great deal on its cultural makeup and the strategic vision that this entails. In some cases, the firm already has the appropriate culture in place rooted in its history and experiences. In other situations, top managers forcibly alter its traditional mission, values, and goals—responding perhaps to recent traumatic events—in ways that may be more or less welcoming of new ideas. We find as well that individual champions or intrapreneurs manage to affect change in an organizational culture through their influence or force of argument so that their favorite projects are more readily accepted at the highest levels of the corporation. Between the late 1980s and the early 2000s, Nucor's corporate culture did indeed change in important ways. In the end, Nucor embraced both technologies but, if they flourished within the company, they did so under two very different cultural environments.

2.6.1 Commercializing the Thin Slab Process: Nucor's "Internalized Static" Culture and Technology Selection

As Nucor contemplated scaling up the thin slab process, its top managers asked themselves if this plant would fit comfortably into the company's cultural milieu.[26] Iverson had created Nucor in his image, that is, he instilled throughout the company an aggressive, highly competitive character. Smaller companies in an industry tend not risk directly challenging the big outfits in their central markets thinking there is little chance they could survive their blistering counterattacks. Instead, they try to gain markets by nipping at the smaller, less important customers that the big, established firms wouldn't necessarily care about or even miss. But Nucor rejected this guerilla-type strategy; Iverson wanted to directly challenge Big Steel in its most valued markets.

In addition to pushing the idea throughout the organization that aggressive behavior is to be valued, Nucor under Iverson increasingly exhibited a highly internalized culture. The company reflected this self-contained orientation in a number of ways, from supplying its own inputs through its minimills to minimizing reliance on strategic partners. Nucor's story of origin was simple—that the firm was unique, sui generis, even exceptional. Its singular abilities—which could not be replicated within other organizations—positioned the company to be the scourge of the complacent integrated mills and, at the same time, to lead the United States back to industrial greatness. In this telling, Nucor was both David (vs. Goliath) and Moses rolled into one. This almost biblical belief in its own greatness, a confidence supported by its continued growth and market success, led Nucor's management onto a dangerous path; it accepted German demands for a nonexclusive license, meaning that SMS could license its thin slab patents technology to any other company anytime it wanted and anywhere in the world without having to ask Nucor's permission. Nucor did not fight this demand because it wanted to build its thin slab plant as soon as possible and because it felt there would be little risk in agreeing to such a provision. After all, the Nucor team reasoned, the company would construct the new facility in a way that could not be easily (if at all) be replicated by any other firm. Let SMS license to whomever it wanted, Nucor executives told themselves; no licensee would be able to match Nucor's distinct process—a process incorporating thousands, even tens of thousands—of small but vital innovations of varying types. Many of these would be ad hoc, spur of the moment changes made on the plant floor most of which would not even be recorded. Creating a complete working process that outperformed all others from these numerous and hands on improvements is just the thing at which Nucor excelled; no one else will be able to come close in technical and operational excellence or so Nucor management thought. This sort of aggressive self-reliance did indeed help drive the company's success with this landmark technology. If it had known or at least suspected that other companies would soon be able to compete effectively with Iverson's new process, it would likely have put its money into other, more promising projects. In fact, eventual and unexpected replication of the thin slab process by other minimills in the United States forced Nucor to seriously question its belief in its "not invented here" mentality. Doing so prepared it to undertake the next major leap into thin strip casting.

2.6.2 Commercializing the Thin Strip Process: Nucor Creates a Dynamic Expansionist Culture

In the first 5 years after the Crawfordsville thin slab plant had begun operations, this enthusiasm for the project and this belief in Nucor's exceptionalism seemed to be well founded.[27] The facility was the most efficient steel plant in the industry. Whereas the large integrated mills required approximately four man hours to make a ton of their steel, Nucor's thin slab process demanded only 45 man-minutes (or 0.75 man-hours). This greater productivity meant that Nucor held between $50 and $75 per ton advantage over Big Steel which translated into a 25% greater cost savings. (By 1996, Nucor had increased productivity in its thin slab plant to only 36 min per ton. or 30% savings) Moreover, the smaller competitors—other minimill operations—seemed, as had been predicted, unable to quickly catch up to Nucor even when they themselves licensed out the SMS technology.

However, by 1995, or about 6 years after the first thin slab plant began operations, Nucor had a rude awakening; competing minimills started to rapidly advance in thin slab production and become competitive with it. Other plants either soaked up know-how from former Nucor employees who had changed jobs or had worked out for themselves how to successfully take SMS' prototype design and create their own version of the process. Once they had, Nucor's profits at Crawfordsville nosedived, and Iverson and his managers had to reconsider the situation. These external pressures radically altered Nucor's cultural orientation in two ways. First, the company was determined to become more dynamic technologically. It would support a new process that was more flexible than the thin slab caster turned out to be. It wanted a new plant that could turn out different products for a variety of high-value markets. In this way, the new technology would not be so dependent on one market and so would be far more resilient to competitive attacks.

Secondly, Nucor had to do more than build a novel minimill and expect it could remain king of the mountain for very long. It had to reach out and take more control over what went on in the industry, especially how any new process it developed was dispersed and how much advanced steel was made and who made it. Clearly, a sea change had to be made to the company's cultural thinking. It had to be more on its guard in protecting its technologies and had to find a way to secure more capacity to supply the greater number of markets it hoped to dominate. It needed, in other words, to become more outwardly oriented. It began to license its most important technologies to other steel companies and to demand an exclusive licensing agreement from the original patent holder so that other companies could not, without legal repercussions, "copy the process the way they did with thin slab casting."[28] Nucor pressed for and got an exclusive licensing arrangement with Castrip LLC and then proceeded to set up its own licensing agreements with the small- and medium-sized minimill operations. The process, being very pliable, could be adjusted to the particular requirements of Nucor's clients. It thus provided them an affordable and efficient way to integrate backward to control their supply of coiled steel and bypass the high prices demanded by Big Steel. In return, Nucor, now in total control over its own creation, spread its process widely to increase capacity to meet demand and to receive royalties and fees from all licensees. In stark contrast to its experience with the earlier thin slab process, it exercised greater control over the thin strip casting technology and in turn over the strategies of its minimill competitors who licensed it.

In geopolitical terms, we can say that Nucor had transformed itself from a company with an internal, isolationist mentality to one with an external, empire-building culture. The CASTRIP process was the appropriate technology for the Nucor of the late 1990s. Its flexibility assured that Nucor could be the first mover in a multiplicity of prime steel-consuming markets. Then too, with such a nimble tool, it could tailor design its process for the particular needs of each licensee in terms of volume of production, type and specification of product, and so forth. This custom fitting of process with client meant that the licensee was able to exploit the technology to the fullest, which was a good thing for Nucor since it captured a percentage of each client's revenues attributable to its CASTRIP operations. It seems reasonable to assume that, if both the thin slab and thin strip process had come along at the same time and the expansionist Nucor of the late 1990s was faced with deciding where to best position its scarce resources, it would select to focus on the more elastic and wide-ranging CASTRIP process over the less adaptable thin slab technology.

To this day, Nucor and its dynamic expansionism continues to lead the industry in new and more productive CASTRIP plants and in developing ever more advanced ultrathin steels and microalloys to expand and broaden present markets and enter into new applications. As of 2016, Nucor has set its sights on an improved, second-generation CASTRIP process to increase productivity and extend the range and quality of ultrathin steel and its family of microalloys. With this and future advancements Nucor hopes to further expand its influence over the 21st century global steel industry.

REFERENCES

1. As reported in Mangels. John (December 9, 2001), "US Steel's Hot Commodity" *Cleveland Plain Dealer*.
2. For review of developments in steel minimills, see Stubbles, J. (2006), "The Minimill Story," AISTech Iron and Steel Technology Conference, Warrendale, October Issue, p. 33 and Barnett, D. and Crandall, R. (1986), *Up from the Ashes: The Rise of the Steel Minimill in the United States*, Washington, DC: Brookings Institute.
3. Preston, R. (1991), *American Steel: Hot Metal Men and the Resurrection of the Rust Belt*, New York, New York: Prentice Hall Press, pp. 79–81.
4. Ibid., pp. 91–94. See also Boyd, B.K. and Grove, S. (2000), "Nucor Corporation and the U.S. Steel Industry" in Hitt, M.A., Duane Ireland, R., and Hoskisson, R.E., *Strategic Management: Competitiveness and Globalization*, 4 edn, Cincinnati, Ohio: South-Western College Publishing.
5. Stubbles (2006), pp. 25–34. See also Boyd and Grove (2000).
6. Mangels. John (December 9, 2001), "US Steel's Hot Commodity" *Cleveland Plain Dealer*.
7. Blejde, W., Fisher, F., Schueren, M., and McQuillis, G. (2010), "The Latest Developments with the Castrip® Process," The Tenth International Conference of Steel Rolling, Beijing, China, September 15–17, 2010.
8. Deng, B. (2015), "New Lightweight Steel Could Improve Cars' Fuel Efficiency," *Scientific American* (Online). http://www.scientificamerican.com/article/new-lightweight-steel-could-improve-cars-fuel-efficiency/. Accessed November 26, 2015.
9. Iverson, K. (1998), *Plain Talk: Lessons from a Business Maverick*, New York, New York: John Wiley & Sons, Inc.. Also see Preston (1991), pp. 71–90.
10. Iverson (1998), pp. 59–61.
11. Ibid., pp. 143–163.
12. Rodengen, J. (1997), *The Legend of Nucor Corporation*, Fort Lauderdale, Florida: Quality Books, Inc., p. 55.
13. In *Plain Talk*, Iverson devotes a full chapter to the importance of risk-taking in business. See Iverson (1998), pp. 143–163.
14. The following narrative is based on a number of sources including Preston (1991), pp. 91–276, and Iverson (1998), pp. 150–157.
15. Waugh, J. and Varcoe, D. (2000), "Commercialization of Strip Casting," South East Asia Iron and Steel Institute, 2000 Vietnam Seminar, Hanoi, Vietnam. See also Ge, S., Isac, M., and Guthrie, R.I.L. (2012), "Progress of Strip Casting Technology for Steel: Historical Developments" *ISIJ International*, *52*, 12: 2109–2122 and Blejde et al. (2010).

16. For a lively discussion on the scaling of thin slab casting, see Preston, American Steel. For additional sources for both thin slab and thin strip casting technologies, see Govindarajan, V. (2000), Nucor Corporation, Tuck School of Business, Dartmouth College, pp. 11–13. Boyd and Grove (2000), "Nucor Corporation and the U.S. Steel Industry," has some excellent material on Nucor's technology.

17. Preston (1991), p. 93.

18. For a discussion on the rise of Nucor's culture, refer to Preston (1991), pp. 71–90. See also Iverson (1998) *Plain Talk* and Govindarajan (2000), pp. 4–11.

19. Preston (1991), pp. 103–114.

20. Ibid., p. 101.

21. Iverson (1998), pp. 79–99.

22. Govindarajan (2000), p. 12.

23. Iverson (1998), pp. 143–163.

24. Waugh and Varcoe (2000); Ge et al. (2012); and Blejde et al. (2010).

25. Ge et al. (2012), p. 2118 and Boyd and Grove (2000).

26. Preston (1991), American Steel and Iverson (1998), *Plain Talk*, pp. 79–99; 150–157. See also Govindarajan (2000), Nucor Corporation, Tuck School of Business, Dartmouth College, pp. 11–13.

27. Waugh and Varcoe (2000). See also Ge et al. (2012) and Blejde et al. (2010).

28. Mangels (December 9, 2001), "U.S. steel's hot commodity," *The Cleveland Plain Dealer*.

3

HIGH-PRESSURE TECHNOLOGY AND DUPONT'S SYNTHETIC FIBER REVOLUTION

For some time I have been hoping that it might be possible to [prove the existence of macromolecules] from the synthetic side. The idea would be to build up some very large molecules by simple and definite reactions in such a way as there could be no doubt as to their structures. This idea is no doubt a little fantastic and one might run up against insuperable difficulties

Wallace Carothers, 1927

The story of superpolymers, notably nylon and its family of synthetics, is very different from that of the advanced steels. These "miracle fibers" were the product of a large, established, and highly bureaucratic corporation. They were created in the laboratory and as part of revolutionary scientific investigation. The champions of this technology did not occupy the highest levels of DuPont's hierarchy. Rather, they were forward-looking middle managers who had to convince a deeply conservative Executive Committee—whose members tended to be at a distance from what went on in the laboratory and pilot plant—to provide the resources to pursue this technology. They did so, not simply by persuading top management of the possible benefits of success but, even more importantly, by offering compelling reasons at critical points in the innovation process why attempting this endeavor would not pose serious risks to the company.

That DuPont's eventual success in synthetic fibers is a landmark achievement in advanced materials cannot be overemphasized. Nylon itself of course became and so remains a singularly important material. Today it is not only a fiber but the basis for an important plastic for many structural applications and a possible polymer for the coming 3-D printing revolution. It has become an essential matrix in modern composites.

Advanced.Materials Innovation: Managing Global Technology in the 21st century, First Edition.
Sanford L. Moskowitz.
© 2016 John Wiley & Sons, Inc. Published 2016 by John Wiley & Sons, Inc.

When combined with such reinforcing fibers as glass or carbon fibers, the resulting material is extremely dense and strong, finding increasing application as a substitute for metal in under-the-hood automotive components. Recent research is looking into the creation of advanced composites made of nylon and nanomaterials that have superior mechanical and thermal properties and that many in the field believe will revolutionize synthetic fabrics and provide structural materials for aerospace systems, electrical and thermal conductors for energy applications, and delivery systems and other vital technologies involving the biotechnology sector.

The advanced materials process behind the success of nylon is high-pressure catalytic technology, which was critical in the making of a crucial intermediate without which DuPont would not have proceeded with nylon development. This process itself has continued to play an important role in 21st-century materials development, notably in the manufacture of nanotubes and other nanomaterials. The following chapter explores the rise and impact of this significant process, its importance to the creation of the first superpolymers, and less positively, its role in DuPont's decision to relegate its R&D to more incremental innovation while leaving other companies to pioneer the next generation of breakthrough synthetics.

3.1 BACKGROUND: THE HIGH-PRESSURE PROCESS AND ADVANCED MATERIALS

By the 1930s, the synthesis of chemical products using very high pressures and selected catalysts to accelerate reactions led to a true revolution in advanced materials. Never before could so many materials be made in such large volume and at such low costs. Some of the most important chemicals known to man that previously could only be turned out in limited quantities and with great difficulty could now flow continuously from large-scale chemical and petrochemical plants. These included sulfuric and nitric acids, ammonia, and methanol. In a number of very important cases, totally new materials never before produced commercially at all—nylon, polyethylene, and other macromolecular fibers, plastics, and composites—entered into and greatly influenced US manufacturing and, through technology transfer, the global economy.

3.1.1 The Nature of High-Pressure Synthesis

High-pressure processes have been important in creating advanced materials because new and more complex materials can be created when the basic components or intermediates are combined under high pressures, often under high temperatures, over catalysts.[1] Catalysts—often composed of metals, alloys, and their oxides—are themselves advanced materials. They accelerate reactions that otherwise would not be commercially viable because they would take too long to carry out. Catalysts also heighten the economy of any given process to the point of making it an acceptable economic risk by minimizing unwanted side reactions and optimizing yields for the desired products. However, high-pressure work was not for the faint of heart. It was dangerous business, with explosions, even on the laboratory scale, a common—and often deadly—occurrence. Even if one could avoid such catastrophes, frustrations aplenty awaited the chemist as he or she hiked up the pressures.

Chemists and engineers had to take into account many factors as they experimented and scaled at high-pressure processes. These included the chemical and physical makeup of the catalyst, pressure control, thermal conduction (heat transfer), and fluid flow. The catalyst of course had to have the proper chemical composition in order to make the desired reactions work, and just this in itself took many hours of empirical work in the laboratory. It also had to have the proper physical composition so that it could withstand the high temperatures and pressure applied. Controlling these massive pressures brought immense challenges into chemical manufacture. The reactor had to have the proper design and be made of very tough alloys. Gauges, valves, and other mechanical components also had to be specially designed to work properly while the reaction progressed. Heat control was particularly problematic. Often, very high heats were applied to urge the reaction along. But even if this was not the case, the reactions themselves were usually exothermic, which meant they gave off massive amounts of heat on their own. If the chemist or engineer could not contain the thermal buildup, the high internal temperatures could damage the catalyst and even cause a runaway reaction to occur, not unlike an uncontrolled nuclear power plant. Consequently, the reactor system had to have a way to dissipate heat buildup and do it quickly, one of the most difficult challenges in any high-pressure work. In addition to all of this, the flow of the gaseous materials had to be carefully monitored and controlled. The reactant gases had to be properly blended and then propelled through the system so that the mixture flowed smoothly and uniformly over and through the catalyst. Disruptions to this current meant low yields, poor quality product, or both. It could also mean local buildup of pressures or heat not uncommonly resulting in catastrophic failure of the system.

In any high-pressure system these variables had to be in perfect balance with one another. Any change in one could throw off balance one or more of the others causing system breakdown. But there was no practical way to foretell how change to one variable would affect any of the others in any particular experiment. Furthermore, even if no changes were made to any of the factors within the laboratory, it was nearly impossible to envisage how scaling up a process from laboratory scale to pilot plant to semicommercial facility and then to fully operating plant would impact any of the critical forces involved.

One has only to read through technical articles and research reports, memoranda, and other communications coming out of DuPont's Ammonia Department—the epicenter of the company's high-pressure operations—during the 1920s and 1930s or of such companies as Sun Oil, which was the first to apply catalytic technology to petroleum refining, to understand the difficulties faced by the scientists and engineers trying to design and scale up high-pressure process.[2] Each problem had to be faced head-on, often entailing fundamental research and extended development. Then once in place, that improvement forced change on another operation, which again demanded separate R&D efforts and so requiring further time and additional costs. Attempting to scale up the process would throw the entire system out of whack and require going back to the laboratory to research and develop a new catalyst, material of construction, component, or entire mechanical system.

Time and costs continued to mount as researches worked their way through the growing problems. In short, high-pressure work was risky, not just to life and limb but also—and more often—to a company's bottom line. A pioneering company could expect one delay after another and the cascading of problems without any guarantee of success. One problem or bottleneck might be the one that can't be solved forcing the abandonment of the project, even after spending considerable capital—financial, physical, and human. Failure could also hurt—even ruin—a company's reputation both with important customers and aggressive competitors further eroding its position in the industry.

And yet, under certain conditions, high-pressure technology has been a very powerful tool. As with any tool, how and when it should be used must be clearly understood. On the other hand, those applying it incorrectly and in the wrong context court disaster. High-pressure catalytic processes are extremely effective when focused on producing large volumes of a single, standardized product, such as ammonia, methanol, or a nylon intermediate. Just like Henry Ford's assembly line, a high-pressure process had built-in rigidities: once a company went through the expense of getting a process right to make a product, it was very good at making that one thing cheaply, and lots of it. But ask it to shift over to produce another, different thing, and it simply can't do it easily and certainly not without entailing great costs: for one must start again practically from scratch, and the solutions and routines that worked before often do so this time around.

The rigidity—or "stickiness"—of high-pressure technology proved to be quite a challenge for those companies that decided to go into the field. The Germans, as so often the case, were the first movers in the technology, but just as typically, the Americans soon pounced on the opportunity and took the process to its furthest limit.

3.1.2   DuPont: High-Pressure Synthesis and Its Road to Advanced Fibers

DuPont began its existence as a single product operation. It was established in 1802 on the Brandywine River near Wilmington, Delaware.[3] Its founder, Eleuthere Irenee du Pont (1771–1834), was a French émigré and protégé of the great chemist—and victim of the French Revolution—Antoine Lavoisier. DuPont's claim to fame at the time was black powder—a mixture of saltpeter, charcoal, and sulfur—for the military and also for farmers, mining companies, frontiersmen, and other civilian customers for clearing land and blasting rock and earth.

Beginning with the Civil War, DuPont's main customer was the US military, supplying the Union army with high-quality gun powder. By the 1880s, DuPont began manufacturing a new type of explosive—dynamite—which was initially invented by the Swedish chemist Alfred Nobel. By the late 19th century, DuPont added a third member to their explosives family, smokeless powder.

3.1.2.1 DuPont's Diversification Strategy   DuPont soon expanded the range of its product lines beyond explosives. It entered into this program of diversification for two main reasons. First, the US military began to build up its own explosives-making capability, thus threatening to deprive DuPont of its main source of revenue. Second, because of stricter federal antitrust legislations, DuPont faced the growing possibility of government action against its virtual monopoly of the explosives industry. DuPont could parry both threats by moving into other product areas. The company's approach—as was typical of many chemical companies—was what management strategists call concentric diversification, that is, it considered only those product technologies that were linked by a common thread, one that reflected DuPont's existing technical competency. Specifically, it explored those products that were based on the two most important substances used in explosives production: the element nitrogen and the naturally occurring organic compound cellulose. Through research and strategic acquisitions, by the 1920s, the company had become a leader in products based on nitrocellulose, including lacquers, synthetic leather, paints, rubber, packaging (cellophane), plastics (viscoloid), fibers, and fabrics (rayon). By the early 1930s, over 90% of DuPont's revenues came from these newer products.

While these products shared nitrocellulose as their essential intermediate, they had little else in common. They had different physical properties and so served vastly dissimilar markets (an important point in avoiding government antitrust actions). And they were made by completely different processes. Knowledge of how to make one did not readily translate into a greater expertise in the manufacture of others. DuPont did not create these materials internally from scratch using a common pool of knowledge, skills, and technology. Rather, the company obtained the technology for each product by another route: it acquired the leading company within each product area: Fabrikoid Company (synthetic leather), Fairfield Rubber Company (synthetic rubber), Harrison Brothers (paints and coatings), and so forth. It is true that in most of these cases—Fabrikoid's artificial leather process is one good example—DuPont employed its own growing research capability to greatly improve the process they had purchased. But in each case, they worked the product area as a distinct research problem; what they learned and how they improved the technology had little linkage to other products and their processes. In short, DuPont's diversification resulted in an expansion in the number of discrete, semiautonomous product fiefdoms increasingly jealous of—and in competition for resources with—the other of DuPont's product groups.

However, after World War I, there was a movement afoot, led by the more scientifically trained managers, to place DuPont's production on a more rational basis. Under their influence, DuPont embarked on something new—an actual process that would be the foundation for many products, including what would turn out to be the basis for the most important one of all, nylon.

3.1.2.2 *DuPont Enters Upon—and Struggles with—High-Pressure Synthesis*
Prior to the early 1920s, DuPont had had little experience in high-pressure catalytic technology.[4] The closest it had come was in its work on the synthesis of dyes, which required the use of elevated pressures. Following World War I, DuPont management supported the pursuit of high-pressure chemicals such as synthetic ammonia because,

if successful, it would be a route to a continuous and cheap supply of nitrogen, the most important raw material for all of its various products.

By the early 1920s, two men at DuPont began to steer the company toward the high-pressure route: Fin Sparre, Director of the Development Department at DuPont and a relative newcomer to the company, and the physical chemist Roger Williams who was appointed director of the newly formed Ammonia Department. Scouting around for a technology to develop, they directed DuPont to a promising French high-pressure process—a variation of Germany's pioneering Haber technology—called the Claude process. DuPont formed a holding company jointly with the Claude interests in France and a separate agreement with one of the largest ammonia companies in the United States at the time, National Ammonia Company. The Claude interests agreed to supply this joint venture, called the Lazote Company, the process and one-quarter of the capital in the new venture. DuPont was responsible for further developing the process. National Ammonia would be the sales arm of the venture. Sparre became the president of Lazote and Williams its technical director. The plant was built at a site in Belle, West Virginia, under the direction of the Ammonia Department.

DuPont focused a great deal of its time on finding appropriate catalysts. This shows us an important example of how a company must often reach outside for critical knowledge it does not have internally, particularly during the research phase. While DuPont certainly knew a great deal about catalysts in general, its understanding of catalytic activity under very high pressures was sketchy at best. A government laboratory known as the Fixed Nitrogen Research Lab (FNRL) did a great deal of work on high-pressure ammonia reactions and processes during World War I. After the war, DuPont attracted many of the lead chemists at FNRL to work for Lazote.[5] Then there was the question of the technology itself. The high pressures and temperatures required presented a series of problems. Constant replacement of valves and other components, problems with temperature control, mixing of gases, effect of corrosive chemicals, and many other difficulties continued to plague the project. One of the biggest—and most expensive—challenges was the unwanted formation of carbon monoxide, which had to be removed because it destroyed the effectiveness of—or poisoned—the ammonia catalyst.[6]

While Williams and his team found solutions to this and the host of other problems that came their way, they spent considerable time and money doing so. The bottom line was that the Claude process caused so many difficulties for DuPont, it never achieved the results—technical or commercial—that DuPont had hoped. By the early 1930s, with the depression devastating the industrial sector, the world was awash in unwanted ammonia. Even the hope of the Ammonia venture to supply DuPont's other departments with nitrogen turned to disappointment given the large amounts of cheap ammonia that could be secured at any time from many suppliers in the United States and globally. Losses mounted to the tune of $400,000 per year, and the Ammonia Department had to sharply curtail R&D expenditures and lay off a major portion of its staff. From 1931 to 1933, Ammonia Department personnel shrunk from over 120 to less than 50 scientists, technicians, and managers.[7]

3.1.2.3 Roger Williams and the First-Generation High-Pressure Chemicals

What happened next is typical of what we find so often in large corporations: the company basically washes its hands of a project or department that is underperforming but lets it survive in a truncated form.[8] This tactic reduces the risks it poses to the company financially but keeps it alive with the hope that it can produce needed revenue in the near future. If, under these reduced circumstances, the remaining group of researchers can rouse itself and start showing profits, the department can continue to operate; if it cannot, it will eventually be allowed to die out or simply ordered to shut down. As sometimes occurs under these pressured conditions, an individual champions the project and effectively employs the limited resources he or she has in hand to completely turn the situation around and create an R&D initiative that eventually persuades top management to embrace the revised technology and integrate it into the company's strategic planning. The champion might be a bench scientist who convinces his immediate superiors to let him continue with the research as long as he does so under severe restrictions of men and money. The champion could also be a senior executive who believes in the program or as is true in this case is a middle manager who runs the department that is under siege and has the wherewithal, like a Henry V before the Battle of Agincourt, to stir his depleted troops to action to accomplish great things. That manager was Roger Williams, officially the technical director of DuPont's Ammonia project. While trained as a scientist, Williams had a remarkable, innate managerial capability that was recognized early by DuPont executives. Ndiaye, in his book *Nylon and Bombs*, tells us that at this time " ... despite his youth, [Williams'] qualities as an organizer and leader of men already seem to have been evident ... everyone [at DuPont] had the greatest admiration for Williams."[9] This company-wide respect was clearly well deserved, for Williams' ability to rouse the hobbled and dispirited group comprising the Ammonia Department was a tour de force of technical insight and managerial agility and inspiration. The fact that he was an excellent physical chemist played an important role as well since it gave him an understanding that the high-pressure process that his department commanded had applications far beyond merely ammonia. Williams was able to convey this to his small, informally organized technical group. He used his team to show DuPont that one process did not necessarily have to make only a single mass-produced product; the same equipment could also turn out many chemicals. So even if one or two chemicals lose their markets, there are others with very different demand curves that can bring in cash to the company. This use of economies of scope strategy went far to allay the fears of corporate executives that high-pressure technology was nothing more than a money pit with uncertain means to recoup investment. It was Williams then who helped convert the culture of DuPont from a static (one product per process) to a more dynamic culture (many products from one process), something that would prove extremely important to DuPont in its future strategies.

As Hounshell tells us, under Williams' leadership, the group's search for new products and processes based on high-pressure catalysis established a coherent and well-oiled team, one that " ... developed the spirit de corps of a military unit that had gone through many campaigns together."[10] Although working with diminished resources, Williams and his researchers worked extremely efficiently with the little they had. Williams understood that his department needed to start producing important new materials to survive. They looked for new applications for the products they had and extended their high-pressure process that they had to synthesize new products for commercial use. In the former category, they developed a methanol antifreeze for automobiles as well as a urea-based fertilizer from ammonia. They also moved into the synthesis of new materials made of acrylic, which eventually became DuPont's famous Lucite plastic. By the late 1930s, the Ammonia Department had invested upward of $50 million in its Belle, West Virginia, plant, which turned out over 80 products. It had become DuPont's largest department.[11]

However, soon the department experienced more problems. This initial spurt in the production of these "first-generation" high-pressure chemicals did not last long. Diseconomies of scale began to kick in. It became more and more difficult to extract additional products from the increasingly troublesome and expensive high-pressure technology. In order to keep the product flow going, higher and higher pressures (and temperatures) had to be applied and more advanced catalysts found, meaning it cost DuPont more to extract each new product that generally could not even sell at premium prices. Attempts to extend the process to the making of truly revolutionary products such as advanced plastics did not pan out; they failed technically as well as economically.[12]

The fact was that the high-pressure process at DuPont had reached its breaking point and could not be coaxed any further into making important new products for DuPont. Nevertheless, the department was able to exploit the products it did have so that if not exactly flush with cash, it could at least begin paying for itself. By the mid-1930s Williams and his crew could breathe a bit easier, for it was safe from extinction at least for the time being. But they were still very far from being a major asset for the company. The department continued to live on the edge, and any downturn in their fortunes could prove fatal to their existence. They needed an opportunity—a bit of luck really—that would show the company, and the world, just how important high-pressure technology was to the new materials revolution that was brewing. And luck would soon come their way. While they were not able to synthesize a major new product on their own, they proved their worth many times over by being the creators of the most important input of the nylon revolution. Even more, they were a critical component of a new model—a methodology really—for DuPont that prescribed how the company would proceed on creating new and more advanced synthetic fibers beyond nylon.

3.2 DUPONT'S NYLON REVOLUTION

It was a good thing for DuPont and the world that Roger Williams had what it took to keep the Ammonia Department going. The department became the vital component of what would become DuPont's most important undertaking. With the coming of Williams to the Ammonia Department, DuPont had taken its diversification strategy to a new, more dynamic level. Rather than simply acquiring technical competency in different products by purchasing companies, Williams and his team showed that new synthetics could be obtained internally using a single process. Whereas the nature of Nucor's culture hinged on whether it was internally or externally oriented, for DuPont, it depended on whether it was static or dynamic in nature, that is, whether it concentrated on only a few large volume products or whether it embraced a continual stream of new synthetics even to the point of sacrificing older, known products for new, rapidly growing ones. This dynamic ethos at DuPont resulted from external events (threat of antitrust action and the military deciding to make its own explosives) and the example of internally created synthetics set by Roger Williams and the Ammonia Department. Through the 1930s and well into the 1960s, DuPont and its dynamic culture would continue to exploit its internal diversification strategy by creating a series of new synthetic fibers. DuPont management accepted each of these as important to the company, even when they ended up taking the market away from existing products. The first and, to many, most revolutionary of these new fibers was nylon. DuPont's nylon project was a complex affair requiring the services of three very different champions to achieve final success.

3.2.1 Charles Stine and DuPont's Central Research Department

If dynamism characterizes DuPont's culture in the 20th century, it took two very different forms over time.[13] Until the 1930s, DuPont had a culture that might be characterized as "dynamic externalism." As discussed, due to environmental pressures, the company wanted to diversify its product line, and it did so by acquiring other companies. But Charles Stine, a respected DuPont chemist, had other ideas. He championed what we can call "internal dynamism," that is, internally generated diversification or the application of basic science to in-house invention of new products. He ultimately succeeded in compelling DuPont to accept his way of thinking and with stunning results.

Until high-pressure technology came on the scene in the 1930s, DuPont had shied away from fundamental chemical research. Rather the company had its chemists working on solving immediate production problems associated with the making of explosives, its most important product group at the time. DuPont really had very little incentive to do otherwise since it relied on the government as its most important customer. It was for this very reason that the company became known, rightly or wrongly, as a war profiteer during the First World War. The need to diversify into new product areas compelled DuPont to acquire existing firms and take over their products and markets. In pursuing this road to firm growth, DuPont executives decided to decentralize its research and development in order to serve the growing number of departments formed from the absorbed companies. That DuPont management would revert to forcing university-trained chemists to merely supporting existing plant processes was in fact the company's default position. Unless it could be convinced that allowing its chemists to do anything more innovative would result—and quickly—in an improved bottom line, it had little interest in giving its scientists and engineers any more creative free rein.

DuPont would finally change its stance and create a strongly centralized R&D program focused on fundamental research into new materials only when a forceful, persuasive, and respected visionary appeared who was able to convince top management that pursuing such a path was in the company's best strategic interests. That prophet-salesman did indeed appear at DuPont in the 1920s in the form of Charles Stine, DuPont's director of the Chemical Department. Stine did not aim his sights on any specific product. He believed, instead, in pursuing a general program of R&D, with a focus on fundamental research independent of the day-to-day concerns of plant managers that would lead to new materials as yet to be determined. Stine, with a Ph.D. in chemistry, believed strongly in the role of science in industrial research. He applied this scientific approach throughout his career at DuPont, starting with explosives manufacture in 1907. He was highly critical of the empiricism with which DuPont approached problems. Finding solutions to any particular problem took time and money and generally could not be used more than once, as different problems required very different solutions. Stine reasoned this did not need to be the case. If only one could find the common principles linking the different product technologies, then that person could transfer the techniques, skills, and concepts used to solve one problem to attack other bottlenecks faced in different product areas. Scientific understanding would, Stine contended, supply this common link, not only in addressing current plant issues but, more importantly, in creating entirely novel families of products that would open up new and promising markets for DuPont. This was not by any means a totally new idea for DuPont since its experience with the Ammonia Department showed that understanding the principles behind a group of products—in this case, high-pressure catalysis—could extend the reach of the company through internal growth and do so in a way that was at least self-supporting. What Stine wanted to do was to expand this approach to make not only novel but truly breakthrough synthetics that would replace existing products over a wide range of industrial applications. He believed such pioneering work would indeed propel DuPont into the front ranks of chemical innovators and, as a result of its increased competitive power, fuel greater growth and profitability the likes of which it had never before experienced.

The area of scientific research that Stine believed united DuPont's disparate products—paints, plastics, fibers, and films—was one of the most fundamental of all: the study of chemical bonding, or how atoms combined to form molecules and how the latter came together to make polymeric materials. For Stine then, fundamental work in this field of polymerization was the key to placing DuPont research on a rational scientific basis and to building the common platform for the synthesis of entirely new families of advanced materials. At the same time, he was a realist; he knew that DuPont's conservative leadership would not be willing to throw money at what it considered risky ventures with pay offs, at best, in the distant future. This meant that the chemists who worked for him could not just be let free to investigate any area of polymerization they desired. They had to address problems and opportunities that interested—or at least appeared to interest—in some way the industrial departments. If their work was to digress too widely into the realm of the abstract, blue-sky research, Stine and his group would lose the support of departmental directors and then of the Executive Committee itself.

Stine understood from the start that, to remain a viable part of the company, his research department would have to walk a fine line between rigorous scientific research—his personal preference—and at least the appearance of pursuing practical commercial application. The agreement he made with DuPont's Executive Committee to fund fundamental chemical research clearly played to the Committee's fears with industrial research. Stine agreed that his department would concentrate on working with the existing industrial departments to improve the economics of their processes. The Executive Committee did give Stine leave to pursue new. product development but only in those areas that would use DuPont's existing technology base. The Committee further reduced the risk associated with undertaking fundamental research by attempting to keep Stine and his team on a very short leash: they reserved the right to review the performance of the department annually to determine how much—if any—money it would allocate to Central Research. Luckily, the timing was right. In the year of decision, 1926, the company did very well financially, and in any case, initial allocation barely put dent in DuPont's budget. In that year, the amount allocated to R&D ($300,000) accounted for only 0.8% of its 1926 income of $38 million.

Stine, anxious for DuPont to take on really radical R&D, must have chaffed more than a little under such rigid restrictions. But he accepted the fact that to get DuPont to fund a fundamental research unit at all was a major achievement. He also understood he had quite a bit of "wiggle room" to work with and with which he could nudge the new department in the direction he wanted it to go. The trick for him was that in so doing, he did not alienate the Executive Committee. In terms of project selection, Stine preferred fields that had strong theoretical possibilities but also assuaged top management with their potential applications. One of the areas selected, high-pressure catalysis, with applications in nitrate chemicals, satisfied both requirements. The study of colloids with their use in the making of paints was another. Once the fields to be studied were determined, finding the right people to direct the various projects became a critical consideration. Stine wisely avoided doing the obvious: going after established academic specialists in these areas, for they would be theoretical prima donnas who would eschew practical applications as beneath their talents. A department stocked with these sorts would antagonize the Executive Committee and eventually sink the entire venture. Instead, Stine approached young, eager chemistry graduates from top schools. Reasoned he could more easily influence and shape these freshly minted candidates, especially in getting them to understand the importance of wedding their theoretical interests to practical industrial problems.

3.2.2 Stine Finds His Star Scientist: Wallace Carothers

For Stine, no chemical area held more promise than polymerization, for, more than any of the other project areas, this one could catapult DuPont into a whole universe of new and revolutionary products.[14] So finding just the right person to lead this particular group was top priority. After some initial difficulties in filling the position, Stine hit upon the perfect candidate. Wallace Hume Carothers (Figure 3.1) was a rising star in the field of organic chemistry. A Midwesterner by birth, Carothers was an instructor in chemistry at Harvard when first approached by DuPont. At Harvard, Carothers was a dedicated researcher but less than enthusiastic about his teaching responsibilities. It would seem then that DuPont would be just the right place for him at this time. If he worked for the company he would have more time, money, and manpower than he ever had at his disposal to work on the type of chemistry that interested him the most. However, a major concern for Carothers, and one that made him at first play hard to get, was his fear that DuPont would not give him the freedom he wanted to pursue the scientific portion of his work, especially if it did not have apparent and immediate market potential. And for a brilliant organic chemist hungry to make a name for himself, the times could not have been more exciting. A debate was raging in the world of science and Europe over what certain groups of materials—rubber, plastics, and coatings—were actually made of. One group of scientists believed these substances were made up of many smaller molecules linked together by some sort of intermolecular force, in much the same way as colloidal materials. The opposing view believed that they were, in fact, composed of single, very long molecules (the "single-molecule" theory). While evidence existed that supported both theories, the German chemist Hermann Staudinger had recently conducted a brilliant set of experiments that pointed toward the single-molecule theory as the correct model. What Carothers wanted to do was extend Staudinger's work and prove once and for all the existence of single long-chained molecules by synthesizing them one step at a time using well-known chemical reactions.[15]

Stine understood Carothers' desire to carry out these scientific investigations, but he also perceived the practical application of this work: if Carothers was right, then it

Figure 3.1   Wallace Carothers (on the left) on a fishing trip in 1925, with his friend and fellow organic chemist Carl Marvel. Picture was taken a few years prior to Carothers beginning his pioneering work at DuPont. (Source: Courtesy of the Chemical Heritage Foundation, Philadelphia, Pennsylvania.)

would be possible to engineer entirely new and commercially important macromolecular materials. Moreover, because of Carothers, DuPont would be in the forefront of this field and well positioned to control the critical patents that would allow the company to dominate these new markets. Of course, the fact that Stine could not predict when such new materials would be commercially available, and even what these materials might be, posed a big problem: how to get the Executive Committee to go along with this type of fundamental—and uncertain—research. While Stine already had the funds to support Carothers for 1 year, the Committee could remove funding the next year if, in its annual review of the department, it rejected Carothers' work as too distant from the needs of the company. Stine proved his mettle as a research director by skillfully acting as a buffer between Carothers and the Executive Committee. He continued to urge Carothers to pursue his polymerization studies and to publish those results in prestigious academic journals, in order to build up his own—and DuPont's—reputation in the field. At the same time, he translated Carothers's scientific progress in a language that the Executive Committee could understand and support, in particular, that Carothers was getting closer to finding commercially viable materials with potentially large markets that, by dint of its patent position, DuPont could control. In these reports to the Committee, he also implied that Carothers, as well as the other project leaders under his command, are being molded into "DuPont" people who will contribute substantially to DuPont profits in the years to come.

3.2.3   Carothers and Nylon

Carothers began working at DuPont in 1928 as leader of the polymerization group.[16] He and his team worked in a special facility called Purity Hall, a part of Central Research, dedicated to the most fundamental types of research.

3.2.3.1   Nylon: Research Phase   Very quickly, Carothers began pushing his chemical research into new directions. Carothers' genius was that he understood intellectually and, it seems, intuitively how to take simple organic molecules and, through the design of self-sustaining reactions, form long-chained molecules. No one else had done this before. Slowly and methodically, Carothers experimentally built up the case for the single-molecule theory of polymers. By 1930, Carothers and his team discovered in the laboratory the most important supermolecules up to that time: neoprene rubber and nylon fiber.

But the economic turmoil of the depression began to slow down this work. Predictably, the depression compelled reduced funding for Carothers that forced a serious reduction in his team; in 1930 half of the 16 people working for him had departed. DuPont management allowed Carothers to continue his work in part because of its deep respect for the chemist. But even Carothers could proceed only if the impact of his group on the company's resources was minimal. DuPont might have had a high regard for Carothers, but the company was certainly not ready to take a risk on him, especially in such financially precarious times. A further difficulty for Carothers was a change in leadership in the department. A large company like DuPont offers many opportunities for talented chemical individuals. Such a prospect opened up for

Stine in 1930 with a chance to become a member of the Executive Committee. He jumped at the opportunity and, by so doing, removed himself as head of the Chemical Department and as the major force in the nylon enterprise. The new director of the Chemical Department, Elmer Bolton, was a very different kind of manager. Also a Ph.D. chemist, Bolton did not raise science to the supreme level as his predecessor. Whereas Stine made a gentleman's agreement with Carothers that he should continue to pursue his scientific work without worrying about immediate application, Bolton was more of a "50/50" man: one-half of the work should be scientific in nature, but only if it can be shown that science bore directly on DuPont's commercial interests. This meant that Bolton actually wanted his researchers to have closer daily contact with the industrial departments in order to find out, and begin working on, the practical problems that needed to be solved. This of course was the "carrot" that Stine had initially dangled in front of the Executive Committee to get his Central Research established. Once he got his funding, Stine didn't push Carothers very hard on doing this; Bolton, now in charge, did expect his chief scientists to honor this commitment.

While Bolton's more goal-oriented, "pay-your-way" approach to R&D could have scuttled Carothers' scientific research, the Executive Committee at least continued to fund this work. It did so because it had, under agreement with Stine, set aside $25,000 monthly specifically earmarked for fundamental research. Such an amount had little impact on DuPont's budget, and it didn't even have to be tapped during the bountiful years of the late 1920s. With the beginning of the depression, a rather large pool of money was available to keep Carothers' fundamental research program going for a couple of years. By 1933, Carothers had completed the bulk of his studies on polymerization. His research proved convincingly that the "single-molecule" theory was correct. This was a monumental achievement. But, while his international scientific reputation had grown, he had still not produced any commercial products for DuPont. With completion of his scientific work, Carothers was somewhat rudderless and needed direction on what to pursue next. He even had thoughts of leaving DuPont and going back to academia. By this time there was not much money left to keep Purity Hall operating.

It is at this point that Bolton begins playing a critical role in the nylon story. We here see an example of "consecutive championing" at work, which often happens in the life of an innovation within a large corporation. Stine's role as champion was to create the context from which nylon could emerge and set the project in motion. However he did not have a direct influence on the conversion of this scientific discovery into a commercial product. Carothers took over for Stine as scientific champion, which is the person who takes the first important steps of actually inventing the new material in the laboratory. Bolton and Carothers at first worked together as joint champions for a few years, until the latter made his great discovery. Then, with the chemist's early death, Bolton becomes sole champion of the project and leader of nylon's development, scale-up, and commercialization phases.

Joint Champions: Bolton, Carothers, and Nylon Like Roger Williams over at the Ammonia Department, Bolton marshalled his troops at Purity Hall to use the technology they had built up to commercialize new products. Most importantly,

Bolton energized Carothers to take what he had learned in his polymerization studies and apply that to finding new and practical fibers. The biggest hurdle was that Carothers' laboratory fiber had too high a melting point to be commercially spun as fiber. Bolton challenged Carothers to apply his knowledge of the synthesis and structure of the polymers to find a polyamide (an organic molecule containing nitrogen) fiber with appropriate physical properties. Carothers took up the problem and began seriously looking at ways to proceed. Bolton supported the project by making sure he had Carothers' back: as the reserve funds dried up, Bolton went to the Executive Committee to assure it that a new and important fiber was soon to be created, that success was close at hand (even if he knew it really wasn't), and that the company would not have to spend very much money on the project. By easing the Committee's mind as to the risks involved, Bolton kept the money coming and so provided shelter for the research effort.

In the spring of 1934, Carothers and his team found their fiber. Success in doing so depended a great deal on an important piece of technical information that came from outside DuPont. A few years earlier—around 1930—they had reached a roadblock in synthesizing macromolecules. They could only make relatively short polymers with fairly small molecular weights, too small in fact to be useful fibers. The problem appeared to be the formation of water during the reaction, which, as it built up, slowed the forward reaction and blocked further polymerization. The key to keeping the reaction proceeding and creating longer molecules was to remove the excess water. It is at this point that Carothers recalled something he heard about within his scientific network, specifically, at a conference he attended a few years earlier—a device called a molecular still. He began thinking that if the polymerization reaction could be carried out in the apparatus, the still could remove the water as it formed thus allowing the reaction to proceed. Experiments showed that the still could make polymers two to three times longer than before. This success energized the project, giving Carothers and senior executives at DuPont a belief that victory was at hand. It reduced the level of perceived risk in the endeavor and was an important reason executives allowed the project to continue. With the still, Carothers found his fiber, a lustrous strong polymeric filament that would soon be known as nylon. His laboratory work moreover gave Bolton insight into how the pilot plant should be built.

Sadly, Carothers' mental state began to seriously deteriorate soon after his discovery. Professional and personal reasons have been cited by biographers and historians. Suffice it to say here, Carothers ended his life in a Philadelphia hotel on April 29, 1937, by taking cyanide he kept in a vile (and which reportedly he carried with him for years). This tragic and premature end to the life of a troubled but brilliant scientist—and likely future Nobel Prize winner—did not slow DuPont down in proceeding to commercialize the important discovery. The cold reality was that, by the time of his death, Carothers had probably contributed all he could to the project. From here on, emphasis would be on the engineering and market development of Carothers' process, areas in which this purest of scientists had little interest in pursuing and no evident capability even if he wanted to stay involved. DuPont now had its nylon and the company had no intention of squandering this unique opportunity to become the world leader in advanced industrial polymers.

By all accounts, Bolton was an exceptional manager who, like Stine, knew how to unite people and organizations to work together for a common purpose. Also like Stine, he was extremely clever in convincing the Executive Committee (now with Stine as one of the members) to go along with his nylon program. In order to gain further support in the company for his project, and give the nylon enterprise further momentum, he convinced other managers in the company to support his efforts. The nylon effort now came under the management of the Chemical and Rayon Departments, the first under the direction of Bolton and his assistants Ernest Benger and Crawford Greenewalt and the second under its general manager Leonard Yerkes and his assistant George Hoff, who served as liaison between the two departments. This combined and powerful management team now had the explicit task of developing, scaling, and commercializing nylon. Bolton remained the first among equals in this arrangement and, as such, took the lead in the championing efforts.

3.2.3.2 Nylon: Development, Scale-Up, and Commercialization Caution continued to rule the nylon project. Most importantly, Bolton and his crew had to show upper management that nylon "had legs," that Carothers' laboratory invention was not just a scientific curiosity, but that it could be turned into a living, breathing commercial product and in a way that did not break the company financially or antagonize its most important customers. For only then would the Executive Committee be willing to commit serious money toward the project.

Bolton Minimizes Perceived Risks of Development, Scale-Up, and Commercialization
At this point, Bolton had little more than a laboratory sample and a possible route to a pilot plant. But DuPont executives had seen other examples of impressive laboratory work that never amounted to much of anything. During the development phase, Bolton had to take it upon himself to chart out a course that would minimize costs and keep the project on a tight track. Delays, rising costs, and uncertain markets, any or all of such problems, would alert the Executive Committee that nylon was in trouble and might need to be abandoned.

The fact that now another department had joined the nylon effort was important and helped to keep it funded for a while at least. But DuPont's executives, to maintain their interest, needed to feel very sure that the risks involved in further chasing nylon were quite small. They had to believe that the nylon process could be developed, scaled, and commercialized rapidly and with few major problems encountered along the way. Tasks had to proceed rapidly and in a direct, linear manner. Diversions, back-looping, and other such difficulties would only delay final success and sap company resources. In order to meet these strict requirements, Bolton conceived of a simplified approach toward commercialization: (i.) only one process route would be selected for commercialization; there would be no hedging of bets with attempting the more expensive and time-consuming method of undertaking simultaneously different possible avenues, and (ii.) the project would focus on only one market—high-end women's hosiery—rather than attempt to test out different opportunities. If DuPont did pursue a number of markets at the same time, it would have had to be able to make a considerable volume of nylon just so these many potential consumers would have sufficient samples to test out on their machinery. This in turn meant having to build relatively large and expensive semicommercial units, a risky approach that was scotched early on.

Given these strict parameters, Bolton and his team had every reason to be confident in success. Carothers' genius was one reason why. While he was no longer in the picture, his work lived on after him. Carothers eased the way for DuPont on two fronts: firstly, through his work, DuPont held a very strong patent position, since no other companies in the United States or elsewhere were developing any synthetic fiber resembling nylon, and secondly, Carothers' studies laid out how the critical polymerizations were to be carried out. His notes were in fact a detailed blueprint for how to go about creating this complex polymer. Together, Carothers and Bolton actually created the basic design for a pilot plant, the prototype for eventual nylon production. Carothers' laboratory was the site of a minipilot plant that could make 100 pounds of nylon per week. Bolton then expanded this laboratory setup into an actual 500 pound per week pilot facility, which in turn " ... was critical to getting [the] Seaford [nylon plant] up and running in record time."[17]

But between the laboratory work and the industrial plant, concerns arose within DuPont as to whether the company could commercialize such complex technology on a large scale. Nylon polymerization is, after all, a catalytic process that is vastly more complicated than the technologies DuPont traditionally worked, such as explosives manufacture and production of nitrocellulose synthetics. The success of the nylon enterprise from this point on depended less on bringing in knowledge of organic chemistry and much more on creatively applying highly advanced chemical engineering practice. The polymerization reactor itself, within which the nylon was formed, and the new and complex melt spinning process, which made the nylon fiber, both involved very intricate, complicated engineering problems. But the most urgent concern was that there was as yet no commercial method for making the key nylon intermediate called hexamethylene diamine (HDA). This problem, if not solved, could have very well been the deal breaker. If another more convenient intermediate had to be found, it would likely alter the properties of the fiber so that alternative polymerization processes and markets would have to be explored. The "one process–one market" strategy would then be shot, and DuPont would find itself in the position it was trying to avoid, leading to delays, rising costs, and unacceptable technical and market uncertainties.

Nylon and High-Pressure Synthesis   It is at this point that the Ammonia Department and its high-pressure technology proved its full value to the company. The impressive work that Roger Williams did with his high-pressure team in the Ammonia Department was well known throughout the company. It was generally acknowledged that this was where the most technically talented individuals worked and that the department could pretty much solve any problem that came its way. Bolton understood that close association with the Ammonia crew would help ease concerns of top management about the nylon project. Through his efforts, Bolton brought Ammonia close into the nylon camp, a smart move for two reasons: it compelled the company to look favorably on the nylon project and it actually increased the venture's chances for success. The Ammonia Department drove the nylon project on three fronts. In the first place, it was the source of DuPont's most skilled technical personnel. The Ammonia Department was where the company cut its teeth on the most sophisticated chemical engineering problems that could be encountered in an advanced materials company. As Hounshell and Smith remind us:

> Much of the early chemical engineering research at DuPont was done ... in connection with the development of high-pressure processes to produce ammonia and nitric acid in the mid-1920s. With success in those areas, the company would soon initiate more general chemical engineering studies.[15]

The lessons of chemical engineering applied to many products, not just those created by high-pressure technology. But high-pressure work, because it handed technical personnel the most difficult problems to solve that involved large-scale chemical production, formed a cadre of highly trained chemical engineers that were supremely confident in their ability to handle a wide range of difficulties. No one knew this more than the men responsible for managing the nylon project. This knowledge gave them ammunition with which to approach the Executive Committee with optimistic forecasts on the project's success.

The second way in which the Ammonia Department drove nylon was in finding a way to make a crucial input. The decision to commercialize nylon hinged upon a successful process to make HDA, and DuPont management had every confidence in the department's ability to come up with the technology to make it. The Ammonia Department did so, and fairly quickly. It applied sophisticated chemical engineering methods to devise a reaction system carried out in the vapor phase using a unique catalyst that converted adipic acid to the vital nylon intermediate, adiponitrile, which could then be converted into HDA by conventional technology.

The third means by which DuPont's chemical engineers eased nylon along was in showing to upper management that it would be relatively simple to construct profitable commercial plants that were smaller than initially planned. This meant less investment needed to be spent in equipment and machinery than originally thought necessary. So with the risks of failure—as perceived by management—reduced to a minimum, the company could begin to sink important money in building a commercial facility. The belief was that problems, such as those involving the polymerization and spinning operation, would continue to surface and that needed to be addressed but that could be solved in a methodical—and more or less predictable—fashion using the skills and expertise of chemical engineers, many of whom worked—or had previously been trained in—in high-pressure methods.

This confidence in the Ammonia Department–nylon partnership turned out to be well founded. Development and commercialization of nylon proceeded, more or less, as planned. The processes initially targeted were the ones scaled up commercially; the market aimed for—women's hosiery—quickly came into focus as nylon's most important (its use as a strategically critical material during World War II was an added bonus for the new polymer). DuPont's technical expertise, especially in high-pressure work, was a major factor in the rapidity with which nylon was commercialized.

Bolton was instrumental in making sure there was close, virtually seamless communication between Carothers' laboratory, the pilot plant facility, the Ammonia Department, and final commercial plant at Seaford, Delaware. The fact that the chemical engineers spoke the same language as the construction team and that DuPont was organizationally structured to facilitate quick and easy communication between these departments greatly helped the innovation process along. Whereas it was typical for a major new material to take a decade or more to move from discovery to a commercial product, nylon—the most important synthetic of the 20th century—went from laboratory curiosity to marketplace in the space of only 5 years. That initial Seaford facility employed 850 people and turned out four million pounds of nylon per year. By the mid-1990s, that plant, still in operation, had 1600 personnel and a capacity of 400 million pounds of Carothers' miracle fiber.[18]

The nylon project was a turning point in the history of DuPont. Its remarkable success gave DuPont executives great confidence—and more willingness—in undertaking new types of advanced materials. The revenue it brought into the company provided the funds for new research and development. Nylon's success piqued the interest of the textile industry in synthetics in general and most particularly in the new materials that DuPont might develop. Textile manufacturers were favorably disposed toward DuPont research. If not exactly a captive audience, they believed that DuPont was the one to watch in cutting-edge synthetic fibers and, as a result, were ready and willing to embrace the company's future innovations. As importantly, DuPont executives believed that the nylon project furnished them a road map for future success in synthetic innovation. The "nylon model" of innovation was simplicity itself: apply the scientific principles stemming from Carothers' research to create new fibers in the laboratory; then select one set of processes to develop and commercialize and one well-defined market to target; finally, depend on DuPont's expertise in advanced chemical engineering—stemming from its unrivaled superiority in high-pressure catalytic work—to solve the many problems that inevitably arise when scaling up a complex technology.

DuPont as a whole embraced the nylon model as the guide for future success. With this proven script in hand, top management believed there was little risk in moving ahead into new fiber technology. More so than was the case with nylon, DuPont's big plunge into the next wave of synthetic fibers—Orlon and Dacron—was instigated from above, for the Executive Committee now felt that DuPont's next nylon had to be just around the corner. For DuPont, success in exploiting these opportunities was as close to a sure thing one could hope for in this business. To the Executive Committee, this belief in their minds became a virtual certainty. It is for this reason that money and people came pouring into the Fibers Department in the years following World War II.

But the world of innovation is rarely so simple. DuPont let this great success go to their collective heads and greatly overestimated the power of economies of scope wielded by their much vaunted nylon process. The fact is DuPont was very lucky that things went so smoothly in the quest for nylon. If, for example, their initial target market did not work out as planned, the project would most likely have been in big trouble. A new objective could very well have demanded a different type of nylon with other sorts of properties, which then would require different intermediates. But it is not at all likely that the Ammonia Department could have easily or cheaply come up with another process to meet this new demand. As discussed, it became progressively more difficult to tease new and important molecules out of the increasingly intractable technology. And nylon intermediates are extremely complex. That the high-pressure engineers came up with its HDA process so quickly is impressive but in the same sense as a hole in one: amazing but not likely repeated in the near future. To exacerbate this situation, the reactor system used to create the bulk nylon polymer and the melt spinning technology that takes the polymer and spins it into fiber—both of which are very sensitive to any changes in the chemical composition of the nylon—would have had to be revamped as well.

The bottom line is that DuPont's nylon process is very inelastic and because of this rigidity, shifting market possibilities would have increased costs and caused extensive delays—that is, increased perceived risks—to the point that funds would most likely have been cut off and the project terminated. This is exactly what happened when DuPont attempted to project its "nylon model" onto other fiber projects. It really did not apply to other new synthetics, and the costs for mistakenly assuming so were high, in fact much higher than DuPont executives, besotted with the nylon triumph, had imagined. When they realized that the path would not be as smooth as it was with nylon—that the nylon model could not be more generally applied—they became quite sheepish toward new polymer development and sharply pulled the company back from further pioneering adventures.[19] But before this realization took hold, DuPont did have one last hurrah with breakthrough fibers as it pushed ahead with pursuing nylon's children soon to be known to the world as Orlon and Dacron.

3.3 NYLON'S CHILDREN: ORLON AND DACRON

The discovery of nylon greatly excited the world. It was a big hit at the 1939 World's Fair in New York and immediately became a big seller for DuPont. The major chemical companies started to eagerly search for the next big synthetic fiber. While companies like Union Carbide and American Viscose found their own synthetics—such as the Vinyon polymers—DuPont once again took the lead in the industry with the commercialization of the next two most important synthetic fibers: Orlon acrylic fiber and Dacron polyester fiber. In fact, the Rayon Department—renamed Textile Fibers Department—took an even more important role in the Orlon and Dacron projects than it did in nylon. A new champion emerged for these technologies from this department. More than any single person at DuPont, W. Hale Charch, Director of the Textile Fibers' Pioneering Research Laboratory, was most responsible for their eventual success. Charch, another of DuPont's Ph.D. chemists, believed nearly as deeply in the importance of scientific research as Stine conceived it. But at the same time, he was very practical in his thinking and often encouraged his chemists to use empirical methods in finding new materials rather than in being bogged down in speculative theory. Like both Stine and Bolton, he had worked at DuPont for many years before entering into synthetic fiber work. Like them as well, he was deeply respected throughout his department and the company for his prior accomplishments, notably his earlier invention and development of one of DuPont's major products, cellophane, in the mid-1920s. Charch turned out to be one of the most aggressive and committed champions of new materials that we will encounter in these case histories. He approached this job with something approaching religious fervor. It is a good thing he did or else DuPont might never have succeeded in finding the next "big thing" in synthetic polymers.

3.3.1 Orlon

Orlon came to light from research conducted within DuPont's Textile Fibers Department in the early 1940s.[20] However, it was only after World War II that the department was able to capitalize on this early discovery. Orlon would become DuPont's second major synthetic fiber.

3.3.1.1 Orlon: Research Phase   Orlon was somewhat of an accidental discovery. Its inventor, DuPont researcher Ray Houtz, was actually working on a project to improve the strength and resilience of one of DuPont's existing products, rayon. At the urging of his boss, W. Hale Charch, Houtz attempted to combine cellulose, a basic component of rayon, with a synthetic material, vinyl polymer. This resulted in a new type of synthetic, a polyacrylonitrile (PAN), that showed good chemical stability and that could be spun into a fiber. Houtz however became intrigued with the possibility of taking his research and discovering an entirely new type of polymer material. His attention then shifted away from cellulose and he began looking seriously at the possibility of polyacrylonitrile alone being made into a fiber. Charch's enthusiasm for the discovery and the fact that DuPont's corporate management was very intrigued in pursuing new and important fibers along the lines of nylon meant that the Executive Committee immediately began showing an interest in the new material.

One important difficulty with polyacrylonitrile was that, unlike nylon, it could not be spun into fiber simply by melting it and then drawing fibers from the melt. The reason was that it was more unstable than nylon at the higher temperatures needed for melting. A solvent had to be found that could dissolve polyacrylonitrile, forming a solution at lower temperatures. An understanding of the molecular structure of the polymer and cut-and-dried empirical testing of possible solvents led Houtz to hit upon the right liquid: dimethylformamide (DMF). Houtz then employed well-known techniques in cellulose fiber production to spin polyacrylonitrile fibers from the PAN–DMF solution.

At this point, the new fiber was poised to enter into the development and commercialization phases. But problems immediately arose that did not come into play with nylon. These obstacles threatened to kill the project even before the new material could get out of the laboratory. In the first place, the PAN fiber, unlike nylon, did not have a clearly defined market. This was because it could be spun as either a wool-type or silklike fiber, each of which had different market applications. In addition, the polymer showed superior resistance to light, chemicals, and bacteria, which opened up additional market opportunities. Since it was impossible to tell at this point which market held the most potential, PAN fibers would have to be supplied to customers in different markets for end-use testing. But this meant having to produce much larger quantities of polymer than was the case with nylon. Unlike nylon, where Carothers' laboratory setup served as a prototype that could be expanded relatively quickly and cheaply into a small pilot plant capable of producing enough polymer for initial market testing, Houtz's new fiber demanded a much larger pilot facility very different in design and operation from his much smaller laboratory apparatus. Not only did Charch's Pioneering Research Laboratory not have the funds to build a pilot plant large enough to turn out sufficient quantities of the PAN material, designing and building one would take time and manpower away from other more immediately profitable projects. Here we find then that market risks and the risks involved in securing a suitable prototype were closely linked. And this was not all. Charch faced two other issues that made DuPont management nervous. One was that PAN fiber could not be dyed given the existing technology. The potential markets for PAN fiber, however, did demand that the fiber have color and thus be dyeable. Another concern was that Carbide and Carbon Chemicals Corp. (Union Carbide) had already come out with its own polyacrylonitrile fiber, Vinyon N, which was priced lower than any PAN-based polymer DuPont produced could ever hope to be.

These three problems, and particularly the market-prototype issue, soured the top managers of the Textile Fibers Department—Leonard A. Yerkes and Ernest Benger—to doing further development work on PAN fiber. However, Charch and his technical staff in the department's Pioneering Research Laboratory believed in the polymer and did what they could to convince their supervisors to continue with the project. Charch directed whatever discretionary funds he could find toward building the pilot plant. He told his superiors that a pilot plant could be built fairly easily because it didn't have to be commercially viable, only good enough to turn out sufficient quantities of the new fiber for market testing. His goal of course was to ease the fears of top management and keep a major fiber project on track. So, he didn't worry about whether the prototype plant operated efficiently, and he was able to cannibalize parts and apparatus from Houtz' laboratory whenever possible. In the end, while not pretty from an engineering standpoint, the makeshift facility did the job; it produced enough PAN fiber—both the wool and silk varieties—so that the Textile Department's Development Section could do its own end-use evaluations and so get a start on market evaluation. Charch dealt with the dyeability problem by selling the government on the advantages of the new fiber's resistance to sunlight, chemicals, and bacteria just at the time the United States entered into World War II. The military purchased significant quantities of the material to make tents, tarpaulin shoelaces, and other supplies needed by soldiers fighting in the Pacific jungles and where the dyeability of the fiber was of no consequence. Finally, Charch addressed the perceived threat of Vinyon N by convincing Yerkes and Benger that it presented no risk to DuPont's fiber. He argued that DuPont's polymer had more potential uses than did Vinyon, especially in the woolen markets. He also noted that Vinyon, because it melted and decomposed at a much lower temperature than DuPont's material, would not be able to match DuPont's PAN fibers in terms of processability and product quality.

3.3.1.2 Orlon: Development Phase   These efforts on the part of Charch saved the project, at least temporarily. Initial market responses were encouraging and the fact that the new material could generate an income from military sales also helped Charch's cause. The development of PAN fiber proceeded slowly but steadily over the next few years so that, by 1944, Charch and his team could produce marginally more polyacrylonitrile. But the laboratory had difficulty generating any more PAN fiber that could be sent to nonmilitary customers so that possible postwar markets could be tested and prioritized according to their future potential. The question of building a pilot or semicommercial plant arose once again so did the dyeability problem as the military cut back its purchases of the new fiber, and appealing to commercial markets once again had to be dealt with.

By the spring of 1946, the project reached a critical stage. The continued commercialization of PAN fiber was now in serious question. The development of the new fiber seemed to have gone awry as far as the Executive Committee was concerned because it had not adhered closely enough to the nylon model, the script for successful polymer innovation. No one single market had been able to be identified as *the* most appropriate for the new synthetic (nylon had women's hosiery); and no one single best process route could be identified, as it could for nylon, a result of the fact that the patchwork pilot facility build by Charch before the war—and meant to only work well enough to turn out the minimum amount of polymer needed to sell to test markets—did not point to a particularly superior commercial way to make the new material. To make matters worse, engineers wanted to make whatever route was chosen in the end a fully continuous process. This greatly increased the costs of scale-up and, even worse, veered into uncharted and therefore risky technical territory—nylon never had to go through such hoops.

The time had come to either get the project back on track—the nylon track—or cut it loose. Upper management in the Textile Fibers Department pursued the former course by appointing G. Preston Hoff as special coordinator—and de facto leader—of the PAN fiber development effort. We first met George Hoff when he came aboard the nylon project a decade earlier as assistant to General Manager Leonard Yerkes and as liaison between the Rayon and Chemical Departments. With the entrance of Hoff, general management could breathe easier. Being a force in nylon research and the man behind the triumphant commercialization of nylon fiber, Hoff was just the person who could bring the nylon model to the rescue of a troubled PAN fiber project. As an effective, no-nonsense leader, Hoff fiercely believed in the nylon approach and he did not waste any time to put his plan into action. He directed that the market with the greatest potential be explored. He also took a lesson from nylon by concentrating on a single workable process. Whether it embodied the best process technology possible was less important than it satisfied the minimum satisfactory conditions for commercial production. Improvements and fine-tuning could come later; initially it was vitally important to get the plant to turn out acceptable volumes of fiber at reasonable cost. An important reason for this was, quite simply, a very high probability that DuPont's chemical engineers—familiar with such complex technology as high-pressure synthesis and very successful in translating this knowledge into the famous Seaford nylon plant—could reach this minimum level quickly; and it was very important to show DuPont's top management that the process was doable on the larger than pilot plant scale. It was for these reasons that Hoff decided to discontinue work on a fully continuous process for the new fiber. Such embellishments to the basic process would seriously slow down—and increase the costs of—development, and for what? The Executive Committee wasn't demanding it and nylon's great success a few years earlier never had to depend on it. If the nylon didn't need it, Hoff reasoned, Orlon—as the new fiber came to be called—could certainly do without it as well.

The strategy of forcing the Orlon project back onto the nylon track appeared at first to be a sound one. The textile industry began to show real interest in the fiber and the process technology continued to advance. The Executive Committee, no doubt feeling great relief, continued to fund the project. In the spring of 1946, it authorized money for the design and construction of a pilot plant to be followed by a full-scale commercial unit. But technical problems continued to plague development. The most critical of these was the troublesome dye problem. While DuPont explored different possible avenues, the solution remained elusive. But, once again, the example of the nylon experience managed to calm management's nerves. The Fibers Department told the Executive Committee that a solution was just around the corner. After all, nylon also had a major difficulty—lack of a process to size the fiber—when its Seaford plant was approved for construction and that DuPont's chemical engineers came up with the answer in time for the facility to open on schedule. There is every reason to believe, they argued, that the same would happen with Orlon and its dye problem. This optimism along with the fact that DuPont customers in the textile industry were lining up to buy the new fiber convinced the Committee to push ahead with the commercial plant. But mounting process difficulties and quality problems over the next few years began driving customers away, and by 1951, DuPont management was on the verge once again of abandoning Orlon fiber.

3.3.1.3 Orlon: Scale-Up and Commercialization   DuPont would surely have given up on Orlon altogether if it had not been for the determination and foresight of its great champion. Because Charch took it upon himself to have the Fiber Department's Pioneering Research Laboratory investigate the physical characteristics and market potential of synthetic fibers as wool substitutes, he offered the project a lifeline—rather than Orlon as a single fiber (which DuPont called Type 41 Orlon), DuPont would explore it as a staple material (Type 42 Orlon). With one failed Orlon plant in the books and anxious customers becoming rapidly disillusioned with the prospect of acrylic fibers—and with the reputation of the company's fiber program as a whole on the line—DuPont strongly felt it had to get a commercial product out quickly. This meant building a large plant capable of turning out 30 million pounds per year of Orlon 42 staple. The building of the plant itself went relatively smoothly. As previously mentioned, close communication between the chemical engineers who worked on the pilot plants and the engineering team responsible for plant construction hastened the process. But, with the plant on line, DuPont had this large facility capable of turning out a product that DuPont didn't know very much about. The company was now was clearly entering into uncharted territory, one that was very different from the more familiar one inhabited by nylon. In deciding on this change so late in the game, " ... DuPont would be making a highly unorthodox move in that not a single test of the Type 42 staple had been carried out in a customer's mill. For DuPont, this was a drastic move. But [DuPont] was in a position that demanded such action."[21] But an unexpected development saved Orlon: the explosion in the demand for knit ware, particularly bulky sweaters, and then woven fabrics created a rapidly growing market for the fiber. By the end of the 1950s, Orlon sales reached nearly 100 million pounds per year. DuPont narrowly escaped the abyss and had succeeded in its second major synthetic material.

3.3.2 Dacron

The story of Dacron closely parallels that of Orlon. DuPont developed, scaled, and commercialized both materials more or less simultaneously in the late 1940s and early 1950s. There was even technical spillover so that experiences from one helped solve problems encountered in the other. For example, finding ways to build resiliency into Dacron applied to Orlon as well. There is another similarity in the innovation of these two materials: while both materials were very different from nylon in their chemical and physical properties, top managers of the Textile Fibers Department insisted on overlaying the "nylon model" onto them. They did so because this model defined the boundaries of their comfort zone when pushing forward in a completely new technology. It was terra firma and held out the hope of a sure, swift, and successful outcome. To the true believers, it meant that money spent on experiments and production plants would not be wasted, customers expecting the next great fiber from DuPont would not be disappointed and (even worse) alienated, and (maybe most importantly from a personal standpoint) managers who put their careers on the line for the fiber would not be held back from promotion. Adhering to the nylon model was also an internal strategy that helped win over the Executive Committee—who also saw it as a virtually risk-free avenue—to continue approving (and funding) the project. In both cases as well, reality reared its ugly head more than once revealing the inconvenient truth that these new materials were not just like nylon, that pursuing them would not be easy, and that sticking with their development pulled DuPont deeper into the unchartered terrain of radical innovation where risk and uncertainty are the lingua franca of the land. In both cases continually reverting back to the nylon paradigm following these periods of crisis helped to maintain support from the top. However, the often thankless effort of the champion for both fibers—W. Hale Charch—to throw the cold water of reality in the faces of his superiors by pushing for approaches to development and commercialization that took into account the important differences between Orlon and Dacron on the one hand and nylon on the other—by offering an alternative way to go just when there appeared to be none—was no less important in keeping these new fibers alive when they could have easily faced a swift and ignominious death.

3.3.2.1 *Dacron: Research Phase*   Dacron is a polymer based on polyethylene (more specifically it is what chemists call a polyethylene terephthalate or PET). Like nylon (a polyamide) and Orlon (a polyacrylonitrile), Dacron fiber originated with the polymerization work of Carothers. In 1934, a member of Carothers' group at Purity Hall, Edgar Spanagel, prepared laboratory amounts of PET polymer as part

of his scientific work into new fibers. Spanagel attempted to generate interest in his discovery but work on nylon took precedence. Besides, chemists did not think that polyesters had a future as commercial materials. By 1944, DuPont showed renewed interest in PET, in part because it had learned of the British firm ICI's work on its version of PET, which it called Terylene. The news particularly inspired Charch and one of his research chemists at Pioneering Research, Emmette Izard. With Charch's encouragement, Izard continued the work of Spanagel on PET and discovered that the chemical would indeed make a promising polymer. By 1946, using processes similarly described by Carothers in his work 16 years earlier, Izard and his team came out with a polyester fiber with superior properties, particularly an acceptable viscosity and high melting point. This meant that polyester fiber, like nylon, could be melt spun by which fibers are drawn from molten material, rather than from a solution. This simplified the process immensely since a separate solvent did not have to be found. It also meant, to the relief of senior executives, that Dacron, in the laboratory at least, was not so different from their poster child for synthetic fiber success, nylon. The perceived risk of building a pilot plant was considerably lower than it was for Orlon as a large Dacron pilot plant could simply be made in the existing nylon plant at Seaford thus reducing considerably the costs and time involved.

3.3.2.2 Dacron: Development The polyester fiber, demonstrating impressive and commercially promising properties and showing even closer kinship to nylon than Orlon, impressed both the top managers of the Textile Fibers Department and the Executive Committee. The deal was sealed when, on Charch's recommendation, the development work was transferred to the nylon research laboratory, once again, under G. Preston Hoff. This meant that DuPont's "go-to guy" for nylon, Hoff, would be applying the nylon model to both polyacrylonitrile (Orlon) and polyester fiber development. Hoff and his colleagues once again concentrated on a single market for polyethylene fiber, in this case, tire cord. But it didn't take long before problems arose in the performance of the fiber in tires under test conditions thus scotched this market possibility, at least for the immediate future. As with Orlon, the one-market model that worked so well for nylon appeared unsuitable to polyester fiber.

But Charch and his team at Pioneering had continued to conduct research on the fiber, much as they had for Orlon, and they came out with the same results: the new polyester fiber—as with Orlon—would make a superior wool substitute in clothing and textiles. In fact, Pioneering was able to synthesize a new type of polyester polymer with excellent wool-like properties. But now a decision had to be reached, ultimately by the Executive Committee: whether to move ahead with developing and scaling up the process for making this "wool-like" polyester. One of the most important considerations in the minds of the members of the Executive Committee was whether this new fiber would compete against its two other synthetic fibers, nylon and Orlon. The Committee was not easily satisfied on this issue. It rejected the first report the Textile Fibers Department gave it on the subject demanding more details and elaboration of the data. It had to be absolutely certain in its collective mind on this score before moving forward.

A second concern involved time and money. The nylon project succeeded so well because the material found its market quickly and without incessant delays and escalating costs. With the difficulties that Orlon had in meeting this criteria possibly fresh in their minds, Committee members demanded that this new fiber better stick more closely to the nylon scheme or else. The Committee wanted the Textile Fibers Department to find a way to produce enough polyester fiber to send to different customers in the textile industry for testing and evaluation. It wanted the department, in fact, to skip entirely the pilot plant stage and move directly into semicommercial production of the fiber—and it wanted this done as soon and expeditiously as possible and in a way that would allow the plant to be scaled up to a commercial unit in a short period of time and with minimum expense. Committee members were no doubt relieved to learn that this new fiber could be polymerized, spun, and drawn on slightly modified process equipment used to make nylon. This discovery fit very nicely within their comfort zone. DuPont then built a semicommercial plant at its Seaford, Delaware, nylon works. Doing so saved considerable time and money and produced enough fiber for market testing and development.

3.3.2.3 *Dacron: Scale-Up and Commercialization*   Success here prompted executives at the department and corporate level to push ahead with a full-scale plant, basically along the same lines as the Seaford works only larger, to be located in North Carolina. Just as occurred at a critical point with nylon and Orlon, DuPont was now at a crossroads, a "yes" or "no" decision on whether to spend considerable resources on a commercial unit. While market tests for the new fiber, called Dacron, were favorable and while the new fiber could be produced using some aspects of the familiar and vaunted nylon technology, risks existed at this late date. As with Orlon, DuPont chemists had not yet solved Dacron's resistance to dyeing. There was also the question of whether the production process was to be continuous or batch. Nylon didn't have to be processed continuously to succeed as a fiber. For Dacron, however, continuous processing greatly improved the properties of the polymer and so clearly had to be part of the production plant design if it was to become a major synthetic material.

Because the Executive Committee felt time was of the essence, it had approved the commercial plant even before these two major problems—dyeing and continuous processing—had been resolved. One could argue that top executives were finally willing to take a great risk—a leap of faith—that all would turn out well in time for the opening of the new facility. While this is certainly true as far as it goes, it should be noted that, as with Orlon, this risk taking occurred very late in the innovation process. As the case with Orlon, there were good reasons to take the leap now, and these echo those that compelled DuPont to push ahead with the Orlon staple plant. For one thing, a great deal of time and company resources had already been spent on Dacron. The careers of managers and executives were on the line, some of whom had friends and allies on the Committee. By this time, customers in the textile industry had seen, tested, and evaluated large quantities of the new material. Dacron was no longer a secret, internal project. DuPont's reputation in the wider world was once again at risk. If it did not follow through on Dacron, the company left itself exposed to the whispers of the marketplace and possible alienation of important clients and its standing as a major innovator in synthetics. And then, of course, top management could turn to the nylon experience. As already noted in our story on Orlon, significant problems had still not been solved even as the first nylon plant started to go up. But, through its engineering genius honed by years in high-pressure technology, DuPont took care of these issues just in time for inclusion in the first production unit. This same argument pushed through the Orlon plant and the Textile Department and, Executive Committee had every reason to believe lightning would strike once again.

In fact, this turned out to be the case. Whether taking wild risk or making rational decision under pressure, the resolve to pursue commercialization paid off. Not only did DuPont's technical group, now under increased pressure to succeed, find ways to improve dyeing and to design continuous production into the Dacron fiber process, it also addressed and solved other problems, notably "pilling" ("the development of little snarled balls of staple on the surface of woven fabrics") that came to light as production ramped up. After a shaky start, by the late 1950s, DuPont's Dacron, in the form of staple and blends, had become the company's third most important and enduring synthetic fiber.

REFERENCES

1. A number of technical books describe the nature of high-pressure reactors. See, for example, Holland, C. (1979), *Fundamentals of Chemical Reaction Engineering*, New York, New York: Prentice Hall. Articles in engineering journals also trace the history of high-pressure technology. See, for example, May 1930, "Pressure-Synthesis Operations of the DuPont Ammonia Corporation" *Industrial and Engineering Chemistry*, 22(5): 433–437.

2. See, for example, Krase, N. (1930), "High Pressure, High Temperature Technology—The Chemical Engineer's Newest and Most Effective Tool" *Chemical and Metallurgical Engineering*, 37: 532–539. Further evidence comes from an interview of one of DuPont's most important engineers and executives Crawford Greenewalt; a transcript is found at DuPont's Hagley Library that recounts the difficulties in carrying out high-pressure R&D (Hagley Library, Greenwalt Interview, November 8, 1982, pp. 18–19).

3. Hounshell, D. and Smith, J. (1988), *Science and Corporate Strategy: DuPont R&D, 1902–1980*, Cambridge, UK: Cambridge University Press, pp. 11–146. See also Ndiaye, P. (2007), *Nylon and Bombs: DuPont and the March of Modern America*, Baltimore, Maryland: The Johns Hopkins University Press, pp. 5–70.

4. Hounshell and Smith (1988), pp. 183–189; 275–285. See also 1930, "Pressure Synthesis Operations of the DuPont Ammonia Corporation," pp. 433–437.

5. Hounshell and Smith (1988), pp. 183–186.

6. Ibid., pp. 186–188.

7. Ibid., p. 187.

8. This narrative of Roger Williams and DuPont's high-pressure work comes from a variety of sources including Ndiaye, pp. 78–90, and Hounshell and Smith, pp. 183–189. The DuPont archives at the Hagley Library also have material that was used in this section. Particularly important are transcripts of interviews of the Ammonia Department manager John C. Woodhouse conducted in 1982.

9. Ndiaye (2007), p. 78.

10. Hounshell and Smith (1988), p. 188.

11. Hounshell and Smith (1988), pp. 188–189.

12. John C. Woodhouse recalls, for example, the difficulties the Ammonia Department had with developing such important new materials as polyethylene and the methacrylates (Hagley Library, DuPont Records, Woodhouse Interview, October 14, 1982, pp. 45–46).

13. The following discussion on Stine's career is taken from Hounshell and Smith (1988), pp. 223–248.

14. This description of Carothers' background and the competing theories of polymerization is found in Hounshell and Smith (1988), pp. 228–233 and Furukawa, Y. (1998), *Inventing Polymer Science: Staudinger, Carothers, and the Emergence of Macromolecular Chemistry*, Philadelphia, Pennsylvania: University of Pennsylvania Press.

15. Hounshell and Smith (1988), p. 276.

16. Much of the following history of nylon research, development, scale-up, and commercialization can be found in Hounshell and Smith (1988), pp. 236–248; 257–273 and Ndiaye (2007), pp. 71–105.

17. McAllister, J. (1995), *The First Nylon Plant*, Washington, DC: American Chemical Society, Division of the History of Chemistry and the Office of Public Outreach, p. 2.

18. McAllister, J. (1995), *The First Nylon Plant*, Washington, DC: American Chemical Society Division of the History of Chemistry and the Office of Public Outreach.

19. John C. Woodhouse discusses DuPont's growing conservatism toward polymer innovation because of rising costs in his 1982 interview (Hagley Library, DuPont Records, Woodhouse Interview, October 14, 1982, pp. 44–47). See also Hounshell and Smith (1988), pp. 372–383; 423–444.

20. The following discussion on Orlon and Dacron is based on Hounshell, D. and Smith, J. (1988), *Science and Corporate Strategy: DuPont R&D, 1902–1980*, Cambridge, UK: Cambridge University Press, pp. 384–420.

21. Hounshell and Smith (1980), p. 406. The risks in scaling up the Orlon plant involved increasing expenses of building chemical plants. The rising costs of plant construction can be observed from documents on the history of DuPont's engineering department at the Hagley Library (see DuPont Records, Series II, Part 2, Accession 1274, Box 111).

4

LOW-TEMPERATURE (INTERFACIAL) POLYMERIZATION

Dupont's Specialty Fibers Versus General Electric's Polycarbonate Revolution

I kept cooking it … Eventually I raised the temperature … The stirrer kept stalling, and I kept twisting it. Finally it wouldn't stir anymore. ..When I [pulled] the thermometer out, I [pulled] a long fiber out with it. Don Sargent [then] came in. He … knew something about polymers. He was kind of impressed with the fiber …

Daniel W. Fox, 1986

In retrospect, the difficulties experienced with Orlon and Dacron had shown how risky these endeavors really were. At the end of the day, DuPont got their fibers but also saw that the bill for this success was very high—much higher and more onerous than it had been for nylon. Even more importantly, these two fibers had taken DuPont to the precipice, an experience DuPont had no desire to repeat. By the late 1950s, DuPont executives were asking themselves if this was the way of the future: longer periods for development and commercialization of new fibers resulting in escalating costs and constant delays in reaching markets with ultimate and humiliating failure a very possible outcome. By this time, intensification in competition for synthetic fibers from other chemical companies and growing evidence that a major new fiber along the lines of nylon, Orlon, and Dacron was not likely to emerge from any company anytime soon further eroded the confidence of DuPont in the technologies it had been using. With further innovation looking more and more risky, DuPont management

Advanced Materials Innovation: Managing Global Technology in the 21st century, First Edition. Sanford L. Moskowitz.
© 2016 John Wiley & Sons, Inc. Published 2016 by John Wiley & Sons, Inc.

in the Textile Fibers Department and on the Executive Committee felt increasingly uncomfortable about further pursuing new mass market synthetic fibers.

A new process called interfacial—or low-temperature—polymerization soon came along that offered DuPont a chance to create new products without the threat of the risks that plagued Orlon and Dacron. While the company succeeded in exploiting this technology in designing new polymers, these synthetics never achieved breakthrough status. DuPont executives consoled themselves with the belief that this lesser type of innovation was now the new normal for the chemical industry. But in fact, DuPont completely missed out on a new polymer revolution—the polycarbonates—that relied on the very process that the company itself had invented. As it was, another large, established company but one that had virtually no experience in advanced materials ended up trumping the great DuPont on its own turf.

4.1 INTRODUCTION AND BACKGROUND

Since the 1960s, interfacial polymerization has become an important process in the world of advanced polymer materials.[1] It is a highly flexible process in that it can make a wide variety of specialized polymers exhibiting particular properties— thermal resistance or high strength or plasticity—or it can synthesize large volume polymers for general engineering application needed in the transportation and other industries. The technology is used to make polymers with superior thermal properties for a variety of applications from aerospace engineering to tire cords for cars and cables for anchoring oceanic oil-drilling platforms. It has produced new generation of engineering plastics capable of replacing metal in many structural applications. It is important in biomedical technology, such as in the making of biodegradable sutures and artificial limbs and organs, and in the synthesizing of proteins and nucleic acids and encapsulation technologies that allow timed release of drugs in the body. Researchers are also exploring interfacial polymerization as a means of synthesizing nanomaterials, particularly nanofibers and nanocomposite thin films.

The rise of interfacial polymerization is best understood within the context of the history of polymerization in general. Polymerization has been known as a laboratory process since the early 19th century. Chemists used it to synthesize both familiar chemicals and new substances. The first truly synthetic polymer—that is, made totally in the laboratory without the use of naturally occurring materials—that found commercial use was invented by Leo Baekeland in the early 1900s, which he named Bakelite. A phenol-formaldehyde resin, Bakelite found wide use as an insulating material for the growing electrical industry. Baekeland's use of elevated pressures and temperatures to carry out the polymerization process became standard practice in the industry. Other polymers surfaced around the same time for specialty applications including early polyesters (used in paints) and polybutadiene rubber.

Despite these advances, the science of polymerization still remained somewhat of a mystery well into the 1930s. Certainly the work of the German chemist August Kekule in the 1860s began to shed light on the molecular structures of organic compounds, but could not tell scientists what exactly polymers were and how they were formed. Empirical, hunt-and-peck experimentation outpaced general scientific understanding. By the 1920s, the prominent theory of polymer structure was that polymers were essentially composed of many small, discrete molecules held together by a mysterious secondary force. At this time, as noted in the previous chapter, the German organic chemist Hermann Staudinger attacked this model and, in its place, posited a new and revolutionary concept: that polymers, rather than being aggregates of many smaller molecules, are, in reality, very long individual molecules, or macromolecules. While Staudinger's work was a giant leap forward in understanding polymers, it was up to Wallace Carothers to prove once and for all the macromolecular theory of polymers through his work on the synthesis of nylon.

Over the first half of the 20th century, polymerization technology continued to evolve along with polymer science. Union Carbide, an early innovator in industrial polymers, developed plastics made of polyvinyl chloride (PVC), and Dow was the first to commercialize polystyrene. But it was DuPont that stood at the center of the polymer materials revolution during the 1930s and 1940s. Prior to his work on nylon, Carothers and his team at DuPont's Purity Hall succeeded in synthesizing the first important synthetic rubber, neoprene. Hounshell and Smith tell us that at first the process was simply a scaled-up version of the laboratory equipment: "Neoprene was produced by pouring liquid chloroprene into open pans and later into five-gallon, wide-mouth bottles. It was allowed to sit until it had polymerized."[2] But this approach, called bulk polymerization, had its problems: it was very slow, resulted in low polymer yield, and produced variable quality material from one batch to the next. DuPont's chemists soon came out with a variant of a German technology that they called "emulsion" polymerization by which liquid chloroprene was poured into water where it polymerized into neoprene rubber. This new technique greatly improved the performance of bulk polymerization on all fronts: the speed of the reaction increased; yields grew by 50%, and, because of the greater control of the polymerization process, the final product was more stable and of a higher quality.

One of the difficulties of the polymerization reaction—whether performed in bulk or in emulsion—was that it generally had to be carried out in a closed autoclave using a catalyst and applying elevated temperatures and pressures. In such a system, a complex relationship exists between process (defined by intermediates used and temperatures, pressures and catalysts applied), product characteristics, and market applications. If the target market shifts, then costly and time-consuming development work has to be undertaken to first find alternative catalysts, and then look for appropriate temperatures and pressures under which the catalyst can operate. Then too, different temperatures force changes in pressures and vice versa. It takes time and money to find the proper balance between the operating variables and embody this new balanced system into pilot, semicommercial, and then commercial plants. To put it another way, such technology is "sticky"—it does not adjust easily to market changes. Nylon, Orlon, and Dacron, all of which required either bulk or emulsion polymerization technology, faced this difficulty, as we have seen.

By the 1940s, polymer chemists began asking whether the polymerization process could be simplified so that, for example, it didn't require the use of elevated temperatures and pressures. Under such less rigorous conditions, it could respond to shifting markets much more rapidly and efficiently. A new, radically different, and simpler type of polymerization process would emerge in the years following World War II that allowed the synthesis of certain polymers without the need for applying or controlling high heats or pressure. The process, called interfacial polymerization, was quite revolutionary in concept. It involves forming two distinct layers of liquid, much like what happens when oil is brought into contact with water, and having the polymerization occur at room temperature at the interface separating the layers. The technique produces high molecular weight polymers at room temperature and moderate pressures, the first time such materials could be made under such mild conditions. This ability meant an unprecedented gain in production efficiencies and product quality as well as large savings in equipment, time, and energy.

Both DuPont and General Electric (GE) discovered the process more or less at the same time, although independently. The discovery "opened up a whole new world of polymers to researchers . . . ."[3] While DuPont made good use of the process in its synthesis of two important but relatively minor specialty fibers, spandex and Kevlar, GE—a relative newcomer in the polymer field—much more successfully exploited the process to introduce a true revolution in man-made plastics, the polycarbonates. How this new kid on the polymer block managed to beat the great DuPont at its own game is the subject of the remainder of this chapter.

4.2 DUPONT AND SPECIALTY FIBERS

DuPont's three revolutionary synthetic fibers were extremely important to the company and more particularly the company's Textile Fibers Department.[4] By 1960, accounting for approximately 40% of DuPont's total revenues, it was the largest, most successful department in the company. The burning question facing the department was, what should they do next to follow up on these successes? Thanks to Carothers and the work that followed him, DuPont had gained an unprecedented knowledge of the molecular structure of fibers and how these structures influenced their physical properties. Why not then use this knowledge and expertise to design a number of high-performance fibers based on Carothers' polymerization studies for particular, well-defined market applications? One might demonstrate superior elasticity, another heat resistance, a third exceptional electrical properties, a fourth ability to absorb liquids, and so forth. Also, because DuPont would assign each fiber to a particular market segment, failure to develop and commercialize any one polymer would not have as great a negative impact on the company in terms of anticipated revenue or reputation in the industry as would have been the case if either nylon, Orlon, or Dacron had flopped. And such specialized fibers, even though they each would serve limited markets, could prove quite profitable since they were expected to have significantly higher value per pound than did DuPont's three major fibers. As the case with DuPont's earlier fibers, this search for high-value specialized fibers accorded well with the company's diversification strategy. The goal was to find a process that could produce such focused polymers without the problems that plagued Orlon and Dacron.

4.2.1 Lycra Spandex and the Block Copolymers

Hale Charch once again took the lead in DuPont's foray into these new polymers. His firm belief that other "nylons" were just around the corner and that DuPont enjoyed an enviable advantage to lead this new revolution fueled his optimism. As he wrote as early as 1941:

> It can be stated with almost complete certainty … that polyamide polymers are not the last super-polymers which the world is going to see in the form of threads and fibers.[5]

Charch saw interfacial polymerization as the magic bullet to a growing family of new, powerful polymers without having to face the uncertainties of past innovations. But first, the process itself had to be perfected. The major battlefield in the evolution of this technology was the research phase. Here is where the chief, potentially show-stopping problems surfaced. An important reason why DuPont succeeded here was vital scientific information obtained from outside the company. DuPont's lead scientist in this effort was Emerson Wittbecker. He tapped the knowledge gained about the process by German chemists who worked on interfacial polymerization prior to World War II and was contained in postwar Allied reports on the state of German chemical research. At Charch's instigation, Wittbecker took German science and scaled the process up to show how it could operate on a pilot basis. As Hounshell and Smith describe it, "the Interfacial process was a general one that applied to many types of polymerizations. [The process] could carry them out at room temps thus saving time and energy."[6]

Charch immediately saw this as an important discovery that could lead to just what he was looking for: precisely engineered structures for specialized purposes using a highly elastic process that worked at less rigorous operating conditions. The question now was which polymers to make.

Charch and his chemists stuck close to what they knew and tried to squeeze new materials out of Carothers' legacy in polycondensation technology. Their greatest success in this endeavor was in developing the first major man-made elastic fiber, which DuPont called Lycra spandex. First commercialized in 1962, Lycra eventually became one of the most profitable of all DuPont's creations in terms of returns on investment, although it always has been a specialized fiber in terms of volume. Even so, it took nearly 20 years from initial research to first commercialization.

Lycra was invented at DuPont by Joseph Shivers, who had a Ph.D. in organic chemistry from Duke University. Soon after joining DuPont in the late 1940s, he began work to make a new type of elastic fiber as a replacement for rubber in garments. DuPont management did not initially support the project and ordered it shelved. Shivers however continued to work on the new fiber in his spare time using whatever equipment he could bootleg. By the 1950s, he had made significant progress. By modifying the molecular structure of Dacron, he created a highly flexible fiber that could withstand heat. At this point, Charch began working closely with Shivers and his team to perfect and develop the new fiber. A remaining problem was learning how to construct a molecule that would have the proper elastic properties—not only stretching but also returning to original shape—that would not degrade over time. This involved engineering a rather complex molecule, composed of alternating "hard" and "soft" segments. It took quite a bit of time before Church and Shivers figured out what these elastic molecules should be made of. One of the problems leading to delays was in obtaining critical information from an "outside" department that could supply an important sense of direction to the project. In the case of Lycra, researchers in Pioneering needed more information and insight into the structure and manipulation of complex molecules. However, they wanted to find things out for themselves believing only they could truly understand how polymer fibers worked. For their part, the chemists in DuPont's Organic Chemistry Department (Orchem) coveted their research and detailed understanding of molecular structures and had little incentive to share what they knew with another department, especially Fibers, which would likely take the information and apply it to further their own interests, with little benefits coming to Orchem. Eventually, Church was able to establish a bridge between the two departments and arrange to bring in experts from Orchem to talk to his chemists in the Pioneering labs "in the hope of stimulating some different thinking about elastomer fibers." This infusion of knowledge from Orchem was an important turning point for the project. The tactic worked and "gave Pioneering researchers the chemistry they needed to make a good synthetic elastomeric fiber."[7]

With this new insight the project received a much needed shot in the arm. With ultimate success now looking very likely and with Church continuing to support the effort, DuPont's management kept the effort going. Happily, things went extremely well once development began for "development of this fiber worked out far more smoothly than that of either Orlon or Dacron; Indeed the Lycra venture appeared to progress as smoothly as had nylon two decades earlier."[8] The pilot plant was fairly easy to build and closely shadowed the laboratory setup. Of course there were the usual issues, especially those associated with scale-up, but these were very familiar now to Church and he knew they could be solved. But even better than nylon, the interfacial process was very "giving" and so could be scaled up fairly quickly and make various types of elastic fibers for different customers for testing and analysis. Those buyers interested in the material didn't even have to wait for the first pilot plant to be built. Church, remembering the case of nylon and the ease with which its laboratory apparatus could be made into a pilot facility, directed the laboratory to expand its experimental unit to the tune of 150 lb/month. This facility could even accommodate a continuous polymerization process, something that could not even be done for nylon in its early years.

Based on the better than expected progress made by the Lycra team, Church and the senior executives of the Textile Fibers Department could report to the Executive Committee that Lycra would eventually break even in 5 years, a very positive projection for a major new fiber. As importantly, they believed that total costs for the first plant would not exceed $30 million, a relatively low amount for a new manufacturing facility. They felt very confident for two reasons: a new and powerful process and the fact that the new material was rooted in Carothers' nylon chemistry. These advantages also went far to lower the risk level perceived by members of the Executive Committee. The close communication between the chemical engineers at Pioneering and the engineering team on the plant construction side also helped the forward momentum of the project. As with nylon, great progress was made in a relatively short amount of time. As it turned out, it took less than 2 years to start making profits for DuPont. By 1976, Lycra accounted for approximately one-fifth of all earnings of the Fibers Department. Furthermore, as had been anticipated, the interfacial process that made Lycra could make other fibers as well, thus helping to cover initial research and development (R&D) costs through the benefits of economies of scope. Research into these elastomers and into the role of interfacial polymerization in producing them led to what has become known as block copolymers, a complex field that continues to be mined by chemical companies and advanced material firms for commercial possibilities.

4.2.2 Kevlar and the Aramids

A second, closely related line of research pursued by DuPont Fibers was in a group of materials known as the aramids. Like nylon, the aramids are polyamides but with a difference: whereas the nylon molecule is a linear, open polyamide chain, the aramid molecule is a closed-ring (aromatic) polyamide. Kevlar is one of the better known of the aramids. It is a material that possesses great strength. Five times stronger than steel and with superior rust and corrosion resistance, Kevlar has found specialized applications in underwater cables, and bulletproof vests and within high-performance composite materials. In recent years, firms looking to develop nanocomposites turn to Kevlar as the material they need to challenge in the marketplace. Kevlar is also one of the first man-made liquid crystal polymers.

DuPont began researching Kevlar in the mid-1960s, or about the time Lycra was coming into its own as a commercial product. Spirits were thus high at Pioneering labs that a new fiber was just around the corner. Kevlar was discovered by organic chemist Stephanie Kwolek. With the support of Hale Charch, she began looking for a new type of fiber that could be used in making lightweight, strong tires. Her initial results were less than promising and she came close to discarding her cloudy, seemingly useless solution. But she decided to test it by spinning a fiber from the substance. To her surprise, the fiber was indeed very strong and did not break. Her accidental discovery showed great promise. At this point, Kwolek handed the material off to Charch for development.

But Kevlar was quite difficult to tame. It took nearly 15 years and considerable resources before the company brought it to the market. Part of the reason for these delays and mounting expense was in finding the right markets for it once the initial application that DuPont targeted, tire cords, fell through. This meant separately developing it for many different specialized applications rather than a single large well-defined market. Precommercial costs for Kevlar were therefore very high. The interfacial process itself, being very elastic, could be adjusted fairly easily to make different Kevlar-based polymers for the shifting markets. Charch believed in Kevlar, because like nylon and Lycra, it was made from the very familiar process (polycondensation) that Carothers perfected at DuPont. Also, the success of Lycra was still fresh and confirmed the excellent performance of interfacial polymerization. If the market for Kevlar was still uncertain, not to worry, the extremely pliable process behind it could be adapted as needed to satisfy future market possibilities. Thus the "comfortability factor" was quite high and allowed Charch to be most optimistic when reporting to the Executive Committee. But, just as attempting to use nylon as a model for Orlon and Dacron had major problems, so closely comparing Kevlar with Lycra held out only a false promise. Making the polymer itself was not the only part of the production process that made Kevlar marketable. The complexity of Kevlar's structure meant that new methods of spinning it into fiber also had to be developed for each particular market application. This greatly slowed the development phase and Kevlar did not enter markets until the early 1970s. By 1980, *Fortune* magazine proclaimed DuPont's venture into Kevlar a "miracle in search of a market."[9] The material never did reach a sizeable market and has yet to become a major product.

DuPont explored the use of interfacial polymerization in creating other types of the so-called aramid fibers. Nomex, for example, introduced DuPont to the field of synthetic paper and spun-bonded materials for industrial sheet products. An aromatic polyamide, nylon-like material, Nomex was one of the more difficult materials for DuPont to commercialize. Even the low-temperature polymerization process—generally an agreeably accommodating technology—proved to be very intractable in this case. In order to make the process work, researchers had to significantly modify it. Even so, making a stable and useful aromatic polyamide was frustratingly elusive. Nomex, like Kevlar, took a long time—nearly two decades—before commercialization took place and, like Kevlar, has never had a major market presence.

DuPont's foray into spun-bound aramid polymers continued with work carried out in the 1960s on three promising products: Tyvek, Reemay, and Typar. The focus of this line of research depended more on mechanical rather than chemical innovation. These three products were very costly to get to market and disappointed once they got there. Taken together, by 1980, all of the specialty fiber and spun-bonded products barely reached 200 million pounds compared to nylon, Dacron, and Orlon that together had production in the tens of billions of pounds. While the price per pound for the specialty fibers exceeded that for the three big fibers, the small markets involved and the exorbitant R&D costs required forced DuPont to declare the specialty fibers program less than successful.

Overall, from the late 1930s and the market entrance of nylon to the 1980s, DuPont Fibers came out with progressively less successful synthetic products. The following table, which rates these innovations on the scale of one (lowest score) to five, clearly reflects (Table 4.1) this declining trend. These ratings are based on the results obtained by Willard F. Mueller in his classic 1959 article on the origins of DuPont's basic inventions. I have modified these ratings to reflect more recent research carried out by Hounshell and Smith in their history of DuPont's R&D Department. These new numbers are derived by considering four factors: (i) technological leap of innovation, (ii) market growth of innovation over the first decade since introduction, (iii) profits from innovation over first decade since introduction, and (iv) current standing of polymer today (2016).

This continuing deterioration in DuPont Fibers' innovative output came with a very high price tag. The fact that this department that clearly the first important super polymers could spend so much for so long for such meager results reflects the power it wielded in the company following its success with its three great early fibers. DuPont management supported the department because by the 1960s it provided the bulk of the company's total profits. Placing confidence on the future growth of the department seemed to be a good bet. Accordingly, what Fibers wanted to pursue, it got to pursue. DuPont Fibers management no doubt believed it was milking its new low-temperature polymerization technology for all it was worth. The thinking at DuPont was that if Fibers couldn't make more of the process than it did, then no company could. By the 1970s, DuPont's growth now rested not on radical new products but on improving its current technology. It extended its manufacturing facilities, including making nylon production fully base and continuous. It also improved the properties of its existing product portfolio in order to meet the needs of the ever-changing fashion industry. This too turned out to be expensive, pushing up R&D costs substantially without creating any major new product. But the reality was that, without introducing important new synthetic fibers, the fortunes of the department—and of DuPont as a whole—declined. The Fibers Department was no longer the engine of growth it had been in the past, in large part because it continued to spend a lot of money on promising fibers that in the end did not perform as well as expected. By the early 1980s, once DuPont's powerhouse operation, it had become the smallest department in the company.

Ultimately, DuPont Fibers worked its low-temperature polymerization process into the ground. Strangely, the company failed to apply this technology where it would have had the greatest impact and, in fact, made the company the center of yet another synthetics revolution: the polycarbonate plastics. But the stark separation of DuPont Fibers from its plastics division (Polychemicals)—a divide born of the company's increasing decentralization and Fibers' surging power and attendant

TABLE 4.1   Relative Importance of DuPont Fibers: 1938-1972[10]

Fiber Innovation	Chemical Composition	Year Introduced	Relative Importance (5 Denotes Greatest Importance)
Nylon	Polyamide	1938	5
Orlon	Polyacrylonitrile	1951	4
Dacron	Polyester	1953	4
Reemay	Polyester	1960	1
Typar	Polypropylene	1960	1
Lycra	Polyurethane	1962	3
Nomex	Polyamide (aromatic)	1967	1
Tyvek	Polyethylene	1967	1
Qiana	Polyamide	1968	<1
Kevlar	Polyamide (aromatic)	1972	2

belief that it really didn't need any help from any other group—meant that the vital connection between interfacial polymerization and the polycarbonate plastics never materialized. While the great DuPont missed out on one of the most important advanced material discoveries of the era, another large company, and one not even primarily in the chemical business, took that very same process and refashioned it to create breakthrough polymers that would become one of the most important materials to ever emerge from a laboratory.

4.3 GENERAL ELECTRIC AND THE POLYCARBONATES

GE established one of the first R&D organizations in the United States.[11] This is no surprise considering GE was founded by Thomas Edison, who ran the famous Menlo Park research complex in New Jersey in the late 19th and early 20th centuries. GE's R&D laboratory was similar to what would become AT&T's Bell Labs in the importance it gave to scientific research. Both labs garnered Nobel Prizes for their scientific work such as Bell's Clinton Davisson for his work on the quantum nature of electrons and GE's Irving Langmuir for his research on surface phenomenon and atomic structure. Both labs respected the work of their top scientists and rewarded their researchers who published in leading scientific journals. However, the two R&D organizations differed significantly on one point: the degree of intercourse their scientists should have with the manufacturing arm of the business. Whereas scientists and engineers at Bell often worked separately from AT&T's business and production activity, those at GE were expected to pursue research that had well-defined and measured practical ends. This meant a closer collaboration between GE's Central Research Laboratory (CRL), located in Schenectady, New York, and its various operating units around the country. While Bell Labs had very much a university feel to it, GE's laboratory was a true center of industrial research.[12]

While much scholarly attention has been focused on GE's central labs, the company also had other laboratories working directly for the various operating divisions. These facilities did not conduct fundamental research but rather worked on practical problems that turned up on a day-to-day basis. One of these divisional laboratories, GE's Steam Turbine Laboratory, also located in Schenectady, housed a small group of scientists who focused on new materials that could be used in the latest turbine units. This Materials and Process Group was particularly concerned with developing new polymer materials for electrical applications. It regularly sent its new Ph.D. to CRL to become intimately acquainted with the more long-term issues it was working personnel on and to investigate to what extent this work might be applied to improving power equipment performance.[13]

4.3.1 The Polycarbonates: Research Phase

Unlike DuPont and its nylon, GE did not discover the polycarbonates. It was in fact first synthesized in 1898 by the German chemist Alfred Einhorn working out of the University of Munich. However, nothing was done with it commercially until the early 1950s. At that time, the German company Bayer patented the first linear polycarbonate under the brand name Makrolon. Yet, no chemical company in the United States pursued this material. It would be GE that would bring the polycarbonates to America. That story begins with some difficulties it was having with its steam turbines.

In the 1950s, the Steam Turbine Group faced the problem of finding advanced insulating coatings for wires that could withstand high temperatures. GE's management pushed this investigation because the company's reputation as the premier producer of electrical components and equipment was on the line. Finding the newest and most advanced polymeric materials to coat wires economically was not only a point of pride with top management but a necessity if they were to hold on to their customers in an increasingly competitive market. While management set the general direction, to find a superior polymer that can be quickly and cheaply applied to wire, there were a number of possible materials to investigate. Whether it approved the R&D of any particular polymer depended on how promising the new material appeared and, in general, the perceived risks involved in following this path.

In 1953, Daniel Fox, a young recruit to GE, a newly minted Ph.D., and organic chemist from the University of Oklahoma, stumbled upon one of the great chemical discoveries of the century. Initially wooed by DuPont, he decided against working there mostly because he was put off by their rigidity and hierarchy and, as he put it, its "long chain of command."[14] GE's informal, fluid organizational structure appealed to him more. Assigned permanently to the Steam Turbine Lab's Materials and Process group, he was lent out to GE's CRL to find new insulating materials for high-performance electrical wire. Fox, by his own admission, was no theoretician, especially when it came to understanding polymers. More akin to Edison than Langmuir, he looked at problems from the practical point of view and tried different routes using informed intuition and trial-and-error testing based on a trained knowledge of how molecules are put together. He grew increasingly interested in the work the CRL was doing in polymeric films and enamels as insulating materials for wire.

While toiling on the problem of coating polymers, Fox brought into the research outside knowledge he had acquired years before he ever set foot in GE. He began thinking of a compound he knew well from his graduate school days. He had in fact synthesized the chemical called guaiacol carbonate a number of times and had studied its properties. At that time, his interest in the compound was as an antiradiation agent. But he remembered that it also was a good insulator that did not degrade when wet. While the compound itself did not have the right physical properties to serve as a coating, he thought it might be just the right starting point for the synthesis of a carbonate-based polymer possessing excellent insulating properties. Further experimentation showed that a second compound—bisphenol A—was needed to carry out the polymerization and, by good fortune, GE's laboratory happened to have this chemical in stock. When Fox mixed the two chemicals, he was able to pull out of the solution "a long fiber of the polymer."[15]

Fox appears not to have understood the importance of what he had done—he did not think it was the breakthrough insulator he was looking for. But some of his colleagues in CRL, a number of whom had worked in other parts of GE and so had studied how polymers behaved in different types of practical settings, understood the structure and behavior of polymers intimately. They saw immediately that Fox had invented a totally new and potentially important material and, with Fox, began studying its structure. They determined that the polymer was a long chain of phenyl groups—six-sided molecules—connected to two methyl groups (simpler molecules composed of one carbon and three hydrogen atoms), with one methyl molecule pointing up and another pointing down. The two methyl groups and the six-sided phenol groups "anchored" one other so that the molecule as a whole was quite rigid. This meant that the polymer was remarkably stiff, strong, and durable. It also meant excellent thermal resistance and the possibility that the polymer could serve as a superior insulator, something in which GE—as well as other electrical companies—would be very interested. And this was not all: the rigid molecular structure made the polymer quite transparent with the ability to transmit light nearly as well as glass. Thus it had superior optical properties. If a flexible enough process could be found, it could turn out different versions of the polymer, each emphasizing a particular characteristic—or series of characteristics—for targeted applications for GE's own use and for outside markets.

If these more experienced polymer chemists in CRL had not been there to set Fox straight on his achievement and its great promise, he, being not very impressed at all with his experiment, may very well have missed its significance, consigned his new material to the garbage, dismantled his apparatus, and moved on to the next experiment.[16] But now a group of scientists and engineers, excited by this discovery, began forming around Fox in the hopes of working with him on the project. That critical piece of knowledge that Fox had dredged up from his memory bank had set the ball rolling and created project momentum—a new type of important polymer now seemed to be a realistic possibility.

4.3.2   The Polycarbonates: Development and Scale-Up

Researchers knew well that creating a small sample of a new material in the laboratory and recognizing its potential significance was only the first, certainly important, step on the road to creating a commercial product. GE management, while possibly intrigued with the novel polymer, was certainly not ready to embrace it. In fact, that dismal reality known as competition for resources put further work on polycarbonates on the back burner. CRL had made much further progress on a different type of coating material—enamels. This is what took precedence now. Enamels research absorbed the men, money, and equipment that could have been employed on the polycarbonates. Fox reluctantly accepted this reality: "I was told that I really had to leave the polycarbonates sit for a while until we got the wire enamel out the door." The new and intriguing material would have to wait its turn.[17]

At this point, Fox moved back to the Steam Turbine Lab to work on a variety of issues directly related to turbine operations. A small sample of the polycarbonate that Fox had made back in CRL was shunted off into a corner and left there. One of the advantages of GE's more flexible organizational structure was the opportunity of entrepreneurial personnel to create an entirely new department focused on pioneering ventures. An engineer and marketing man, Zay Jeffries, spearheaded the creation of this venture group in the 1940s as a result of his championing of GE's earlier materials breakthrough, the polysilicones. It was Jeffries and his venture group who created the favorable context—an enclave of radical innovation within GE—where groundbreaking projects like the silicones and polycarbonates could incubate and develop into business enterprises. Here we have an excellent example of innovation in a large company coming from a department outside of the regular business. This internal venture outfit—initially called the Chemical Engineering Group and then simply the Chemical Department—could explore new materials and processes that were of potential interest to GE and that might even be sold in the open market. Operating out of Pittsfield, Massachusetts, far from GE's headquarter and central laboratory, it had the freedom to delve into less immediate, more long-term projects. Nevertheless, there were restrictions. GE management made it very clear to the founders of the new Chemical Department that this was no carte blanche arrangement. They had to proceed judiciously and with great caution, or not at all. In other words, their projects had to steer very clear of "blue-sky" research, for such endeavors would not be supported from above. Rather, GE expected the venture group to search inside the company for ideas or processes that were already there and that did not require a lot of money or a long time frame to start making money.[18]

The director of the department at this time, Alphonse Pechukus, was very anxious to demonstrate the importance of his team to the company's bottom line. Pechukus, a well-known chemist, had become an influential manager at GE (and, when he left the company, a corporate consultant). He is considered by many colleagues the real champion of the GE's polymers and the major force in directing polycarbonate development and commercialization. He knew that the most likely place to find good opportunities was CRL in Schenectady. As usual, CRL was only too happy to share the fruits of its labors with other scientists and engineers at GE. This was after all its mandate and so its effectiveness—and the career advancement of those who worked there—depended on its connecting with and transferring its knowledge, skills, and technology to other departments. This technology transfer process proved critical to the future of GE's creation of a viable polycarbonate technology. Researchers there, still excited about Fox's polycarbonates, showed Pechukus the laboratory sample made earlier by Fox and his team and recounted how it came to be synthesized. Greatly impressed with its mechanical and thermal properties and pleasantly surprised that it was not too brittle—a characteristic he had always associated with the polycarbonates—he decided that his department would take on the task of developing the material. He also knew that his work would have to be done more or less out of view of the upper management since they wanted to focus on enamels and not on polycarbonates. Pechukus would have to tread lightly and in a way that did not draw attention to his project from corporate executives.

One of his first decisions in this endeavor was one of his most important: luring Fox—the actual inventor of the polycarbonates—away from the Steam Turbine Lab to head up the research team on polycarbonates. Pechukus and Fox took an "under-the-radar" strategy for the project. Working by stealth—making progress using little money—was particularly important. While Pechukus may have been able to siphon funds and even equipment from other projects to the polycarbonate effort, doing so might very well have called attention to their work before they had anything substantial to show GE's administration. This could easily have resulted in the endeavor being killed outright. What top management needed to see—and as rapidly as possible, before the enamels became the dominant coating material, was a viable and flexible process able to make enough polymer to send to interested "customers" within the company for their various applications. As one of GE's main researchers on the earlier silicones relates: "if you are first in the business, go ahead and take any process you have [even if less than perfect] in order to get into production and make the products available to customers. Get your market position and in the meantime go ahead with the development process and try to improve the economics of the process that you have. But get your market position first ... Polycarbonates benefitted vastly from our experience and our learning and know-how [in silicones] in market development."[19] So researchers had to get some process, or any process that would work well enough to pump out polycarbonates of reasonable quality in quantities that could be sent to the molders for testing. The quickest and cheapest way to get a workable technology was to modify a well-known and proven polymerization process. A common and generally effective approach was to carry out the polymerization when the reactants are in a molten state. Fox and Pechukus had every confidence in this "melt polymerization" technique. While they understood it would likely present problems during scale-up, for now, on a smaller scale and in the short term, it worked well enough to move the project forward in a positive direction. And so Fox and his crew built up the flawed but temporarily acceptable technology as best they could to somewhere between a pilot and semicommercial plant. Their work paid off: in short order they had built a five-million-pound-per-year facility. They did so without spending much money or man-hours and so did not bring unwanted attention from inside the company on the project. This output sufficed to begin sending out samples to a limited number of users in the company and to outside plastic molders for evaluation.

At this point, the project still was being pursued as a secret "underground" venture without much interest shown by GE's top management, which is not surprising since they still did not know very much about what Pechukus and Fox were up to. This is exactly what Pechukus wanted: he had no intention of approaching corporate executives for support until the new material could demonstrate that it was superior to enamel as an insulator, that it could be made cheaper, and that equipment engineers within GE and plastic molders outside of the company wanted to use it. As expected, molders who had been testing out the material complained to Pechukus that the polycarbonate sent to them could not be worked easily. They needed a more tractable polymer. They also complained that the polycarbonate they received was suitable for only a limited range of markets; they wanted to test out different types of polycarbonates for a broader selection of customers. The process in place did not easily adjust to these different demands. Even as engineers worked the pilot plant, Fox and his fellow chemists, under the direction and encouragement of Pechukus, were back in the laboratory in hot pursuit of a superior and more permanent solution to the polycarbonates. Experimentation soon moved away from modifying or "piggybacking" off of melt polymerization to a completely new approach: having polymerization take place at room temperatures at the interface between two solvents. It is interesting that the team appears to have come upon interfacial, low-temperature polymerization around the same time as DuPont Fibers. GE's discovery came rather undramatically through pure empirical—hunt-and-peck—experimentation. If there was no thunderclap associated with GE's breakthrough, the process dramatically saved the day for the polycarbonate project. It is, as we noted, a highly supple process that, in the hands of competent chemical engineers, can economically create many varieties of a polymer for different purposes (the problems that came with Kevlar and its process stemmed not from interfacial polymerization but the fiber spinning process, which was not an issue for the polycarbonate plastics). The impressive economies of scope achieved by the process meant that GE could create many types of polycarbonates for uses both internally and outside markets and thus more readily recapture money spent on developing and scaling up the new material. This and the fact that, unlike high-pressure technology, the new process scaled relatively easily and did not require expensive equipment creating dangerously rigorous and energy-intensive operating conditions did much to push production and market acceptance of the new polymer. In short, interfacial polymerization was tailor-made for what Pechukus and his project needed: a cheap, effective way to expand production and flexibly respond to the demands of a growing number of clients.

With the solving of the technical issues and demonstrating a major market for the polycarbonates—and doing so under corporate's radar—Pechukus was ready to present his case to GE's top managers. They too were impressed. But the project was not yet out of the woods The question now turned on whether the new material fit into what senior executives believed to be GE's corporate strategic context. If it did not, the company could very well delay its market entrance or even reject it outright as a commercially acceptable product. In fact, despite the convincing arguments put forth by Pechukus and Fox and the obvious promise of the new material, GE did indeed come dangerously close to blocking the commercialization of the polycarbonates. That it didn't depended on Pechukus guiding and patiently but forcefully nudging GE into significantly reorienting the strategic direction of its R&D program and the role of that program within the company as a whole.

4.3.3   The Polycarbonates: Commercialization Phase—GE Research Shifts from an Internally Directed to Externally Oriented Culture

As noted in previous cases of innovation, the commercialization of a growing technology—even one that is on the way to being scaled—is not preordained. The persistent question is whether actually commercializing a new technology appears to be too risky to a company's power structure. In other words, does its introduction into the market correspond with what the organization at the time believes its strategic imperative to be? This strategic vision is often—in fact usually—strongly linked to the company's corporate culture, which, in turn, emerges from its historical context, as often modified by more recent events and urgent necessities. As was true of Nucor, the main issue swirling around GE's cultural climate at this time was the question of whether GE's corporate powers perceived their company to be internally or externally oriented. In the case of GE's R&D, we find an organization traditionally bound by an internally centered way of thinking—it was used to strongly protect the company's patent position against all external competitors and only creating technologies for, and transferring them to, other divisions and departments within the company. For GE, sharing its homegrown knowledge with other companies posed a great risk. The improvement to its electrical products that came from the work of its CRL and divisional research departments could just as well benefit its competitors should they get wind of its discoveries. The result of course would undermine GE's competitiveness in the industry and threaten its leadership position. The problem was that remaining isolated this way worked against taking advantage of the new polymer materials. They had patent issues that could only be resolved satisfactorily by shared licensing arrangements, and their cost effectiveness required large, continuous manufacturing plants that turned out mass-produced polymer in a volume requiring a market demand extending far beyond the limited needs of GE's divisions. The first issue compelled GE to find ways to negotiate with, rather than struggle against, competitors, and the second to "get out more," listen to what external customers wanted, and work with them to satisfy these needs.

4.3.3.1 The Patent Issue
The patent problem for GE did not just involve defending its intellectual property position as a competitive tactic. GE certainly had plenty of experience when it came to leveraging its expertise in electrical technology. As the leading company in this industry, it protected its patents aggressively in the courts and through the use of patent "add-ons" in order to extend the life of its monopoly on important products and systems. But polymers were a very different situation. In any patent war, GE would not be the leader but rather the newcomer—the David—who opposed the large established chemical Goliaths in the United States and Germany. GE was intruding on their turf now, and this meant employing a more defensive—even conciliatory—tactic in order to survive. In this case, GE found itself confronting none other than the legendary German chemical company Bayer Chemical, which appears to have independently discovered its own process for making polycarbonates at about the same time as GE. Fox tells us of his and GE's shock at suddenly finding this out: "we [GE] were in total ignorance of their work until we saw [it] published in a German publication (Angewandte Chemie) ... This shook us up. [GE's management was] reluctant to continue [with the project] not knowing whether or not we would be able to [succeed against the Germans]."[20]

Rather than going it alone and attempt head-to-head conflict with the leviathans of the international chemical industry—a very risky proposition indeed, it used its R&D organization to work with GE executives to negotiate deals with the competition that gave both sides what they wanted. In this way, GE avoided costly court battles where they could easily lose. Pechukus played a direct role in helping devise these strategies for management. He helped management understand that, as it did with its silicones, GE could leverage its polycarbonate technology to fashion a cooperative arrangement that would work in the company's favor. As Fox recounts:

Therefore … it would make sense for us to agree that if they [Bayer] prevailed, they would license us and if we prevailed, we would license them. We would cross license any part of the process with some sort of financial compensation to the person who held the patents. We would not exchange any information or any technology. That is the arrangement which we established and essentially have maintained until today.[21]

By this agreement, GE no longer had to worry about which company won the final patent. It would continue to develop its technology, not have to give away any secrets, and sell in its most important markets: GE in the US and Bayer in Europe. This arrangement eliminated the risk from litigation from the established, deep-pocketed chemical companies.

4.3.3.2 The Customer Issue   There was another problem that tested the company's R&D organization's ability to reach out and handle customer concerns. The customer problem stemmed from the physical properties of the polycarbonates. While the interfacial (low-temperature) polymerization process could relatively easily turn out a wide variety of polycarbonate plastics with various physical attributes desired by different markets, that polymer family as a whole had one common failing: the difficulty of processing these polymers using existing molding equipment. The mechanical requirements demanded by the polycarbonate family was simply too high. Fox in fact laments: "We literally had the situation where we had a product but no customers that could utilize it, [and GE] couldn't take [the] chance of waiting and hoping the molders would 'get it' ."[22] The previous experience with the polysilicones and the need for chemists and chemical engineers to work closely with processors (see Chapter 5), were critical in easing the fears of management when faced with these risks. From the silicones, GE learned how effective their technical staff could be when they reached outside of the company and worked with customers to train them in the use of the new materials. Pechukus and his department reminded GE of how successful the silicones had been in reaching out and actively working to shape their thinking about the new material and that the same sort of marketing campaign needed to be done with the polycarbonates. As Fox remembers it: " … the silicone business … had established the whole concept of marketing in support of research … In effect you had to go out there and create markets for your materials … We had to train and teach people how to mold this material, and how to change their molding machines to accommodate it. We had to do a real evangelistic kind of program here educating the industry … this is one of GE's major contributions. We paved the way for the high-performance, high temperature plastics which are common today."[23] This "education of the marketplace" by R&D scientists and engineers actually generated markets for polycarbonates, as they had done for the silicones. Doing so removed the last barrier—the final perceived risk—that could have held the entire enterprise hostage.

Commercialization of the polycarbonates proceeded very quickly as the company more fully embraced the new technology, which, like the silicones a decade earlier, became a major fixture in GE's Plastics Division. Other projects—considered less a sure thing—were given less attention and fewer resources. GE began production of the polycarbonates in 1960, or only about 6 years after Fox's first discovery, under the trade name Lexan. As it turned out, markets initially predicted for the polycarbonates—electrical film and electrical insulation—did not materialize. But the fluidity of GE's R&D organization, the flexibility of the low-temperature polymerization process, and the continued efforts of R&D's technical service outreach to customers overcame this misguided forecast, and the polycarbonates soon found profitable demand as a material for use in components for electrical ranges, wind lights on supersonic jets, transistor radio parts, electrical motor parts, and photographic film. Over the next few decades, the market for polycarbonate sheet and film continued to expand across more industries and applications including in compact discs, DVDs, Blu-ray discs, electronic components, displays (replacing glass), smartphones, packaging, automotive and aerospace parts, and appliances. By the turn of the 21st century, both GE and Bayer used this advanced material as a model for the research, development, scaling, and commercialization of new generation of advanced material technologies, such as nanomaterials (see Chapter 13).

REFERENCES

1. This background on polymerization comes from a number of sources including Hounshell, D. and Smith, J. (1988), *Science and Corporate Strategy: DuPont R&D, 1902–1980*, Cambridge, UK: Cambridge University Press, pp. 426–430, and Furukawa, Y. (1998), *Inventing Polymer Science: Staudinger, Carothers, and the Emergence of Macromolecular Chemistry*, Philadelphia, Pennsylvania: University of Pennsylvania Press, pp. 10–82, 111–144.

2. Hounshell and Smith, p. 253.

3. Ibid., p. 425.

4. The discussion on DuPont's specialty fibers is based on Hounshell and Smith, pp. 423–444.

5. Ibid., p. 391.

6. Ibid., p. 426.

7. Ibid., pp. 429–430.

8. Ibid., p. 430.

9. Smith, L. (December 1, 1980), "A Miracle in Search of a Market," *Fortune*, pp. 92–95.

10. Mueller, W. (1962), "The Origins of the Basic Inventions Underlying DuPont's Major Product and Process Innovations, 1920 to 1950" in National Bureau of Economic Research, *The Rate and Direction of Inventive Activity: Economic and Social Factors*, Princeton, New Jersey: Princeton University, pp. 323–358 and Hounshell and Smith, pp. 183–189; 228–274; 384–444; 474–501.

11. The following case history on the polycarbonates derives from an interview of Daniel Fox conducted by the Chemical Heritage Foundation (CHF) in 1986: Daniel W. Fox, Interviewed by Leonard W. Fine and George Wise in Pittsfield, Massachusetts, August 14, 1986 (Philadelphia, Pennsylvania: Chemical Heritage Foundation, Oral History Transcript #0058).

12. Wise, G. (1985), *Willis R. Whitney, General Electric and the Origins of U.S. Industrial Research*, New York: Columbia University Press.

13. CHF Interview of Daniel W. Fox, p. 17.

14. Ibid., p. 16.

15. Ibid., p. 21.

16. Ibid., pp. 21–22.

17. Ibid., p. 22.

18. Ibid., pp. 33–34.

19. Charles E. Reed, Interviewed by Leonard W. Fine and George Wise in New York, New York, July 11, 1986 (Philadelphia, Pennsylvania: Chemical Heritage Foundation, Oral History Transcript #0051), p. 39.

20. CHF Interview of Daniel W. Fox, pp. 30–31.

21. Ibid., p. 31.

22. Ibid., p. 34.

23. Ibid.

5

FLUIDIZATION I

From Advanced Fuels to the Polysilicones

... The interesting thing is that ... a fluid unit ..., on a small scale, could turn out a couple of drums of this material [polysilicones] a day ... [and] several hundred pounds ... in a short period of time. This greatly impressed the G.E. management.

Charles E. Reed, 1986

The next two chapters center on one of the most powerful and encompassing advanced materials processes ever created. It spans many different materials over a variety of industries. It has played a central role in some of the most historic products of the last half century. This chapter tracks its birth in the petroleum refining industry and then follows its path into chemicals and its early conquest of General Electric's (GE) first major polymer, the polysilicones.

5.1 BACKGROUND: FLUIDIZATION AND ADVANCED FUELS

In the late 1930s, the petroleum refining industry made most of its gasoline by a process called thermal cracking by which oil feedstock was placed in a metal container and subjected to high temperatures and elevated pressures. In this way, the complex compounds within the petroleum broke down into simpler molecules that compose automotive (and aviation) fuel. Distillation technology then separated out and isolated that fraction of the cracked mixture associated with gasoline. Despite great progress

Advanced Materials Innovation: Managing Global Technology in the 21st century, First Edition.
Sanford L. Moskowitz.
© 2016 John Wiley & Sons, Inc. Published 2016 by John Wiley & Sons, Inc.

made in thermal cracking over the years, many problems remained. The process was too slow and not very precise: it was difficult to control the types and quality of gasoline. In the 1920s and 1930s, these limitations were not of major concern because automobile engines were not exactly precision machines. They would not necessarily know "what to do"—that is, operate any better—with high-octane fuel. But engines continued to improve and airplane technology soon demanded gasoline of a very high order indeed. The more farsighted in the refining industry looked ahead a decade or so and saw that their product was not going to be able to keep up with demand if refiners didn't radically improve their cracking technology. The story of how a breakthrough innovation totally revolutionized the US petroleum refining industry begins innocently enough with a visit one day by a French engineer and amateur chemist to the Philadelphia office of the president of one of America's smaller refining companies. That informal discussion would ultimately lead to momentous events in the history of advanced materials.

5.1.1 Sun Oil and the Houdry Process

Through the 1930s, refiners searched for new and better ways to crack gasoline.[1] Sun Oil of Pennsylvania was, like Nucor, a relative small fry within its industry. Sun's President J. Howard Pew was, although politically conservative, a particularly forward-looking executive when it came to technical innovation. He was, after all, an MIT man and more open than many of his colleagues at other companies to new ideas in technology. In the same way as Ken Iverson, he established the context for innovation at Sun, and, also like Iverson, he was not shy about betting money on truly radical projects that, if successful, would catapult him and his company into the big leagues. In short, Pew created the sort of dynamic culture at Sun Oil—similar in a way to that which overtook DuPont in the 1930s—that would be very willing to accept a brand-new technology, even if it replaced existing plants, if it proved its mettle as a superior refining process.

During the 1920s, a wealthy French automotive engineer, businessman, and race car enthusiast, Eugene Houdry, began experimenting with new ways to make high-octane gasoline. He taught himself chemistry, discussed his problem with leading chemists of the day, and then set up his own private laboratory near Paris to conduct experiments using different types of catalysts to convert coal into fuel. By the early 1930s he came to the United States, the world's center for oil and gasoline production, to find a company to support his research and help him turn his pilot facility into a full-scale plant to catalytically crack oil to make gasoline. Although the use of catalysts to make fuel was a totally new concept, Houdry convinced Pew that this indeed is a promising way to gain competitive control of the quality fuel market. In effect, like Ken Iverson at Nucor, Pew agreed to develop and scale an already existing pilot facility on site using foreign technology. However, whereas Iverson trusted his own people to do this work and wanted as little interference as possible from the original inventors, Pew had no chemical expertise at Sun. He needed Houdry to do the bulk of the work. Under Sun's aegis and working closely with the company's mechanical engineering group, Houdry moved himself and his operations to Sun's Marcus Hook site, only 30 miles from Philadelphia, to direct the project there. The fact that Houdry had already developed a working pilot facility prior to coming to Sun and that the refiner had extensive mechanical capability through its large shipbuilding facilities convinced Pew that Houdry would likely succeed in his quest in a reasonable amount of time and without breaking the bank. Houdry himself was not only the inventor but a most aggressive, tireless, and effective champion for his brainchild. He searched for new and improved catalysts, designed pilot and commercial plants, worked with shipbuilding to scale the process, taught Sun's operating crew how to work and Sun's maintain the personnel equipment, hunted for potential markets, helped Sun develop technical service capability to attract licensees, worked out who would license the technology and under what conditions, and engineered important improvements to the technology.

By the mid-1930s, or only a few years after setting up shop at Sun, Houdry had a working unit. An ingenious mechanism, Houdry's catalytic cracking technology was as much mechanical contraption as chemical process. In simplest terms, it consisted of a reactor containing a stationary catalyst bed. The system directed oil vapors to be cracked over the heated catalyst. It removed the gasoline produced, and, since over time carbon deposits formed over the catalyst particles thereby rendering them inactive, it periodically sent a stream of oxygen over the solid to burn off the carbon. Once this "burn-off" cycle was complete, it allowed more oil vapor to enter to be cracked into gasoline.

The Houdry process continued to have its problems, often of a mechanical nature, but it worked, and it shook up the refining industry. Just as Pew had created the context for innovation at Sun, he also firmly established the corporate culture and one that fully embraced this new process. Specifically, he pushed down into the company the belief that new technology, if superior, should replace the old, no matter how much sweat and money had been sunk in the existing plant. This dynamic view of the world accepted, even relished, the working of creative destruction. Even if what is being destroyed is one's own possessions, Pew was ready to cannibalize as necessary, believing that in the end his company would rule the roost. So, in the years prior to World War II, Pew ordered Houdry's plants to take the place of more and more of Sun's older thermal cracking facilities. Sun's competitors came hat in hand to Pew for the privilege of licensing the process. Eventually, Pew acquiesced, but he knew the leverage he held and this privilege did not come cheaply. Sun and its licensees soon were producing significant quantities of catalytic automotive fuel throughout the country.

There is no question that Houdry and Sun did something very important in the world of refining. However, despite its initial success, the Houdry process was not destined to remain the leading refining technology. Essentially, Houdry's technology was a very rigid process, incapable of varying its output very much without causing a whole battalion of problems large and small. As with high-pressure processes, equipment was very expensive to construct and operate. Moreover, there was the problem of one or more of the operating variables—such as the heat control system—failing forcing plant shutdowns. Breakdowns occurred regularly, for instance, when hotspots developed within the catalyst bed, which caused significant havoc to heat balances and the quality of final products. Even when a plant was working smoothly, yields were still less than optimal. There was also the same problem that plagued the older thermal cracking units, that is, it was incapable of successfully processing the heavier oil feedstock and cracking stocks with high sulfur content. These failures seriously limited the economic usefulness of the Houdry process since it seriously restricted the type of oil that could be refined. As automotive engines improved and demanded higher-octane fuel, the Houdry units found it increasingly difficult to meet this demand without extensive and costly modifications having to be made to equipment and plant configuration.

5.1.2 Jersey Standard and the Fluidization Process

By the late 1930s, Standard Oil of New Jersey (later Exxon), a former member of the great Standard Oil trust and largest US refiner, had a problem.[2] It had become very clear very quickly that catalytic cracking was going to be the wave of the future in petroleum refining and that that future was in fact now upon the industry. Jersey Standard had two choices: license the Houdry process from Sun Oil or develop its own catalytic process. While the former might seem to be the lesser risk, this was actually not the case. The fees and royalties demanded by Sun were very high, so high in fact that it would be difficult for Jersey to turn a profit as a Sun licensee. And what would Jersey really be getting for its money? Houdry's technology presented many technical problems for any licensee, and so it seemed it would be buying the rights to a troubled process that would soon be displaced by better cracking methods. To complicate matters, Pew held a special grudge against Jersey Standard. This was further in part due to its size and power as a personal competitor. But there was to Pew a moral issue as well: he attacked Jersey for continuing to do business with Germany's I.G. Farben right up to America's entrance into the war. Pew did not keep his feeling to himself; he personally presented what he believed was strong evidence of Jersey's illegal business activities to the Justice Department. The company of course knew about Pew's campaign against it within the corridors of government. The last thing it needed to do was to put itself into the hands of such a passionate enemy. Finally, and arguably most importantly, fixed-bed technology, even when working smoothly, was then at best only a semicontinuous process that could refine only a limited volume of oil on a daily basis. But Jersey had a much larger volume of oil that had to be processed. What it needed was a fully continuous technology. Thus, fixed-bed cracking did not really serve Jersey's purpose (the first of Jersey's fluid catalytic units had a capacity of 12,750 barrels of oil per calendar day compared to less than 7000 for the fixed-bed process).

The bottom line is that licensing the Houdry fixed-bed process was an extremely risky proposition for Jersey Standard and it did not want any part of it. Developing an entirely new and improved technology had its own risks of course, but Jersey had every reason to believe it would come out ahead in the end. The company already had in place one of the largest and most important research and development (R&D)

organizations in the refining industry that already had created important cracking processes. Through their information and patent exchange sharing agreement with IG Farben, Jersey obtained significant experience in a type of high-pressure catalytic technology called hydrogenation, which could convert cheaper petroleum stocks into gasoline.[3] Their scientists and engineers—far greater in number and professional experience than anything Sun had to offer—could develop their own proprietary and superior catalytic refining process. The person who made this claim and set the wheels in motion to find that cutting-edge technology was an engineer, inventor, businessman, and patent lawyer. Frank A. Howard, president of Standard Oil Development Company (SODC), the R&D arm of Jersey Standard, was the major force in creating the context from which R&D would be carried out. Top management at Jersey Standard worked closely with Howard, for this venture was of the utmost importance to those at the top. Also, like Iverson at Nucor and Pew at Sun, Howard, with approval of corporate leaders at Jersey, set the cultural tone of the place, one that was highly dynamic so that it, as at Sun, demanded the cannibalization of existing plants when faced with a new and exceptional technology.

Howard and his scientists and engineers at SODC believed that only a process in which the catalyst moved—that is, it did not remain stationary within a single reactor vessel—could solve the numerous difficulties that beset Houdry. Houdry himself realized this and was to come out with a cleverly redesigned process using mechanical and ultimately pneumatic means to move the catalyst throughout his reconfigured unit. However, not looking beyond the company for new ideas, Houdry's improvements were simply extensions—albeit creative ones—of his original fixed-bed concept.

For their part Howard and his Jersey research organization had no intention of making the same mistakes as Sun. Howard's shop evolved an "externalist-dynamic" culture: it also wanted to turn out major new cracking processes but was willing and even anxious to bring outside talent and resources into the fold to improve the chances of hitting upon the best possible solution. They vigorously searched for superior scientific and engineering help, and if that took them outside of the walls of their own corporate R&D department, so be it. The immediate question for Howard and Jersey in 1937 now was how to best transport the catalyst from one point to another in a reactor that avoided the numerous problems of the Houdry design. In addition to a rotating screw concept, other mechanical means were considered including "canisters, conveyor chains, grates, blower fans," and other similar devices. None of these was deemed satisfactory due to excessive clogging, equipment failure, and difficulty of scaling up plants.

It became clear to both Howard and Jersey that fundamental research into how solids like catalyst particles moved within various media had to be done. With other more immediate projects on the front burner, Howard could not afford to shunt his limited resources over to what were basically scientific experiments. In order to conserve men and money and also speed results, he decided to outsource this problem to academia. This is the point at which Jersey Standard's close relationship with MIT and its pioneering chemical engineering department—manifested in the hiring of

MIT graduates and the use of MIT professors and laboratories to solve industrial problems—really paid off.

In 1938, two MIT professors, both eminent in the field and consultants to Jersey Standard for a number of years, brought truly advanced thinking to the problem of continuous cracking. Drs. Warren K. Lewis and E.R. Gilliland had been studying transport phenomena and their implications for designing chemical plants by examining how the three major unit processes—heat transfer, fluid flow, and the diffusion of matter—could be melded together to form a mutually interdependent system. Both men convinced Jersey to support empirical research to be carried out within MIT's chemical engineering laboratory. Two graduate students going for their master's degrees separately conducted experiments to investigate the behavior of finely divided clay particles within air currents traveling along vertical tubes.[4] Their findings opened up an entire new way to continuously crack oil vapors. Using specially designed glass equipment to observe and tabulate what was occurring within the laboratory apparatus under different operating conditions, the MIT researchers showed clearly and precisely under what conditions finely divided solid particles in contact with upward flowing vapors would behave exactly like a liquid be compelled to flow in a continuous manner. Moreover, this solid–vapor "liquid" could be made to form a stable, well-defined fluid bed or cloud within which solid and vapor particles interacted closely, continuously hitting and bouncing off one another. In such a design, all the relevant phenomena essential to the proper maintenance of the system—heat control, linear material flow, solid–vapor mixing, diffusion, and bed density—were closely linked together, as if prisoners on a chain gang. None of these forces could move without disturbing the others and in a predictable manner that engineers could easily and quickly compute. So, for example, solving the heat control problem at any point in the system also meant optimizing mixing and fluid flow of the solid–vapor blend. One problem could not be resolved without fixing problems found in other parts of the system; thus no surprises awaited engineers when they altered an operating variable or scaled up from laboratory set-up to commercial plant.

MIT research did more than just provide the concepts of fluidization; it also designed equipment and an entire fluidization system, which, while only on a laboratory scale, served as the basic blueprint for what would eventually become Jersey Standard's commercial unit. All that was needed for translating these laboratory results into a working refining plant was to substitute the catalyst for MIT's clay particles and oil vapor for the upward jets of air. The operating principle of this new fluid catalytic cracking technology was to create a turbulent but stable fluid bed within one reactor where cracking would take place and then use pressure differences to compel the solid–vapor system to flow like a liquid into another chamber with its own fluidized bed—with oxygen replacing oil vapors as the gaseous medium—for the regeneration process. Here, combustion burnt up carbon residue that had accumulated on the catalyst particles, which, now reactivated, flowed back again to the cracking sector to reenter the solid–vapor cloud where cracking continued taking place. In this way, fluid cracking embodied a fully continuous process. The degree to which MIT's results pointed the way toward a workable process is shown by the fact that within only 6 months after receiving the school's research results,

Jersey engineers could successfully crack oil using the fluid bed technique and just a year or so after that, in 1940, they designed, built, and placed into operation a 100-barrel-per-day fluidized pilot plant.

By the end of the 1930s, Howard and the management at Jersey Standard could contemplate a relatively easy path to a full-scale plant, the key word here being "relatively." Neither Howard nor his engineers were fooled into thinking it would be calm seas ahead. They knew the many pitfalls to be avoided and problems to be solved that awaited them as they pushed ahead from prototype to full-scale facility Jersey was not alone in this endeavor. In the late 1930s it entered into a consortium of companies to jointly develop the process. The so-called Catalytic Research Associates (CRA) included Jersey Standard, Indiana Standard, M.W. Kellogg (the engineering firm), and I.G. Farben. Other companies later joined the group, notably Royal Dutch Shell, Texaco, and Universal Oil Products (of course, once the United States entered the war, I.G. Farben could no longer remain a member.) Jersey Standard became first among equals in the CRA group; it conducted and paid for the great bulk of the R&D and plant building took place in its facilities. It was the gatekeeper to the technology. At the same time, the other firms in the consortium helped defray the costs and provided specialized technical aid when needed. Even if these companies were to benefit from fluid technology by being part of CRA, there is little question that Jersey had every right to feel it would be able to control the technology's diffusion and garner the bulk of the revenues from licensing arrangements.

In addition to having a very flexible process and corporate partners sharing the risks, Jersey had a third card up its sleeve in the person of the eminent chemical engineer Eger Murphree as technical leader of the project. Murphree, also an MIT graduate (Ph.D.) and with significant industrial and managerial experience guided many important technology projects for Jersey, was highly respected throughout the company. He also organized a top level group of engineers—he and they (Donald Campbell, Homer Martin, and Charles Tyson) came to be called, with some degree of awe by colleagues, the "Four Horsemen"—to carry out development and scaling. The belief around the company was if any team could make this happen, this is the group of individuals to do it.[5]

Over the course of the project, Murphree was highly effective as the project's champion.[6] In this capacity, he wore many hats. He guided the engineering work that solved the many issues that came up with the scaling-up process. Beyond this, he was the important link between Jersey executives and research, development, and engineering. He translated the results of MIT research on the one side in terms top executives could understand and on the other explained how this work greatly reduced the risks of expanding the technology to a commercial unit. With the start of World War II, he took on important technical and managerial responsibilities as a member of the Executive Committee of the Office of Scientific Research and Development (OSRD). With his extensive government contacts, he impressed on the military the importance of Jersey's fluid process and why and how it was superior to Houdry's machine. He explained not only its great ability to expand production of motor and aviation fuel quickly and with relative ease but that it could be modified to turn out large amounts of butane that could then be converted into butadiene for the production of synthetic rubber.

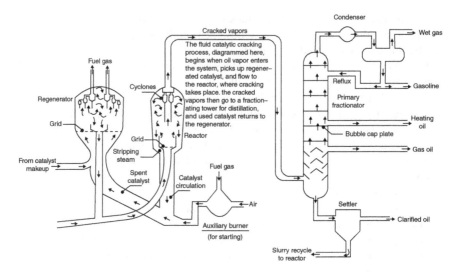

Figure 5.1 Fluid catalytic cracking unit (Source: Spitz[1]. Reproduced with permission of Wiley)

The impressive elasticity of the fluid process is further suggested in the speed with which Jersey could improve the technology so that it operated more efficiently even as it processed the heavier, less expensive oil fractions. Again, the key to the success of the process improvement stage was predictability. Even before the first three fluid plants went on line, Jersey Standard had been working on ways to perfect the process. The close relationship and intertwining of the operating variables permitted scientists and engineers to forecast the benefits of process modifications. The most important of these resulted in what Jersey Standard called the "downflow" design by which catalyst exited the cracking and regeneration vessels from the bottom rather (as in the initial design) from the top. This change afforded impressive design efficiencies and even greater operating flexibility by which the catalyst circulating through the equipment could be easily adjusted " … since the catalyst flowed like a fluid, it could follow contorted paths with relative ease."[7] Figure 5.1 shows a schematic of a typical fluid catalytic cracking plant of the 1940s. On the left it shows the reactor where cracking takes place and the regenerator where carbon is burned off the catalyst particles. Turbulent catalyst beds form in each of the vessels, and the catalyst–vapor mixture flows smoothly between them. The cracked vapors then flow to a distillation unit where the incoming gases are separated into gasoline, heating oil, and gas oil.

The dynamism instilled in the corporate culture by Howard and his R&D organization meant that the new technology received strong support at the very top levels of the company. However, not everyone was happy with this innovation. The traditionalists, the old guard, at Jersey, who were the refinery managers with mechanical training and aptitude but little knowledge or appreciation of the new chemical engineering, fought the policy that meant the closing down of their beloved old thermal units and their replacement by new catalytic crackers. Not only opposed to new ways of doing things, they also tended to be internalists at heart, feeling that they knew how to improve refinery operations on the plant floor and distrustful of outsiders whether in the form of "foreign" influences (such as from I.G. Farben), university research (such as from MIT), or cooperative arrangements with competitors (the CRA). These men—still influential within the company—retained their culture of static internalism and opposed the new dynamic expansionist thinking that was overtaking the organization. They did what they could to slow the process of creative destruction, even once fluid cracking units started to go up. But of course, it was all for naught; the times were changing and there was nothing really they could do about it. As they retired, they were replaced by young chemical engineers, many obtaining their degrees from places like MIT. In the case of fluid cracking, there was no question of delays occurring from the conservatives within the organization since wartime conditions demanded these pioneering plants be built.[3] It is interesting to consider the consequences if the old guard continued to hold sway and Jersey management as a whole had not moved from their static way of thinking to the more vibrant mentality that welcomed new ideas from many quarters. Howard and his group of engineers may have gotten quite far in developing and even scaling up their fluid plant, but Jersey's corporate executives would have likely quashed—or at least severely limited—further commercialization at the expense of the existing refineries. Then the history of this technology—and thus of advanced materials in the 20th and 21st century—may have been very different indeed.

In the years following the war, fluid cracking technology came to dominate the petroleum refining market. Exxon and other companies continued to improve the process, although the basic design of the fluid plant of 2014 closely resembles that of 1942. By 1960, fluid cracking units accounted for approximately 60% of total gasoline produced in the United States. By the first decade of the 21st century, with the rise of globalization, fluid cracking units have spread internationally and account for the great bulk of gasoline produced worldwide.[8]

It is reasonable to ask why Jersey Standard did not enter into the world of petrochemicals at this point (1940s). There were those at Jersey who certainly considered this possibility. However, the idea did not go very far for the simple reason that it did not pass the "relevancy" test—Jersey was a refiner of oil; it did not make chemicals, and doing so was simply not pertinent to what Jersey did. But other companies not dedicated to refining soon began to see fluidization as a very general type of process, one that could be very useful in ways having little to do with the processing of oil. The first important instance of it leaving its home in refining and venturing further afield into advanced chemicals occurred in an unlikely company making a product in which, many believed, it had no business getting involved. This story brings us back to GE, but this time a decade before it ever got involved in the polycarbonates.[9]

5.2   GENERAL ELECTRIC AND THE POLYSILICONES

We first came across GE as an advanced materials company as the innovator of the polycarbonates in the 1950s. But the company's first great triumph in polymers came a decade or so earlier with its discovery, development, and commercialization of the silicones. In fact, GE probably would not have succeeded with the polycarbonates at all if it hadn't been for the lessons learned and experience gained in its struggles with these earlier materials.

5.2.1   The Silicones: Initiation Phase

GE's entry into the silicones hinged on obtaining a crucial piece of information from someone outside of the company, who in fact worked for one of GE's most important rivals and also one of its closest collaborators.[9] The story of GE and the silicones begins in 1938 when the management of the Corning Glass Works in Corning, New York, approached GE's leadership with a proposition. Corning had just synthesized a new material called ethyl phenyl silicone (EPS) and wanted GE to adapt it so that it could be used as a material to make glass fibers in electrical insulation for electrical wiring and machinery. Such a project would clearly benefit both companies with Corning selling the basic chemical and GE buying it for its own uses. GE's top managers assigned the project to the head of the company's Central Research Laboratory William David Coolidge, a powerful executive in his own right.

Coolidge turned the project over to his best chemists, Abraham Marshall and Winton Patnode. Marshall and Patnode, impressed with a young Ph.D. chemist from Cornell, Eugene Rochow, told him about the task at hand hoping to get him interested in working on the problem as well. This was a somewhat unusual move in that Rochow's inorganic chemicals group generally had little communication with Marshall's and Patnode's more prestigious organic polymers team. As Rochow recounts it:

> See, I worked on the sixth Floor [of GE's Research Lab in Schenectady, which focused on mineralogy and ceramics]. All the chemists—including the hotshot guys in physical chemistry, like [Abraham L.] Marshall—were on the fifth floor, and we didn't mix much with them. I was busy, had my own work. They had their work in all sorts of polymers.[10]

Patnode and Marshall must have thought that Rochow would have jumped at the chance to get away from his less interesting projects in ceramics to work on the more scientifically exciting polymer research being conducted on the fifth floor. They were right in their first assumption—Rochow had become fairly bored with ceramics—but underestimated his ambition and independence as a scientist. Rochow was anxious to make a name for himself at GE, and toiling on modifying a material created by another company was not the way to do this. He wondered what the point was of improving a chemical Corning had already made. Corning already having the basic patents, what then would be "in it for GE" beyond simply being a client of Corning's? Rochow was also fascinated with taking what that company had already done and pushing the science to create a totally new type of polymer, one that GE could claim as its own and one, in turn, that would put his name in front of the top executives of the company. However, Rochow was not able to convince Marshall and Patnode to go along with this ambitious project. Neither the money nor the mandate was there to support such a radical plan, which would make failure that much more risky for their own futures in the company.

But Rochow was determined to proceed with or without their consent. The fact that an organizational separation—a type of social and professional firewall—existed between Rochow's group and that of Marshall and Patnode now worked in his favor: he could proceed on his own without having to let the other two know what he was up to. Rochow figured he could let them in on his work when he found something significant. His work had to be done in secret for he meant to enter into forbidden territory. The challenge was to be able to make progress in his research without attracting attention from top management in GE and its research laboratory and even from the organics group working geographically right below him.

5.2.2   The Silicones: Research Phase

Rochow initially knew very well what he was looking for: he believed he could synthesize a novel and superior silicone-based polymer using Corning's material as a starting point. But he soon realized that his best hope for creating something truly important was to part ways with Corning's work and find his own path to a new polymer.

5.2.2.1   Early Research   Like Wallace Carothers at DuPont, Rochow had great insights into the nature of organic molecules and their possible reactions. He understood that Corning's EPS could never itself be polymerized to make any type of stable substance. Its silicon atoms are connected to the carbon skeleton by oxygen so that the material would produce water during reaction that, as Carothers himself discovered in his nylon work, acted to block the formation of long chains. But, Rochow thought, what if there was a way to eliminate the oxygen and attach the silicon directly to the carbon chain? Such carbon–silicon bonds, Rochow emphasizes, " ... was a horse of a different color" because water would now not be generated during polymerization and the silicone polymer could grow very long without hindrance.[11] Furthermore, Rochow understood that polymerization would be further facilitated by replacing the ethyl group (composed of two carbon and five hydrogen atoms) in Corning's EPS with the simpler methyl group (which has only one carbon and three hydrogen atoms) to form methyl phenol silicone (MPS). No person or company had ever done this before, and MPS and the silicone polymer derived from it would be an original GE discovery and a truly radical leap in polymer technology.

But there was a logistical problem that Rochow had to deal with very early in his work. While his research was solidly within the organic chemistry field, the organic chemistry group on the floor below wanted nothing to do with this work. This meant Rochow could not fill out timesheets that were directly linked to the project. Subterfuge had to be used and Rochow's immediate supervisor in the ceramics lab, knowing full well Rochow's capabilities, was willing to help his colleague out in this secret venture. He simply told Rochow to "slip" his timesheets in with other nonorganic-related projects with the belief that no one "will know the difference." Just to be sure, he saw to it that Rochow's bogus timesheets never reached Marshall and Patnode but instead went directly to the front office. Rochow could now continue to conduct his work on this new material within the GE structure. At this early stage in the research, Rochow and his supervisor were the only ones who knew about this secret project to find a new silicone polymer.

Working virtually on his own in the laboratory, Rochow began experimenting with different combinations of substances containing the methyl group and silicon. It did not take him very long to hit upon the right sequence of chemicals to use. He successfully synthesized methyl silicone by combining the compound methylmagnesium bromide with silicon tetrachloride. Upon further heating, he was able coax the methyl silicone to form a viscous polymer that had some important characteristics that could pique the interest of other technical personnel at GE. It turned out to be an exceptionally stable insulating material, even better in this regard than Corning's EPS. Rochow leveraged this benefit as a way to find a wider range of support for his research within the company. He arranged a demonstration of the insulating qualities of his methyl silicone at an internal colloquium of GE engineers. He took two identical copper coils, impregnating one of them with a common phenolic varnish—at that time the standard insulating coating—and the other coil with his methyl silicone. He then slowly increased the voltage on both coils. After a certain point, the prevailing material began to smoke and burst into flame while the methyl silicone remained intact and kept on working (i.e., acting as insulator). Extremely impressed, the engineers wanted to know more about this new material and its structure, preparation method, possible costs involved in making it on a large scale, other potential applications, and so forth. But first of all, they wanted to know if Rochow could make enough of it now to give to them for further testing on various electrical equipment. At this point, Rochow, appearing to breathe a little easier, reflected, "It was a long struggle, but I knew methyl silicone had a future."[12]

5.2.2.2 *Later Research*   While Rochow's new material managed to attract the attention and even support of a number of GE engineers, it still was an underground technology. He had much work to do before he was ready to unveil his new discovery to Marshall and Patnode.

The main concern now was whether Rochow's laboratory procedure had any hope of becoming a commercially viable process. The issue centered on magnesium and its critical importance in Rochow's synthesis of methyl silicone. The question was how to obtain sufficient quantities of magnesium to make enough methylmagnesium bromide for the large-scale production of methyl silicone. But Dow Chemical held a virtual monopoly on magnesium and so could charge GE anything it wanted. Furthermore, Dow and Corning had by this time formed a close working relationship to explore the commercial potential of the silicones (Dow Corning was formally established in 1943). This development had two implications, neither favorable to GE. First, it meant Dow was not just a supplier to GE of a vital input but now one of its competitors in the silicone field. Second, Dow, the major producer of magnesium in the world, had, as Corning's ally, little interest in granting any concessions to GE on this metal.[13]

Clearly, this was unacceptable situation for GE. Magnesium simply could not be a raw material in the silicone process if the company was to pursue silicone polymers, and Rochow would have to find an alternative route to methyl silicone, one that did not involve magnesium. Given the continued uncertainty of Rochow's work, GE remained very hesitant in committing any of its resources to methyl silicone. But Rochow was determined to find a way to methyl silicone without magnesium. As was generally true with Rochow, he did not turn to theory to guide him. In fact, Rochow is quick to admit that there was little theory available to him: "I knew nothing about this mechanism [involved] ... It's all mysterious."[14] He used a combination of basic chemical knowledge and "hit-or-miss" experimentation. Because he still had no budget at his disposal, he had to use chemicals that were near at hand. He knocked around the laboratory's storeroom to see what he might try. He soon found a route that did not need magnesium and, moreover, was much simpler than his previous process. It involved just the single step of passing methyl chloride gas through a fixed bed of silicon copper alloy to obtain a material virtually identical to methyl silicone, called methylchlorosilane (MCS). Once he demonstrated that MCS as a polymer was as good an insulator as methyl silicone, he knew he had seized on a commercially feasible process. It was at this point that Rochow's work began to gain momentum within GE. He was not ready to approach Marshall and Patnode with his results. They agreed to support the project and suggested to Rochow that he build a small laboratory-scale pilot plant. He did so successfully, demonstrating once and for all the commercial potential of what has come to be known as Rochow's "one-step" silicone process.

5.2.3 The Silicones: Development Phase

At this point, GE's top executives in the research organization understood that Rochow was really onto something that could be commercialized.[15] The fact that GE's electrical engineers who had been at Rochow's earlier demonstration demanded more of this material for their insulation needs and that the company's leading organic chemists—who were highly influential within the upper echelons of GE's research lab—sanctioned Rochow's work meant that Rochow's new material and process would finally be given some corporate resources for development. The project graduated from a secretive, underground effort to an official project increasingly scrutinized by senior executives in research and in corporate headquarters. From here on in, project leaders would have to prove to these high-level managers that the venture should be given resources at each step of the way. Failure to do so at any stage would doom the project to oblivion.

5.2.3.1 Early Development   It is ironic that now, when the company appeared to be embracing Rochow's achievement, he no longer would be working on it. As in the case with Carothers and nylon, the brilliant research chemist was no engineer, and engineering—specifically chemical engineering—was what was needed now.

The responsibility for developing and scaling up Rochow's one-step process rested on the shoulders of Charles E. Reed, who came to GE in 1942 with a doctorate in chemical engineering from MIT. It is from Reed that the silicone project received the second critical piece of outside knowledge. The fact that Reed went to MIT turned out to be very significant for the future course of silicone technology at GE. While at MIT, he studied and worked with Warren Lewis and Edwin Gilliland, as well as other bright lights of the chemical engineering profession. Not surprisingly, the institution greatly impressed Reed. He felt that MIT's chemical engineering department was truly pioneering and, as he put it, very much "ahead of their time."[16]

Because of his MIT education, he became exposed, he tells us, to the development of the fluidization concept and technology. Indeed, he was right there as a doctoral candidate at the same time that Lewis' graduate students Chambers and Walker conducted their historic fluidization experiments for Jersey Standard. Reed actually tells us of his luck in being exposed to that development. In addition, because of his experience taking the school's advanced courses in engineering and economics, he fully understood when fluidization would make the most sense economically within a real-world chemical plant. To his way of thinking, a plant process based on the fluidized bed " ... was the way to go about making silicone polymers."[17] He had in fact ruled out a fixed-bed design very early in his research because of the massive heat buildup and blockages to gas flow that occurs during reaction and, because of these issues, the great difficulty of scaling the process up from a laboratory pilot to full commercial plant. As he remembers the problems confronting a fixed-bed setup:

> ... and if the rate of the reaction is too fast, you get a lot of caking in the tube and the tube is blocked and you can't get any more gas through it. The problem of scaling up this from a small tube maybe an inch in diameter, to something where you could produce pounds per day, rather than grams per day, would be a four hundred and fifty fold increase in scale from where Rochow started out.[18]

From his close-hand experience with the fluid process at MIT, he knew both analytically and instinctively that a fluid bed design would minimize these problems and allow for rapid and successful scaling. This was very important for the continued support of the program. He knew that top management would not maintain interest in the project if it was plagued with delays, bottlenecks, and shutdowns. He knew he was playing on very thin ice. He was the first and only chemical engineer ever hired at GE; consequently, unlike the executives at petroleum refining companies, GE's management had very little idea of what chemical engineers actually do, how they work, or how effective they could be. The fact that Reed was proposing to develop a very new and advanced sort of process technology further increased the stakes. Reed tells us that this was a concern when he recalls that the chemists at GE didn't know anything about this technology and, in fact, "they had never heard of fluid bed reactors."

What's more, GE R&D did not at this time have any history with creating new polymers. So, unlike DuPont that harked back on its nylon success to push through

Orlon and other synthetics, GE did not have an example of a successful past material under its belt that could serve as a model for supporting other subsequent polymer projects. For all these reasons, the only criterion that GE management could understand was a regular succession of good results. This meant that Reed had to show to Coolidge and his superiors at GE that this process was moving along smoothly and on schedule. On a personal level, his very career at GE depended on it.

This question of scaling was not just a technical issue as far as the lab's senior managers were concerned. There were economic realities that had to be considered. The continued interest of GE's top executives in polysilicones depended on quickly and smoothly achieving economies of scale in production. Would then the process be able to expand rapidly to meet this internal demand? If it could not, manufacturing costs per pound of resin would skyrocket. Even more seriously, the departments at GE, depending on getting the polymer to improve their products, might very well not receive the quantities they needed to meet production schedules. If the process did not work adequately, the quality of the silicones might suffer, thus compromising the products and components GE turned out for their industrial customers, who might take their business elsewhere. Given the highly competitive environment in the electrical industry in the 1940s, this was a serious consideration indeed. Word of GE's failure and inability to deliver would percolate through the industry, and the reputation of GE research and the company as a whole would be seriously—possibly irretrievably—put into doubt.

GE had much at stake in this project and it was not about to gamble away its resources and reputation on the say-so of one young and untested chemical engineer. Management's concern was very understandable. Reed understood that if he was to keep the project alive, he needed to do what Rochow had done: provide a dramatic and easy-to-understand demonstration of his technology. In this case, a fluid bed reactor was the technology and the company's highest-level executives within GE's research laboratory the audience. The timing was also important, as Reed understood perfectly well: "We knew that we would come to the point when demand for silicones within the company would be large enough to call for a bigger plant."[19] So he and his team built a small fluid bed reactor and showed how smoothly it worked and its great flexibility in turning out significant amounts of polymer, far more than could be delivered by a fixed-bed unit even if it did function at optimal levels. This demonstration had a very big impact on the managers and executives who witnessed it:

> The interesting thing is that I built a fluid bed unit that, on a small scale, could turn out a couple of drums of this material per day. That could [also] produce several hundred pounds of the material in a short period of time. That greatly impressed the GE management.[20]

5.2.3.2 Later Development This presentation marks a turning point for Reed and his fluid bed project. GE's top research management saw with their own eyes the power of the process and turned these men into true believers. They were now convinced that a good, viable process existed. In fact, in this case, perception and reality did not equate, for there were in fact many hurdles to overcome to transform the demonstration setup into a fully realized operation. Reed knew this very well, but he also knew that his audience did not and he made full use of their inexperience:

> They [GE's top research managers] didn't have experience. They had never heard of fluid bed rectors ... It was like taking candy from a baby.[21]

The successful fluid bed demonstration and the ignorance of GE's research management in engineering technology helped relieve the sense of risk that had been building up in the months following Reed's entry into polysilicones and kept resources flowing into Reed's project. Many difficulties did arise as scale-up proceeded, and delays did occur. It turned out that fluid bed processing works extremely well—and is very predictable—past a certain capacity threshold. But until that point had been reached, problems such as difficulties controlling heat buildup, the blocking of tubes and passages with solids, uneven particle flow, and so forth continually plagued the work. If GE's management anticipated these problems, Reed believed, they would not have ventured forth with the fluidized route in the first place. It is also fortunate that Reed's team was able to get its fluid reactor to that critical size beyond which such issues virtually disappeared before management became aware of Reed's subterfuge.

By this time samples of the polysilicones had been sent to different departments within GE for evaluation as to its usefulness in their production lines. The marketing people within these business units also reached out to their current customers in other industries to let them know about the new polymer and its insulating properties and how it would improve the performance and economics of the electrical products and components they sold to them. Demand for the new material grew within GE as outside clients ramped up their own production lines in anticipation of purchasing and incorporating the "polysiliconed" products from GE. Now, with the fluid plant having reached its minimal economic capacity, Reed and his team had at their disposal a very flexible machine that could rapidly be driven to meet this growing demand:

> I remember [Reed informs us] during the course of that plant design they started off at one capacity and I remember one weekend making a rough calculation on the overall plant and P said 'It will never go unless we double it'. So we doubled the size of that plant on that weekend.[22]

But the process was also pliable in terms of the polymer it turned out. Many different varieties of polysilicone material existed, depending on the operating conditions under which it was made. For example, a polysilicone could be made in the form of an oil, rubber, elastomer, or resins. A fixed-bed reactor, being rather inelastic, could not easily shift from one variant to the next without significant delays and costs. The fluid bed reactor was a very different animal, especially when working at full throttle. Reed could make different polysilicone products through simple and quick modifications to one variable or another without having to close the unit down for retooling.

Fluidization embodied the very essence of flexible manufacture in advanced material production. This proved very important when departments within GE, at the instigation of their customers, demanded polysilicones of one type or another, each having its own unique set of properties appropriate for specific applications. In the case of the polysilicones:

> The job was to get a cadre of people who would take [this new product] to customers and say ... "what are your problems? What can you use them for?" And then go back to research and development ... and say, "Now here's this opportunity. How can you change our product or adapt our product to the use that this potential customer has told us about?"[23]

Departments could make the products their customers wanted because Reed's fluid bed reactor could customize the polysilicones to their particular requirements and do so quickly and cheaply.

5.2.4 The Silicones: Commercialization Phase

In the previous chapter, we saw how GE accepted the realities of commercialization of its polycarbonates even though by doing so it had to shift its view of R&D from being an internally oriented activity to one that is more outwardly directed, that is, more directly in communication with final customers. We explained that this change was greatly eased by GE's previous experience with the silicones. But, we may ask, how did the silicones themselves manage to force GE toward this major realignment when for them there was no previous model—no precedent—that could be called upon to reduce the risks of uncertainty? It is in such a case that we have another example of how a large corporation can reverse its own cultural traditions through the power of persuasion of an able champion. As with the polycarbonates, this conversion involved patent strategies and customer relations.

By the mid-1940s, shortly after the war, GE had to face certain realities involving the future commercialization of its polysilicones. Despite the success the company had with development and scale-up of the material thus far, GE's top managers had to deal with important—and potentially project-ending—risks as it considered how, and even whether, to commercialize the polymer. Guiding the company through the commercialization process was the third champion in our story, Dr. Zay Jeffries, a metallurgical engineer and business consultant who obtained his bachelor's degree from Cleveland's Case School of Applied Science (now Case Western Reserve University's School of Engineering) in 1914 and a few years later his doctor of science degree from Harvard. Jeffries had worked for the aluminum industry before finding his way to GE. He was particularly adept at business negotiations and setting up and operating business units. He brought the tungsten carbide manufacturing process from Germany to the United States and established the Corralloy Company as a subsidiary of GE. He also consulted for GE's Lamp Division. Jeffries was the primary force guiding GE in commercializing its polysilicones.

5.2.4.1 Patents   The first issue that had to be addressed was the delicate patent situation. Dow, spurred on by its relationship with Corning, was also working on its own polysilicone technology. Now that GE was contemplating commercializing its process, Dow threatened to haul the company to court on charges of patent infringement. Whatever the merits of the case, such a move could keep GE from using polysilicone to improve its products for years and cost the company millions of dollars in legal fees to boot. This was an unacceptable risk and it had to be dealt with or GE might very well have to abandon its new material. GE was fortunate that Dow also felt the same way and that Jeffries could convince Dow to come to an agreement with GE on the matter. Reed was quite impressed with the way Jeffries reached out to a competitor to defuse this potentially explosive situation:

> It never came to the courts ... Jeffries said that instead of letting the lawyers profit from all of this, we ought to cross license ... We had complete cross licensing as far as the companies were concerned ... from a business point of view the whole thing was resolved in 1946.[24]

This strategic cooperation taught GE the value of striking patent deals with competitors—a decade later this lesson, as we have seen, was recalled when GE had to deal with Bayer on the polycarbonates.

5.2.4.2 Internal Use Versus External Customers   But there was another problem that needed attention. GE's culture simply did not accept the notion of the company pursuing advanced materials as a commercial venture in which other companies would buy the chemicals that GE made. Up to now, GE considered its chemical research purely as subsidiary to the business as a whole—it simply provided materials to be used to improve GE's electrical equipment and components, particularly transformers. The chemicals and polymers rarely left the company.

Reed, the chemical engineer, was all too aware of this disdain for, even antagonism against, GE's chemical work within the company: "the chemical business and its people were treated as poor relatives and a drain on the company because the transformer business was very profitable and looked down at this little chemical business ... which was also irritating many of their good customers. ... [in fact] people at GE were dead set against [commercializing chemicals] because their customers were the big chemical companies and [these companies] brought all kinds of pressure to bear on GE not to go into the chemical business."[25]

Attempts to sell these new materials directly to outside customers were perceived as a definite threat within many of GE's departments. The dilemma for the polysilicones was that all the benefits that could be captured through economies of scale could not be realized by just making the product for internal consumption. New demand had to be found to increase output and reduce costs. The quality of the material also depended on full capacity operations. Soon enough competitors, like Dow, who knew very well how to exploit external markets for their materials, would be making better polysilicone polymer at very low costs and would be selling it even to GE. In short, if GE did not begin thinking about approaching outside customers with its polysilicone, the business would die.

In spite of GE's traditional mindset, the rise of the polysilicones had already started to create significant fissures in that cultural shell that had kept its chemical discoveries from reaching the outside world. Managers and scientists in the research lab in particular noticed a disturbing pattern of their company inventing new materials for internal use only and seeing them in time commercialized for the mass market by another company. Reed vents this frustration: " ... GE [made] excellent chemical innovations which they would then turn over to somebody else [to engineer] who would exploit them and profit from them...."[26]

The question of whether to hold its advanced materials captive or to market them widely outside the company was reaching a critical point. The issue was not only alienation of customers in the chemical industry but also maintaining competitive advantage in the quality and price of electrical equipment that GE sold to its customers. Reed has little doubt that the rise of the polysilicones fueled this tension:

> There [was at GE] the problem of captive use vs. [external] sales to third parties, the argument being that we are going to manufacture this internally because it gives [our equipment] an advantage over the competitors. Now, if we could go out and try to sell [our polymers] to our competitors and to the market in general, we will undercut some of our product advantage because our new material gives us an advantage in the final product [i.e., electrical components]. On the other hand, the advantage of the economies of scale by adding those external sales to internal sales is extremely important as far as the cost of manufacturing goes. It is always a thread of contention in a large company like General Electric ... you've got that constant tension. We had it in silicones in the early days.[27]

The silicon polymer project was such a lightning rod for this issue because of its obvious market potential—really the first "star" polymer to come GE's way—and because the fluid bed process was such a perfect mass production machine. The petroleum refiners had found this out a few years earlier and now GE experienced the power of fluidization. The demand for polysilicones was clearly forming, and GE was the only company that had the ideal tool to exploit it. If it was ever going to control the market for an advanced material, now was the time.

Frustration over lost opportunities and realization that GE now had the technological superiority to dominate a new and powerful industry that it itself had created were clearly tipping the scales in favor of capturing external markets. But, however this may have been true, GE had little experience in commercializing new materials. It needed direction from a seasoned marketing man in such matters and from someone GE executives were willing to follow. It is at this point that Jeffries, a man greatly respected at the highest levels of GE's corporate management, really came into his own. He would guide them in the direction they needed—and increasingly desired—to go. His campaign to finally change GE's internal-oriented culture resembled a type of pincer strategy well known to military strategists. He would surround and engulf the enemy (the traditional culture) through two simultaneous and converging waves of attack: create an engineering capability allowing GE for the first time to mass-produce the advanced materials it invents for the larger market and form a business unit for polysilicones and other cutting-edge polymers to focus on initiating new markets for the pioneering materials and to serve as a model for how to integrate new material products into GE's corporate structure. Jeffries wanted the engineering group to be organized initially around Reed's fluid bed reactor and the polysilicones; but he also assigned engineering the ongoing job of searching out other promising new materials discovered internally and turning them into commercial products:

> Jeffries saw that if they were going to continue in silicones they would have to have new technology sponsorship in the Chemical Department itself, so he set up a Chemical Engineering Department ... to continue designing the first full scale silicone plant but also with developing new polymers which were being investigated [within GE's] laborator [ies]....[28]

Jeffries soon convinced GE management to form a polysilicone business unit to perfect the first commercial unit and that was closely tied to the work of the Chemical Engineering Department. This new group emphasized the important connection that has to exist between engineering, business, and marketing in commercializing a new venture. Under Jeffries' tutelage, Reed, by his account, " ... became increasingly aware of the importance of marketing as time went on."[29] This was absolutely essential for while Dow Chemical, GE's most important competitor in the polysilicone field, did not have a technology to match GE's fluid reactor, it knew how to market, and this was a very important advantage, one that certainly impressed Reed:

> Going out and working with customers, developing new applications. Dow had a tremendous amount of experience and we were handicapped by our lack of commercialization experience in comparison to Dow.[30]

Within the next 2 years, thanks to the efforts of Jeffries, the department began to acquire an appreciation for and "ability to tie into the market."[31]

The creation of an "external" culture within GE advanced materials proved absolutely essential for the company's eventual success in polymers. In fact, GE did not turn a profit on the silicones until 1957 or more than a decade after Reed engineered the first commercial plant. The "old" GE would never have tolerated this financial drain imposed by a new and unproven material. The red ink would have just confirmed the skeptics who would have forced shutdown in operations. But Jeffries had forced a change within GE. Top management now was much more patient with their new polymer. Through contact with the engineering and polysilicone departments, GE executives learned what was involved in exploiting advanced material technology within external markets, the benchmarks for progress, and the signs to look for eventual success. They understood that profits from new polymers take time: technology had to be perfected, costs reduced, and customers slowly accumulated and served so they stay within the fold. By the late 1960s, GE was the world leader in silicone resins and has remained so to this day (Dow ended up controlling the less lucrative silicone oils and rubber).

GE's new "external" culture in materials exerted a deep influence on the company and its advanced chemical projects. The engineering–business model setup for the polysilicones continued to operate as GE entered into cutting-edge polymer development. In essence, the silicones became GE's "nylon," not only because these polymers were GE's first important mass-produced synthetic but also because they supplied a model for polymer development and commercialization for future advanced materials. The infrastructure put in place for the polysilicones also served later materials. For example, as noted in the previous chapter, the chemical engineering department, which did not exist before Jeffries and Reed created it for the silicones, supplied experience and technical input to such important polymers as the polycarbonates. By the 1950s, the organization Jeffries created had become the Chemical Department, the venture group that targeted Fox's polycarbonate work as an important material to nurture and bring to commercial fruition. With silicones as well, GE established its first effective marketing program for polymers carried out within a silicone business unit. Moreover, the silicone experience taught middle managers and, through them, top executives what it took to create a successful new polymer. This experience gave them confidence—sometimes overconfidence—in tackling other new materials. We have seen how the close association of an ongoing project to nylon helped push a new polymer along at DuPont by making top executives believe—rightly or wrongly—that proceeding to the next step entailed minimal risks. The same sort of phenomenon took place at GE. If DuPont had its "nylon" model, GE held fast to its "silicone," for this also was a real breakthrough from which the company learned:

> ... a lot,..[such as] how to go to customers and try to identify opportunities that might be able to accommodate the higher prices for a polymer [with] better properties ... we had such an outstanding market development operation ... that we could take on [new materials] even though we [might not have] advantage in monomers or in process chemistry. But what we had was a powerful marketing organization that [could] take [these materials] out and exploit [them].[32]

As nylon did at DuPont, the silicone experience exerted an important psychological impact on the company. In both cases, the champion of subsequent projects that got into trouble with upper management could remind the increasingly skittish superiors of the earlier difficulties they had with the earlier polymer and argue convincingly that it was in the company's long-term interest to stay the course. Such was the case during GE's struggles with its second major polymer family, the polycarbonates. In the early 1960s, the polycarbonate business, now in its commercialization phase, was losing money, and GE's president at the time seriously considered shutting down operations. It was Reed, then vice president and general manager of the Chemicals and Metals Division, who had to go to the president and the board of directors to defend the struggling polycarbonate project and make the case for keeping it going. One of the most important weapons in Reed's arsenal was the company's past success: because of the company's prior triumph in the silicones, GE's management was "willing to [continue to] invest" in developing polycarbonate materials.[33] These achievements in advanced materials, Reed emphasizes, created an internal momentum within GE that helped keep subsequent polymer projects in play even when top management began to doubt their commercial viability:

> We then had the experience with polycarbonates. That gave us the confidence to go into Noryl [polyphenol ether resin] and gave us the confidence to go into Valox [polyolefin terephthalate resin] .... But in any one of these case [sic], we were in for several million dollars before we broke into the black.[34]

By the mid-1970s, GE had become DuPont's most important competitor in polymers. GE's development of the fluid process—a technology DuPont itself had known about but rejected—was an important reason for this success. At around this time, another company would take fluidization under its wing and brilliantly adapt it to create yet another advanced materials revolution.

REFERENCES

1. Enos, J. (1962), *Petroleum Progress and Profits*, Cambridge, Massachusetts, MIT Press, pp. 131–159. See also Spitz, P. (1988), *Petrochemicals: The Rise of an Industry*, New York, New York: John Wiley & Sons, pp. 123–128. An internal document of December 16, 1942, describing the evolution of the fixed-bed process in great detail is found at the Hagley Library and is entitled "A General Description of the Houdry Cracking Process" (Sun Oil Company records, Accession 1317, Series 5, Box 25, pp. 3–10). For an account of the importance of Sun Shipbuilding facilities to the scaling of the Houdry process, see Giebelhaus, A. (1980), *Business and Government in the Oil Industry: A Case Study of Sun Oil, 1876–1945*, Greenwich, Connecticut: JAI Press, pp. 86–87. Internal documents on Sun Shipbuilding can also be found at the Hagley Library, Sun Oil Records, Accession 1317, Series 5, Box 16.

2. Enos (1962), pp. 187–213; See also Spitz (1988), pp. 128–138. Other important sources for the history of the fluid cracking are contained in Reichle, A. (1988), "Early Days of Catalytic Cracking at Exxon," Ketjen Symposium, Scheveningen, Holland, pp. 1–3 and also Jahnig, C., Campbell, D., and Martin, H. (1980), History of Fluidized Solids Development at Exxon, in Grace, J. and Matsen, J. (eds), *Fluidization*, New York, New York: Plenum Publishing Corp. Another useful source is Flank, W., Abraham, M., and Matthews, M. (eds) (2009), *Innovations in Industrial and Engineering Chemistry: A Century of Achievements and Prospects for the New Millennium*, Washington, DC: American Chemical Society, pp. 131–159; 189–248. Internal information on the tense and ultimately futile patent negotiations between Sun Oil and Jersey Standard is found in a December 15, 1937, memorandum housed at the Hagley Library and entitled "Memorandum of Conference between A.E. Pew, Jr., Frank Howard, and Frank Abrams" (Sun Oil company records, Accession 1317, Series 5, Box 25). Another Hagley document titled "Memorandum of Activities of Standard Oil Company of New Jersey against Houdry Process Corporation"—undated but likely from the late 1930s or early 1940s—testifies to the very tense relationship between the two companies (Accession 1317, Series 12, vol/Box 1039).

3. Ibid., pp. 203–204.

4. The details of the research done on fluidization at MIT are included in the master's theses turned into the Chemical Engineering Department by the graduate students charged with conducting their experiments for Exxon (see Chambers, J. (1939), "Flow Characteristics of Air-Fine Particle Mixtures in Vertical Tubes," Master's Thesis, Cambridge, Massachusetts: MIT and Walker, S. (1940), "Flow Characteristics of Fluid-Fine Particle Mixtures," Master's Thesis, Cambridge, Massachusetts: MIT.

5. The "Four Horsemen" nevertheless did face a number of technical problems during scale-up and postcommercial improvement as related in Jahnig, C., Campbell, D., and Martin, H. (1980), *History of Fluidized Solids Development at Exxon*, pp. 10–14.

6. One of the "four horsemen" of fluid catalytic cracking relates to the author that the "... performance [of fluid cracking] confirmed the soundness of Murphree's judgement" (Campbell, D. (April 12, 1994), Letter to S. Moskowitz). See also Spitz (1988), pp. 133–135.

7. Enos (1962), p. 200.

8. Grand View Research Report. (2015), *Global Fluid Catalytic Cracking (FCC) Market Analysis Size and Segment Forecasts to 2022*, San Francisco, California: Grand View Research, Inc.

9. The early part of this story is related in: Eugene G. Rochow, interview by James J. Bohning at Fort Myers, Florida, January 24, 1995 (Philadelphia, PA: Chemical Heritage Foundation, Oral History Transcript #0129).

10. CHF Interview of Eugene Rochow, p. 21.

11. Ibid., p. 24

12. Ibid., p. 26.

13. Ibid., pp. 30–31.

14. Ibid., p. 31.

15. The following narrative of the development, scale-up, and commercialization of GE's silicone project Charles E. Reed, Interview by Leonard Fine and George Wise at New York, New York, July 11, 1986 (Philadelphia, Pennsylvania: Chemical Heritage Foundation, Oral History Transcript #0051).

16. CHF Interview of Charles E. Reed, p. 4.

17. Ibid., p. 5.

18. Ibid., p. 16.

19. Ibid., p. 20.

20. Ibid., p. 18.

21. Ibid.

22. Ibid., p. 27.

23. Ibid., p. 33.

24. Ibid., p. 29.

25. Ibid., pp. 25, 43.

26. Ibid., p. 25

27. Ibid., p. 34.

28. Ibid., p. 24.

29. Ibid., p. 33.

30. Ibid., p. 30.

31. Ibid.

32. Ibid., p. 54

33. Ibid., pp. 43–44.

34. Ibid., p. 44.

6

FLUIDIZATION II

Polyethylene, the Unipol Process, and the Metallocenes

We started a program in 1971–1972. We said "We know gas-phase works. Let's find out what it takes to do this under low pressure and attack this huge market." … Some people were watching us all along because they wanted to figure out, "Is this really going to happen?"

Frederick J. Karol, 1995

In our first chapter, we were introduced to steelmaking processes that made both traditional steel products as well as a new type of material called ultrathin steel. This chapter explores the origins and development of another process, based on fluid catalytic technology, that can make another well-known advance material warhorse—polyethylene—as well as new variations of this material including high-density polyethylene (HDPE), linear low-density polyethylene (LLDPE), very low-density polyethylene (VLDPE), and ultrahigh molecular weight polyethylene (UHMWPE). These materials as a group have superior thermal, mechanical, and chemical properties used for many 21st century applications in manufacturing, infrastructure, packaging, aerospace, automotive, energy, and biotechnology. These polyolefins plastics are so versatile that they accounts for approximately half of all plastics sold in the United States.[1]

One of the most significant advanced materials processes of the late 20th century, one that touched all of these polymers, is called the Unipol process. Over a hundred 21st century Unipol reactors operate turning out tens of millions of tons

Advanced Materials Innovation: Managing Global Technology in the 21st century, First Edition.
Sanford L. Moskowitz.
© 2016 John Wiley & Sons, Inc. Published 2016 by John Wiley & Sons, Inc.

of polyethylene materials globally. The flexibility of this process hinges on the fact that it works on the principles of fluidization. Through adjustment of operating variables—but without having to retool equipment—the process can produce polymers possessing widely different molecular weights and molecular densities. The process can readily switch from making low- to high-density materials covering an impressive range of physical properties.

That Unipol's creator was Union Carbide is intriguing. Here was a company trying desperately to survive in the face of internal struggles. Top management did not plan on pursuing this risky venture at all. The impetus for going after this technology came from the ranks of middle management and the technical and strategic vision of two champions working together. The company achieved great success with Unipol. However, many factors—including the disastrous Bhopal tragedy—sealed the doom of the company. Nevertheless, before its acquisition by Dow, Union Carbide gave the world a true breakthrough innovation, one which continues to shape 21st century advanced materials.

Before delving into the story of Union Carbide and its Unipol process, it would be useful to first consider the efforts of Carbide's major competitor in plastics, DuPont, and try understand its failure to excel in these immensely important materials. This discussion then sets the stage for the story of Union Carbide and its contributions to the Unipol revolution.

6.1   BACKGROUND: POLYETHYLENE AND THE DUPONT PROBLEM

Within the United States, both DuPont and Union Carbide tried their luck at polyethylene. As it turned out, DuPont came in decidedly second in this important race. This was just another chapter in the continued story of the company's inability to capitalize on its built up knowledge and experience in advanced polymers. We have seen how, after its successes with nylon Orlon and Rayon, DuPont increasingly spent considerable time and resources on what turned out to be relatively minor fibers. More promising was the world of plastics. But despite creating a whole new department to tackle this field and despite its army of brilliant scientists and engineers working in synthetic polymers, DuPont failed to capture the most significant of these materials. So, by the 1960s, this company that was supposed to be the leader in introducing the most important advanced materials to the world found itself virtually marginalized while other, supposedly lesser competitors—some, like GE, not even in the chemical industry at all—grabbed the limelight. While these companies found the "next big thing," DuPont continued to spend lavishly on what would turn out to be minor-league synthetics.

6.1.1   DuPont and the Polychemicals Department

The Ammonia Department played a large role in the success of the nylon project.[2] DuPont's experience in high-pressure technology helped push the nylon venture forward in two ways: it reduced management's fear level in dealing with such a new and untested product, and it actually provided important skills and technology in the making of an essential nylon intermediate. The efforts of the Ammonia Department and its genius for high-pressure work *really did* lower the risks of failure in the nylon venture. But a company's core competency can also interfere with its desire to succeed in a new technology space. DuPont's failure to lead the pack in plastics technology is a prime example—and in truth the only major one in all of our case histories—of how a firm's inveterate capability hinders its branching out into new and more promising avenues. The Ammonia Department once again takes center stage, only this time as the scourge of technological progress.

By the 1950s, DuPont's polymer research and development was divided between two departments: the Textile Fibers department, which dominated DuPont as a whole because of nylon, Orlon, and Dacon, and the newly formed Polychemicals Department, responsible for new plastics. Polychemicals had absorbed, and was under the very strong influence of, DuPont's legendary Ammonia department.

High-pressure technology practically defined the company—as noted it was critical for nylon—and led it into some of the most important research and development work during World War II. Executives closely associated with the high-pressure process and with nylon development—particularly the brilliant chemical engineer Crawford Greenewalt—went on to transfer the knowledge, skills, and even technology they had learned in high-pressure synthesis to design, build, and operate the first plant anywhere to make large quantities of plutonium for atomic weapons as part of the Manhattan Project. The great success of this project further burnished the reputation of DuPont's high-pressure expertise in the years following the war, with the technology gaining an even tighter grip on the company and particularly its plastics work. But this was a large part of the problem. Dedication to its past successes blinded DuPont to new developments that did away with the need for high pressures. European research and DuPont's American competitors—particularly, General Electric, Phillips Petroleum, and Union Carbide—led the charge toward these pioneering technologies. The stories of Delrin and polyethylene in particular illustrate the damping effect an old and embedded culture can have on a new organization's innovative spirit.

6.1.2 DuPont and Delrin Plastic

Besides polyethylene, the two most important polymers that came out of DuPont's Polychemicals Department in the 1950s and 1960s were Teflon (polytetrafluoroethylene or PTFE) and Delrin (polyformaldehyde). But these two materials never came close to becoming the major products that DuPont had expected. Former Ammonia Department managers, now powerful executives in Polychemicals with close connections to the Executive Committee and the company's corporate leaders, strongly advocated for pursuing these materials. This was especially true of Delrin. Initial research showed that Delrin had properties similar to nylon plastic. This of course piqued the interest of a number of DuPont managers. Roger Williams was the most influential of these. A member of the DuPont's Executive Committee and a former mover and shaker in the Ammonia Department, Williams' opinions on this counted for much with the managers in Polychem. On his advice, and pointing to the possibility that

Delrin could become DuPont's nylon plastic, Polychem formed a task force to further study the material's promise. Soon, the number of researches on the project grew from a mere six to fifty as they studied the characteristics of the polymer, ways to make it, and possible markets.

Williams' enthusiasm for Delrin is easy to understand. Its eventual success as a new commercial material depended on finding a process to synthesize its critical input, the chemical intermediate formaldehyde, in a state pure enough to make a stable polymer. While no such process existed in the 1950s, this was the sort of problem Williams' old Ammonia Department specialized in. This was, we can recall, just the same type of challenge that faced the nylon project at its critical moment. Williams' technology succeeded in finding a solution then; it would, Williams reasoned, surely find one now. The old ammonia people, now in polychemicals, had found a new mission. Researchers studied the problem using DuPont's Belle, West Virginia facilities, ground zero for its work on nylon intermediates over a decade earlier. Here, the built up expertise and technology could be applied to the problem at hand. However, the solution to the formaldehyde challenge did not come easily and in fact remained elusive. Researchers could not obtain a sufficiently pure material at reasonable cost. This meant that Delrin would have to remain a high-priced polymer for specialized markets. DuPont then had to reconsider its marketing strategy. Delrin, like Teflon, would have to be fairly expensive; it could not compete on price. Like Teflon, Delrin would be a niche product and serve markets not otherwise populated by other plastics. Such a strategy entailed great risks since DuPont would be hard pressed to sell sufficient quantities of the material to compensate for the high R&D expenses associated with the Delrin project. In fact, coming in at $50 million, Delrin was the most expensive R&D project for DuPont up to that time. Consequently, as Hounshell points out, the "size and cost of the initial plant [would have to be] abnormally large" and sales would have to grow at an unprecedented rate."[3] These perceived risks led to delays in moving forward. Finally, in 1957, DuPont authorized construction of its first Delrin plant, which went on line in 1960. By that time, DuPont faced a competitor, Celanese, which developed a better process for which the latter obtained a strong patent position. DuPont, now forced to share 50% of the Delrin market with Celanese, lost money on the material for years. Delrin never became a big seller or source of profit for DuPont. The case of polyethylene is another example of DuPont and polychemicals holding onto an outdated process and eschewing a promising one because it didn't fit within their expertise and comfort zone.

6.1.3   DuPont and Polyethylene

Like nylon, polyethylene has played an extraordinarily important role in 20th and 21st century advanced materials.[3] By the late 1950s, polyethylene became the first 1 billion pound per year plastic. Unlike nylon, this technology did not originate from DuPont or from any US laboratory.

6.1.3.1   European Developments   The German chemist Hans von Pechmann, who was particularly adept at investigating the molecular structures of compounds,

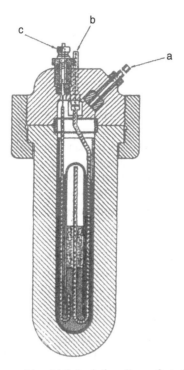

Figure 6.1   Gas reaction vessel in which "polythene" was first observed by ICI researchers (Source: Spitz[3]. Reproduced with permission of Wiley)

made laboratory amounts of a close cousin of polyethylene called polymethylene in the 1890s. No attempt to develop a commercial process for polyethylene was made until the 1930s, and the British, not the Germans, seized on the important advances here. In the 1930s, the British chemical firm Imperial Chemical Industries (ICI) developed great capability in high-pressure synthesis from their own work in synthetic ammonia. Looking to find other applications for their high-pressure facilities, ICI chemists experimented with the creation of new compounds using different types of catalysts. It turned to ethylene as a possible starting point for the making of a new superpolymer. Theoretically, the chemical was a prime candidate for polymerization as researchers demonstrated in the laboratory that ethylene could combine with itself to form very large molecular units. But attempting this feat using commercial equipment was another matter. Explosions under the very high—indeed unprecedented—pressures required occurred regularly. ICI eventually learned how to control the process. It built the first commercial polyethylene plant in 1939 (Figure 6.1).

Soon other companies attempted to create their own proprietary polyethylene processes. By the early 1950s, Europe took another important leap in polyethylene technology with the development of a whole new family of catalysts invented independently by the German chemist Karl Ziegler and the Italian Giulio Natta.

These so-called Ziegler–Natta catalysts for the first time permitted the production of certain types of polyethylene without having to apply very high pressures. A few years after these catalysts appeared, Phillips Petroleum came out with its own proprietary catalyst that was less expensive and easier to work with than the Ziegler–Natta technology.

6.1.3.2 DuPont and the "One Polyethylene" Strategy   DuPont should have had the inside track on being the industry leader in American polyethylene following the war. Through a patents and processes exchange agreement, ICI informed the company of its work in ethylene polymerization as early as 1935. DuPont soon obtained a license to work the process and in 1941 sent representatives to visit the ICI polyethylene plant to learn its operations. DuPont researchers faced even more difficulties in creating and controlling the very high pressures required than they had in developing their ammonia processes in the 1920s. By the early 1940s, the American company devised an improved polymerization reaction and began building commercial reactors. By this time, World War II assured DuPont a ready market for the polymer as polyethylene was needed as lightweight insulation material for radar systems, shortwave radio sets, and high-frequency electrical cables. After the war, technical problems continued to plague the company's ability to turn out low-cost, high-quality material. By the early 1950s, the newly formed Polychemicals Department continued to attack the polyethylene problem using the high-pressure route. Research focused on finding new catalysts that worked best under very high pressures. Equipment and apparatus used in experiments and in pilot plants constantly broke down under the difficult operating conditions. Costs mounted while the department gradually improved catalysts and processing equipment.

Beyond the technology itself was the general strategic approach of the work: researchers continued to manage the polyethylene problem as they had with ammonia, nylon, and the other of DuPont's other high-pressure products, that is, perfecting catalytic processes that mass produced one superior type of product in large volumes. But this strategy completely missed the mark when it came to polyethylene. Unlike these other products, polyethylene was not one material but a number of different polymers with varying properties and applications. Rather than focusing on one process to make a single type of polyethylene for one large market, DuPont should have been spending its research dollars on developing a flexible process capable of making different types of polyethylene for various—and significant—market segments. By pinning all their hopes on the success of only one kind of polyethylene, DuPont had no other material on which to fall back should their one product falter. In fact, this is exactly what happened. Customers—the plastic molders—rejected DuPont's polymer because it was difficult to process on their existing equipment and did not demonstrate sufficiently superior properties to make retooling worth their while. DuPont had to back track, tear their old process apart, and jerry-rig equipment and catalysts to be able to make different sorts of polyethylene. But high-pressure plants, as previously discussed, are fairly inelastic and are not capable of creating different materials quickly or cheaply. It is as if Henry Ford, who was unrivaled for being able to mass produce one type of car for one market and do so very cheaply, was suddenly required to regularly rip his very inflexible plant apart, retool, and redesign it to accommodate the demand for annual model changes. Attempting to do so proved extremely difficult for Ford, and it was as well for DuPont.

6.1.3.3   *DuPont and the High-Density Polyethylene Problem*   One of the variants of polyethylene that was particularly promising is called linear polyethylene, often referred to either as low-pressure or high-density polyethylene (HDPE). In contrast to the traditional forms of polyethylene, which tend to be "branched" in structure and have low density, HDPE has very few branches which makes it highly crystalline and therefore exceptionally strong, tough, and stiff and with many market applications. Such a material could never be made commercially using high-pressure methods—the pressures required to eliminate branching would simply be too great to be used on a large-scale basis. The secret to a practical HDPE process was in developing new and more advanced catalysts that work under much lower pressures. But the sustained influence of what we might describe as a "high-pressure culture" on DuPont compelled Polychemicals to continue to attempt making HDPE through the high-pressure route. Polychemicals once again turned to its Belle facilities to make this happen. Engineers and chemists put ethylene in an autoclave and pushed pressures up to unheard of levels. Explosions, malfunctions, poor conversion rates, and inconsistent quality of the HDPE at last forced DuPont to abandon these efforts. Eventually, Polychemicals in alliance with the Chemical Department hunted for better catalysts that would obviate the need for such rigorous operating conditions.

While DuPont did indeed find the right catalyst, precious time had been lost and DuPont by now faced stiff competition. By this time, the Ziegler–Natta catalysts had been invented and were being licensed around the world. Another player in this field, Phillips Petroleum, had also came out with its own catalyst technology for HDPE that competed with the Europeans. So now DuPont faced "considerable uncertainty" in what markets would be left to it with so many competitors dividing up the spoils. In any case, the company could not rush its HDPE program in hopes of heading off the competition since so much of its R&D funds remained still tied up in pursuing Delrin, that money pit and favorite of the former Ammonia executives. So DuPont continued to drag its feet on HDPE for a number of years and did not even begin to manufacture linear polyethylene until 1960. DuPont ultimately did not capture more than 10% of this market.

But all was not necessarily lost. Another possible approach that would have given DuPont a competitive edge in polyethylene would have been to cut its losses with high pressures and move vigorously to develop a radically new process very different—and vastly superior in terms of costs and product variety—to that which any other companies were using. DuPont was very familiar with fluidization as many of its own chemical engineers were MIT graduates who had studied with Lewis and other leaders in fluid cracking technology. That this new process could also manufacture chemicals was well known through the work of General Electric and the polysilicones. Once again however the strong influence of Ammonia personnel in polychemicals thwarted any attempts to move in this new direction.

6.1.3.4 Dupont and Fluidization   DuPont was, in fact, one of the—if not the—
first chemical company to have seriously investigated moving and fluid bed processes for making chemicals. This is not very surprising since, more than any other
American chemical company, DuPont attracted many chemical engineering graduates and consultants from MIT, the epicenter of fluidization studies. Thus we find
in the DuPont archives a report written by famed chemical engineer and Director
of DuPont's Chemical Engineering Department, Thomas Chilton, to the Executive
Committee in the mid-1930s on work his team had been doing on moving powdered
catalyst beds in the hope of obtaining better heat control in the chemical plant. Here
Chilton describes a series of experiments—quite similar to, but predating, the ones
carried out by Jersey Standard and MIT—using air to force catalyst powder through
tubes of various sizes.[4] It was not until the late 1940s that the company continued these types of investigations, eventually working with the Chemical Department
commercializing a fluid bed process for making an important chemical intermediate, ethylene oxide. But this work came from the Chemical Department. Neither
the Ammonia nor Polychemicals Departments had much use for such research and
failed to pick up on the power of fluid bed technology and particularly, as a way
toward a superior polymerization process. Their core competency was in fixed-bed,
high-pressure catalysis, which was a very different type of process—demanding its
own set of skills, techniques, and designs. In short, fluidization had very little relevance for the scientists and engineers who had cut their professional teeth in the
Ammonia Department and transferred their skills and technological experience to
Polychemicals. Without the support of these departments, the role fluidization could
play at DuPont was severely limited.

After spending so much to develop and commercialize what were, at bottom,
niche materials, DuPont faced many years of red ink on Delrin, Teflon, and its linear
polyethylene. By the 1960s, DuPont's management decided that the Polychemicals
Department should avoid taking on any more large-scale projects with the hope of
developing major new products. Instead, top DuPont executives charged Polychemicals to pursue such incremental, specialty polymers with narrow but assured markets
such as "Vespel" resins and Mylar elastomers.

This strategic shift toward minor innovation reflects DuPont's belief that by the
1960s "the polymer bonanza was dead." But in fact this was not quite true. For by
the latter part of the decade, another revolution in polymers was starting to stir; this
one based on what is arguably the boldest and most creative application of fluid bed
cracking since the early days of fluid catalytic cracking. Once again, a company other
than DuPont, one not so fettered to high-pressure synthesis, managed to turn the world
of polymers on its head in the late 20th century.

6.2   UNION CARBIDE AND THE POLYOLEFINS: THE UNIPOL
PROCESS

While DuPont continues today on as one of the top ten biggest chemical companies
globally, Union Carbide—a major chemical player since the 1920s and for many

years the largest and most diverse producer of synthetic organic chemicals in the world—has not existed as a company since 2001 when it was acquired by Dow Chemical as a wholly owned subsidiary. One may be tempted to conclude that DuPont acted wisely in deciding to avoid taking on the long-ball innovations in favor of the more incremental, base-hit variety and that Carbide—while certainly contributing mightily to global technology by creating an important new process for making polyethylene—did itself in by pursuing this game-changing technology. However, such is not at all the case. For its part, DuPont has suffered from its lack of important new technology: it pretax profits have declined precipitously since 2000, from a little over $6 billion in 2000 to only $949 million in 2012, a result in part of not having introduced a major new material or process in recent years. Discussions in late 2015 of a possible merger with Dow would have been unlikely in the years of DuPont's most technically creative years.[5] As for Carbide, its financial troubles stemmed from causes separate from its R&D initiatives. The Bhopal disaster in the mid-1980s of course took a very high—some believe fatal—toll on the company's resources and reputation. But years before this catastrophe, Carbide had experienced significant business problems. Even as early as the late 1960s, when Carbide was the second largest chemical company in the United States, it suffered declining profits and slower growth compared to its competitors, a result of relying too much on sales of low-margin commodities and inability to develop coherent and effective competitive strategies for its too extensive, overly diverse, and much too unwieldy product portfolio. Because the company made consistently poor strategic decisions, the business press in the late 1960s and early 1970s—including *Business Week* and *Fortune*—attacked the company as a fumbling giant.[6] Over the next 15 years, Carbide struggled toward profitability. Bhopal occurred just when the company finally appeared to be digging itself out of its financial hole. The only real bright spot for Carbide during the 1980s and 1990s was its development of the Unipol process for making an advanced form of polyethylene.

By a combination of operating its own plants and licensing to producers around the world, this important technology brought much needed cash into the company. By most accounts, the Unipol process—and the divesting of nonperforming operations—was the only reason that Carbide could hold out as long as it did. As one Carbide executive concluded in 1995, the Unipol process was the only thing keeping the company "afloat."[7] However, the process by itself just wasn't enough. By the late 1990s, with recession looming, Carbide was ripe for its acquisition by Dow.

However one views the ultimate fate of Union Carbide; there can be little question that its fluid bed Unipol process was a truly major achievement with global impact in the world of advanced materials. Why and how such a large, established—and financially shaky—company chose to do so is the focus of the following discussion.

6.2.1 Union Carbide and Polyethylene: Background

Union Carbide Corporation began in the late 19th century as the Union Carbide Company which at that time concentrated on making the chemical calcium carbide for different industrial applications, such as in steelmaking and in the manufacture of fertilizers.[8] By the early 1900s, a newly formed company Prest-O-Lite began purchasing large amounts of calcium carbide from Union Carbide as the intermediate to make the gas acetylene as a lighting source for the headlamps of bicycles. Soon Prest-O-Lite began selling its acetylene for use in portable lighting systems to automobile manufacturers and in oxyacetylene welding equipment to the construction industry. While even today welding continues to be an important consumer of acetylene, the chemical was soon displaced as a lighting source for automobiles by electric lamps. In an attempt to find new markets for its product, Prest-O-Lite turned to the Mellon Institute of Industrial Research in Pittsburgh to find alternative uses for acetylene. Under the direction of the chemist George Curme, this inquiry led to a more general study of the production and commercial uses of the so-called aliphatic chemicals for which acetylene and ethylene serve as the basic molecular building blocks. Experiments showed that the most efficient way to obtain these intermediates was to thermally crack—that is, break down—ethane, a component of natural gas. Moreover, preliminary investigation indicated that there were a number of new and potentially promising commercial uses of acetylene and ethylene, particularly in the making of industrial alcohols. By 1917, Prest-O-Lite and Union Carbide Company—along with a few other chemical firms—decided to merge to form the Union Carbide and Carbon Company. With their resources now concentrated into one company, these firms could more effectively pursue these lines of inquiry. Union Carbide and Carbon established a subsidiary called Carbide and Carbon Chemicals Corporation to develop and commercialize aliphatic chemicals from natural gas. This organization set up research facilities and built pilot and commercial plants in a place called Charleston, West Virginia, located within the Kanawha Valley, a region in West Virginia containing an abundant supply of ethane-rich natural gas. By the 1920s, Union Carbide had become the first American company to commercially crack ethane to make ethylene. It did so using a proprietary low-pressure (but high temperature) process known as steam cracking. By 1930, the company was the major source of ethylene for the world's chemical industry. With Curme as research director, it also established important markets for the various synthetic chemicals it created out of ethylene.

    With its focus on ethylene, going after polyethylene was a natural step to take for Carbide. Like DuPont, it licensed a high-pressure polyethylene process from ICI in the 1940s (albeit using a special tubular type of reactor of its own design) that made low-density polyethylene (LDPE). During the 1950s and 1960s, as demand for the plastic grew, Carbide expanded its polyethylene capacity. Much of this capacity was high-pressure (low-density) material, and indeed Carbide was one of the most important producers of it in the world. But Carbide also increasingly got involved in low-pressure high-density (HD) polyethylene. Carbide, unlike DuPont, was not saddled with a deeply rooted high-pressure culture. As noted, one of its early and most important creations, steam cracking, did not operate under high pressures. Carbide did however have a much stronger tradition in exploring and exploiting any and all ethylene-based processes that showed promise, whether that technology required high pressures or could operate under less rigorous conditions did not matter; as long as it "ran on" ethylene, it served Carbide and fell within the company's comfort zone. We understand then why the company very quickly licensed a new low-pressure polyethylene process to make HDPE. But as noted, competition from Ziegler–Natta technology was stiff. Through the 1960s and 1970s, Union Carbide continued to improve this process but, despite significant effort, it never could capture a large market for this low-pressure material. Unlike DuPont, which continued to push high-pressure processes beyond the point it should have, Carbide's polyethylene champions had fewer restrictions placed on them to pursue new more pioneering processes.

6.2.2 The Unipol Process: Initiation Phase

The Unipol process did not emerge suddenly; its evolution was firmly rooted within the context of Union Carbide's polyethylene history.[9] While Carbide led the field in high-pressure (low-density) polyethylene, it remained only a minor player in the low-pressure (high-density) polymer. By the late 1960s, Carbide had only 4% of the US market. One the major reasons for Carbide's failure to capture a greater share of the market was its inability to solve problems that continually plagued the process it was licensing for the high-density plastic from Phillips. In particular, Carbide chemists had a difficult time controlling the molecular weight distribution of the final product using Phillips chromium oxide catalyst. This meant they could not predict the physical and chemical properties of the polyethylene they produced with any accuracy. Customers received polyethylenes of varying quality. A further quality problem arose due to the fact that the Phillips system used a slurry—which is a liquid phase—process, and it was very difficult to obtain a polymer with the compact structure desired by clients because the polymer that formed in the solution absorbed the liquid used in the process causing the material to expand.

Faced with these problems, Carbide's top executives supported a program to improve this process. This meant finding new catalysts and looking at different sorts of processes that would minimize polymer swelling. To give this improvement project a shot of creative adrenaline, they put an energetic and highly resourceful man in charge. William H. Joyce had made a name for himself at the highest levels of Carbide as a man who got things done. Unlike many technical personnel within the firm, Joyce had one foot planted in engineering and one in business. In addition to his degree in chemical engineering from Pennsylvania State University, he had an MBA from New York University and would go on to obtain a doctorate in business and finance from that school. Joyce by his very nature was no theoretician but a practical, empirically grounded engineer who liked to "put things together" and had a strong interest in marketing and operations management as related to polymers. Most importantly, he was also a risk taker, a true intrapreneur who had a history of seeing and jumping on opportunities when he came upon them. He pursued these projects because he wanted to show to Carbide's management that he was earning his keep. As he tells it, "I never viewed myself as promising, because I was struggling to earn enough money to pay for all the equipment and everything else I was using [at Carbide]."[10] He also understood that successful business ventures meant advancement. So he aggressively engaged in deal making within the company, but under the radar, until he could come to his superiors with evidence of success. Joyce shows his business sense and deal-making bravado by revealing that, while barely in his 20s and only a couple of years with the company, he was already "... entering into long-term contracts, designing projects and spending capital dollars without authorization."[11]

One of the reasons for Joyce's success in handling these secret agreements was his breadth of vision and his sound strategic thinking. As a person who spanned different fields—engineering, business, and marketing—he held to the importance of reaching out to different departments, and even other firms, for information and insight into a particular problem. He also believed in the value of knowing how the various departments within Carbide fit together—how one could help the other—and the role of outside firms as potential customers. A good example of Joyce's modus operandi surfaced in his early work at Carbide on improving insulating materials for electrical cables for AT&T's manufacturing subsidiary, Western Electric. While Joyce was presented with this engineering problem, he began viewing it more broadly, from a business point of view. He wondered why Carbide never contacted AT&T's research arms, Bell Labs, to discover what the lab was developing and whether this could become a market opportunity for Carbide. So he went to Bell himself and found out it would be interested in Carbide developing an improved polyethylene material as an insulator for a new type of transatlantic cable. Joyce on his own and using bootlegged equipment secretly cobbled together a test amount of the type of material Bell needed, and soon Carbide's top managers found they were involved in a money-making operation that they had never approved. That the venture proved successful saved Joyce's position at Carbide, but Joyce was fully aware that things could have gone far differently: "somebody on high realized how much money we would make, so all the rules were waved aside. It's a good thing or I would have been gone."[12] These sorts of bold moves and, most importantly, the fact that they panned out resulted in Joyce moving up the corporate ladder rather quickly. After only 5 years with the company, he was made group leader. Soon he became technology manager followed by promotion to director of marketing for polymers where he plied his skills in identifying and securing new markets for Carbide's most advanced polymer technologies. Carbide's corporate powers soon moved Joyce into the position of operations manager for the company's low-pressure polyethylene operations with the hope that he would find a way to reinvigorate the program and solve the thorny problems that continued to haunt the Philips process. This task, fully approved by Carbide's power structure, offered Joyce the chance to undertake his most important assignment. It did not take him long before he ignored it and went in his own groundbreaking direction.

Joyce, being the aggressive and ambitious manager he was, immediately went beyond the task he was given by challenging the notion that trying to improve the low-pressure process made any sense. Like Rochow at GE when confronted with Corning's work, Joyce felt that it was a fool's errand to attempt to modify some other company's technology; since Carbide would not be creating anything truly novel, it could not expect to greatly increase its market share no matter how much brilliant engineering it might put into the old process. And the nature of the market had to be considered: there just wasn't that large a demand for low-pressure, HD polyethylene.

Joyce concluded that "All the plans were nuts. It just wasn't going to work ... we had a 4% market share because that was what we deserved ... we really [needed] to forget about all those plans about expansion and work on something [totally new]. New stuff that was going to give us more money. We needed to figure out some way to bring something new to the market."[13]

It did not take long for Joyce to decide on the direction he wanted to go. The big game in the plastics jungle was not high but low density polyethylene (LDPE), the type of polyethylene that Carbide itself had a strong market presence. But, despite the large volumes of the material that Carbide made, it was not much of a money maker. The high-pressure process used to make the material was very costly (due to heavy capital investment and high energy outlays) and inefficient (due to very low conversion rates of inputs into final polymer). He decided to focus on the process rather than the product. Instead of improving an existing process to make HD polyethylene, for which there was a small market, he concluded why not create an entirely new process to make high tonnage LDPE at low unit costs. Now, Joyce felt, this would be something worth pursuing; it would totally revolutionize the polyethylene industry. Carbide would indeed be bringing something new to the table and it would be totally proprietary. Just as Exxon's Frank Howard and Standard Oil Development Corp. rejected licensing Houdry's fixed-bed process in favor of creating a completely new type of catalytic cracking technology, so Joyce and Carbide would now go after their own radically new and disruptive process to make the bulk of the world's most important plastic.

A major difference between Howard's and Joyce's situation in this early stage of innovation was that Howard, as President of SODC, had the backing of his company from the beginning; Joyce did not. Joyce's goal clearly went well beyond his mandate; in fact, Joyce rejected that mandate altogether in favor of one of his own. Carbide management did not give Joyce and his team permission to pursue a low-pressure route to LDPE. The company already had many high-pressure plants operating around the world. It had no desire—nor even a pressing reason—for pursuing a radically novel process that would render obsolete and obliterate all of Carbide's sunk capital in a known and proven technology. Indeed, marketing and other managers strongly opposed veering from the current strategy, as Joyce's most important scientist Frederick Karol was keenly aware: " ... there was a certain amount of resistance [to Unipol]. To come in and bring new systems in was a challenge."[14] Clearly, if Joyce and his crew were going to move ahead in their plans, they would have to tread very carefully and, in effect, stay invisible for as long as possible. There was in fact no organizational context at Carbide for the type of research Joyce wanted to do; he would have to create it.

6.2.3 The Unipol Process: Research Phase

Joyce understood then that he and his team had to demonstrate a certain tactical subtlety and keen psychological subterfuge if they were to proceed on their risky course. If Carbide's top management had gotten wind of what Joyce really wanted to do and where he wanted to take the research, it would have shut him down on the spot.

Even though still a young man, Joyce had been through a number of battles with the corporate hierarchy; as a seasoned veteran and savvy internal entrepreneur, he knew full well it would be useless to attempt to sway his superiors as to the potential benefits of what he was trying to accomplish. He knew that they would consider his plan as "blue-sky" research with an outcome too uncertain. Carbide would never condone it. Outwardly, Joyce and his group continued to work on improving the Phillips Process, as ordered. However, whenever they could, they conducted research on catalysts and equipment aimed at finding a way to make LDPE without the use of high pressures. The composition of the research team itself was of primary importance in order to successfully carry off the charade. Joyce made sure his squad began as a small and tightly knit group and picked a select assemblage of people he knew felt the most excited about the work. As Joyce understood very well, intellectual curiosity and professional pride kept this group going:

> We were pretty excited about what we were learning … we weren't worrying [at this early stage] about the business end … But there was an environment of excitement because there was really an emergence of organometallic chemistry. That's what drove a number of us to work. We thought we were finding out a lot of things.[15]

These researchers were in it for the adventure, and, like Joyce, they were internally motivated to keep going—and to do just about anything to maintain momentum. They worked discretely, economically, and continued to make progress using the very limited resources at their disposal. The primary goal was to avoid attracting too much attention for as long as possible and at least until their work showed commercial promise. As any corporate champion worth his (or her) salt, Joyce understood the importance of taking the initiative in risky projects without obtaining official sanction:

> … the one thing I learned is that you never ask permission. Because when you ask permission, you transfer the responsibility to your boss. People don't like responsibility. They don't like change, let alone being responsible for it … you never ask [if you should proceed]. Then people, all of a sudden, feel it's too late to tell you that you can't do it. They never accepted responsibility for it. I've found that works wonders.[16]

Caution, efficiency, and dedication—and not asking superiors unnecessary questions—characterize this intrepid group of researchers as they took their first important steps toward discovering a game-changing technology. Joyce was very confident that they would succeed. His close familiarity with polymers—and not just an irrational optimism—told him so: "I know from my experience with polyethylene that when you put new branches on there you destroy crystallinity [which was our goal]. So it seemed pretty straightforward that we ought to be able to do that."[17]

During this early part of the research phase, attention focused on finding a brand new and proprietary catalyst that could "put new branches" on polyethylene under low-pressure conditions (branching reduces the polymer's density). It is interesting that Joyce could not find—or could not secure—within the walls of Carbide a specialist in this area to help him. Finding this catalyst was absolutely vital, and without it, Joyce's project could not begin to gain the impetus it needed to get past the research phase. Just as DuPont did with nylon and Jersey Standard with fluid cracking, Joyce looked outside Carbide for help. He found and brought into his group a talented Ph.D. organic chemist from MIT, Frederick Karol, who had been studying advanced catalysts of the type needed for the project. Karol's task was to find the right catalyst that could produce LDPE for making film—the largest single market for polyethylene. The catalyst main job was to insert just the right amount of branching within the polyethylene molecule to imitate the LDPE then made by high-pressure processes. It also had to be inexpensive (<1 cent per pound of polyethylene).

Such a catalyst did not yet exist; Joyce and Karol wanted to be the first ones to find it. Theory played only a small role in this work. Karol's instincts regarding catalysts and what they could and could not do, and hit-or-miss experimentation guided by this intuition were the only ways to find the right catalytic materials. Karol knew from the literature and from his own earlier research that the proper catalyst would be one that contained one or more metal atoms surrounded by certain organic molecules. What these atoms and molecules were remained to be discovered. In addition, these catalysts had to be tested with different promoters—elements and compounds that were mixed into the catalyst to increase its efficacy. This research involved testing many combinations of catalytic materials using laboratory apparatus. A key concern was minimizing impurities contained in the ethylene monomer used in the experiments. There could be no more than a few parts per million of contaminants for satisfactory polymers to be made. When a promising catalyst was found, it was tested in a small (one liter) fixed-bed pilot plant reactor capable of producing a number of pounds of polymer per hour.

Joyce never relied so much on his past experiences at Carbide as he did now. Using the knowledge he had gained and the contacts he had made while Carbide's marketing manager, he drew up a plan to begin establishing new markets for the polyethylene samples his team had made using the most promising catalysts they had tested in the laboratory equipment they had on hand. Joyce realized more research had to be done but felt his crew had progressed far enough along to at least begin to pique the interest of some of the outside clients he knew in his past position. He then went to them and convinced them to purchase quantities of this new low-pressure polyethylene at sharply discounted prices to test out in their processing equipment. He particularly focused on wire and cable applications, a market he got to know quite well through his work with Bell Labs and in his previous position at Carbide. As Joyce recalls it, "We ended up [getting some] business with Western Electric. I got it approved by Bell Laboratories because I knew everybody—I knew how they thought ... and we took all the money from that and spent it on R&D."[18]

6.2.3.1   The Unipol Process: Development and Scale-Up Phases   Joyce was still a long way from being able to come clean with his superiors about what he and his researchers were really up to. As far as senior executives knew, he was still toiling away at improving the Phillips process. But he knew they would start asking questions before too long, and so struggled to build his working pilot plant and have it begin supplying customers as quickly—and silently—as possible. The only way to uncover the most important technical problems was to build up laboratory models into semicommercial units and see what problems surfaced. Identifying and fixing these bottlenecks then became a priority. And this was not all. Customers who began to test the polyethylene samples—including Bell Labs—responded with suggestions of how Joyce and his team could improve the product for better processing on their equipment and for superior performance in final products and components. If these suggestions went unheeded, customers would lose the initial interest that Joyce had generated, or if redesigning the process to create these modified products ended up too costly, the momentum gained would be sacrificed and Joyce would have little to show his bosses that would interest them. And he had to proceed in a way that did not require significant corporate resources, for this would alert top managers of what was going on and he would not only sacrifice his underground advantage but most likely force a shutdown of his work. He was thus in the same position as GE's Pechukas and his polycarbonates. For both men, continuing to work stealthily while building up market acceptance was the ticket to constructing a successful argument when confronting corporate leaders.

One of Joyce's most pressing concerns was understanding what was wrong with the existing polyethylene products as far as the clients were concerned and find ways to modify catalyst systems to meet the client demands.[19] Accordingly, a major problem he and his team faced was that the research phase—that is, discovering new and effective catalysts—never really ended; it continued to bleed into development and scale-up. Every attempt to respond to a client request for significant quantities of a novel type of polyethylene meant relying on Karol and his laboratory to come up with a new and effective catalyst that would inject just the right amount of branching in just the right places within the polyethylene molecule to satisfy that requirement. But this process took precious time. No shortcuts existed to hasten discovery and testing. The large number of possible combinations and permutations tested in Karol's lab continued to extend the development and scale-up phases. While the search for new catalysts plagued other advanced material projects we have examined, including nylon and subsequent synthetic fibers, catalytic cracking, polycarbonates, and silicones, the work on low-temperature polyethylenes was very different. The catalysts involved are far more complex with many more variables to consider and consequently more options to think through and tests to be done. Joyce understood perfectly well that he could not afford to "house" these catalysts in a production process that was rigid and inelastic and that resulted in further delays and escalating costs The commercial process he selected therefore had to be extremely flexible with the capability of accepting any catalyst given to it and rapidly and efficiently turning out different types of polyethylenes.

The center of focus of the project then shifted from chemical research and the laboratory to chemical engineering and the production plant. While the laboratory and early development phases continued to depend on fixed-bed systems, a number of MIT-trained chemical engineers began campaigning for testing low-pressure gas-phase reactors using the fluid bed technique. They apparently were quite convincing. Karol soon championed the process, pointing out that high-pressure reactors "run at [pressures of] twenty thousand to thirty thousand psi [pounds per square inch] and two to three hundred degrees centigrade [and they are] very expensive to build [and modify for different catalysts and products]."[20] By adopting the gas-phase (or fluid bed) approach, which requires pressures of only 100–200 psi and temperatures of less than 90°C, a milk run by comparison, you could " ... get rid of all that [expense and complexity]."[21]

As discussed in the previous chapter, fluidized beds have an important advantage over most fixed-bed systems: the ability to predict their behavior as they scale up and to operate over a wide range of product and product grades without having to shut down and retool. The engineers pushing for the fluidization process convinced Joyce that it would be the quickest, cheapest way to go and offered the greatest degree of process flexibility: "I figured we're going to make with fluid bed because if we could, we could have low cost and make a broader product range."[22] Within the converter section, very fine catalyst powder intermingled with extremely pure ethylene gas to form a turbulent fluidized bed. Within this dynamic, roiling cloud, the ethylene was converted to the type of polyethylene dictated by the particular catalyst used, the resin forming both within and outside of the catalyst particles. As with fluid catalytic cracking and the fluidized polysilicone process, smaller units could quickly and predictably be expanded into large-scale operations.

During the time the team was developing new catalysts and adapting the process for fluid reactors, Joyce himself buttonholed carefully selected and influential executives within Carbide with whom he had formed particularly good relationships while serving as product and marketing manager. He now could lay out his plans, specify the technical and market successes he had achieved and the path he was taking to project completion. He knew these converts to his cause would then spread the message that this research appeared to "have legs" and should be supported to their superiors throughout the organization. It was now critically important for Joyce to demonstrate for senior executives his ability to create LDPE using the new technology. Demonstrating how rapidly laboratory setups could be scaled up to commercial proportions went a long way to impress the powers that be at Carbide and allowed Joyce to meet the demands of customers. The Promethean power and range of fluid technology amazed Joyce and served him well as a tool to bring to bear when discussing the merits of his approach to his superiors:

> One of the features it [gas phase reactor] had was when you make it on a small scale, the product you made was exactly the same as when you made it on big scale so you could scale things up right away [from only a few pounds per year to over a billion pounds annually]. That's an enormous advantage for making new products.[23]

This advantage in fact turned out to give Carbide an important competitive edge in the polyethylenes, an advantage Joyce would make sure management knew about:

> Everyone else [using fixed bed reactors] would do it on the bench, then they would build pilot plant, then a lot of times they would build a small plant. Well, they were using years of time. We were there long before they were anywhere, because we scaled up so quickly.[24]

By the mid-1970s, working closely with plant engineers and aided with improved catalysts coming out of Karol's lab, Joyce's gas-phase process turned out large amounts of low-pressure polyethylene at low unit costs. Different types of this material could be created using the same equipment without recourse to shutdowns or retooling. A family of polymers could simply and predictably be "dialed into" the system by adjusting in situ catalyst composition and operating conditions. Joyce recall that this exceptional plasticity of the gas-phase process—coupled with Carbide's proprietary catalysts—seemed to have the power to produce "an endless stream" of new polymers.[25]

The pliability of this new process meant that Joyce could expand the number of customers willing to purchase his LDPE. The large quantities of polymer produced, even at the semicommercial stage, made the LDPE available to more clients. The very low unit costs of the material, due to the inherent continuity of fluid operations, meant that Joyce could sell his polymer at reduced prices and still realize significant profits. Buyers also knew they could come back to Joyce and his team and ask for—and quickly get—variants of LDPE made to order for their own processing equipment or to satisfy the demands of their own customers.

The impressive tractability of the new low-pressure process is evident in the range of LDPE's it created even before reaching full-scale dimensions. It made polyethylene-jacketed material for telephone cables at "a third of the price of what one could make with high-pressure polyethylene, and with better performance."[26] Soon thereafter, it turned out LDPEs for molding applications. By the early 1970s, the process created an unexpectedly popular polymer called linear low-density polyethylene (LLDPE). Then in 1975, semicommercial fluidized units came out with a very special grade of LLDPE, one that could be turned into polyethylene film, the very epicenter of the world's polymer demand. Shortly thereafter, in the late 1970s, Joyce's technology fashioned a completely different family of advanced materials based on ethylene in the form of synthetic rubber, elastomers, and plastomers; these were also destined to play a major role in advanced polymer markets in the late 20th and into the 21st centuries.

Because of their ability to introduce so many commercially important polymers on a regular basis, the technology became a money-making machine for Joyce and his researchers—and for Carbide. Their still-small group managed to turn a significant profit that they then reinjected back into the project to support further development and scale-up. Historically research and development at Carbide—or anywhere else for that matter—rarely paid for itself " ... the corporation was used to no cash flow coming out [of R&D] ...."[27] Because Joyce could support his research effort with fluid-based processes, he could continue to expand his market base without having to tap corporate resources, thus helping him to keep his project under wraps. But eventually word got around of the activities taking place in Joyce's shop, and senior executives finally sat up and began paying attention to the new technology. But success also instigates resistance from many quarters. Because of the radical nature of the process, some managers at Carbide saw it as a threat to their own expertise and thus their positions in the company. Some failed to honestly appreciate the benefits of the new approach to polyethylene. From them, criticism over the direction of the project did not abate and in fact grew in intensity. Joyce feared these attacks could potentially delay, and even derail, his work. Here again, and quite effectively, the elasticity of fluid beds worked to the project's favor. One executive in particular, rather highly placed, complained that Joyce would not be able to scale his process up in a timely fashion, that " … it was not possible for [the group] to be able to do what [they] were trying to do, that [they] had no chance of getting there."[28] It is quite likely this critic had in mind the well-known difficulties attending the expansion of fixed-bed reactors under high pressure. Joyce and his team responded to this challenge by taking their fluid bed process and readily accomplishing what their detractor had said could not be done. By successfully facing down his critics in such a manner—and having the tool allowing him to do it—Joyce, as he put it, "undermined" them and put an end to potentially dangerous attacks before they had a chance to gain destructive force within the organization.

6.2.4 The Unipol Process: Commercialization Phase

By the late 1970s, Carbide gave Joyce's technology a name: the Unipol process (a combination of *Uni*on Carbide and *Pol*yethylene). By this time, the process had successfully come through development and scale-up and clearly was superior to the existing high-pressure method for making polyethylene both in terms of product range and production economies. The company now had to face a critical decision: to continue to pour hundreds of millions of dollars a year into the project to prepare the Unipol process for full commercial production and release it into the world or to put a stop to the project once and for all and save that money to improve its existing high-pressure polyethylene plants around the world. Whether to fish or cut bait came down to one question: Was Carbide willing to cannibalize—essentially scrap—its existing fleet of high-pressure reactors? Creative destruction coming from outside was one thing, but a company bringing it upon itself was something else entirely; it had control—at least in the near term—whether to do it or not; DuPont did it when it invented nylon, but by the 1930s, the demand for rayon was already in decline. In any case, DuPont's top management did not make the decision to commercialize nylon with the intention of replacing the older fiber; it expected the new synthetic to find its own markets. Carbide faced a far different situation. The polyethylene market was growing and the Unipol process was a radically different way to make this polymer and the family of advanced polyolefins; and no other competitor was even close to developing such a process. The power structure within the company would have to consciously and with intent decide to tear down existing plants and build from the ground up brand new fluidization units.

Never before had a chemical company been faced with such a decision on such a scale. That it had gotten this far is a testament to Karol's talent as a chemist and, even more, Joyce's skill as a visionary and intrapreneur. He had coaxed—even manipulated—Carbide to go along with his project. He had done so by simply continuing to move ahead insistently and in secret. But now top management was privy to the project, and they were now the ones to have to decide on whether to take this final, fateful step. The stakes were very high, to say the least: "We had [our old high-pressure] facilities all over the United States, many, many reactors … Some people were watching us all along because they wanted to figure out 'Is this really going to happen?'."[29]

While Carbide management wasn't sure which way to go on this, Joyce once again moved in to push the company toward his way of thinking. This last campaign was possibly the most challenging. Whereas his previous success rested a great deal on technical achievement and marketing savvy, it now depended on altering the cultural make up of an entire company. Success was contingent on knowing what Carbide managers perceived to be the greatest risks in cannibalizing and finding a way to minimize these risks in their eyes, that is, convince them that these dangers were not so great or that not proceeding to cannibalize would be the far more dangerous course.

In his crusade to reshape Carbide's cultural complexion in order to get the firm to destroy its own working plants to make way for the Unipol process, Joyce exploited a general—and disturbing—trend within the chemical industry, convinced the company that a solution to this problem demanded embracing creative destruction, and articulated a realistic and persuasive strategy for proceeding along these lines. By the late 1970s, Peter Spitz, as historian of the chemical industry reminds us, American industry appeared to be on the decline: "In the 1970s, it became fashionable to characterize such superannuated manufacturing segments as 'smokestack' industries, which were considered to be in a state of [decay] from which they would not recover."[30] High inflation, energy shocks, and growing competition, especially from Japan, took their toll on such industries as steel and automobiles. Chemical manufacturing was also very hard hit. Particularly frustrating to chemical companies was the problem that, even when they made the effort and came out with important new products and processes, they could not capture the profits from them that they once did. In addition to greater competition from Europe and Asia, there were more firms than ever before fighting over the markets for particular chemicals. One of the main reasons for this was the growing role played by engineering consulting firms such as MW Kellogg and Scientific Design. These outfits helped to lower the barriers for entry into high growth products by being able to rapidly design efficient and relatively inexpensive turnkey plants that could be readily purchased and constructed and put into operation by upstart firms wanting a quick and painless way into potentially profitable chemical intermediates and synthetics. Large, established firms, like Carbide, which previously dominated particular chemical markets and could count on receiving healthy returns from their research and development projects over a number of years, now had to share their profits with newcomers who did not have to spend their money on R&D in order to become major players in new technology. These start-ups seriously ate into the total earnings captured by large, established firms from their own innovations thus extending the amount of time it took for the true creators to begin making any sort of profits from new products and processes.[31]

And there was another important and very practical factor to consider. Given the financial problems faced by Carbide, how could it even begin to think about junking its existing high-pressure technology to make way for Unipol. Carbide had plants serving in important markets around the world; it would have to spend many billions of dollars to realize this effort at creative destruction. It simply did not have the funds in reserve to do this and no massive infusion of cash going into Carbide's coffers appeared in the foreseeable future. Even if the money could be found to undertake this mammoth project, the disruptions in production schedules that would necessarily occur during the changeover would threaten Carbide's ability to retain its existing customers, especially when aggressive and efficient start-up firms were looking at opportunities to exploit any and all market openings.[32]

Given these economic realities, it would make perfect sense for Carbide's corporate management to have decided once and for all to cut its losses from the Unipol experience; transfer useful knowledge, techniques, and equipment design from that project to improve its existing polyethylene plants; and shut down the glorious experiment before any full-time commercial units saw the light of day. But Joyce, always the entrepreneur, understood full well, probably better than anyone else in the company, what Unipol meant to Carbide and that it could position the company to regain its place as the most important polymer producer in the world going into the 1980s. In short, he essentially told management that they had in their hands the most important technological innovation that their company ever possessed, that it gave them untold leverage in the world of advanced materials, and that they would be foolish in the extreme to waste this great opportunity and not play it out to its furthest limit. Still, Joyce had to convince his superiors to put aside their short-term concerns and embrace a longer-range strategy. He managed to do this by emphasizing to them the current problems the company was having with its existing polyethylene plants and how the Unipol process could be used to solve these immediate issues. These difficulties were certainly well known to the company leadership. The field was simply too crowded. Too many producers were making too much high-pressure polyethylene for the existing market. Given that its plants embodied old and familiar technology, Carbide could not distinguish itself in the industry. The 1960s were particularly difficult years for the polymer and for Carbide. Joyce argued that embracing Unipol could alleviate these difficulties and bolster the company's resources. The new process—a truly disruptive technology—would unquestionably place the company at the head of the pack. The Unipol process was nothing less than " ... a huge new revolution, because you have high pressure reactors, all over the world, making these materials, and Carbide coming along and saying ... [w]e can do this by a low pressure process, with better economics, in terms of investment and operating costs."[33] With Unipol, Carbide could not only make its polyethylene cheaper than any of its high-pressure competitors, but it could also produce superior synthetics in the form of LLDPE, a major new, advanced material not available from the existing high-pressure technology.

The economics of polyethylene itself was another problem and another reason to scrap current plants in favor of Unipol technology. More than many synthetics, the polyethylene business was cyclical in nature which meant profits that the company made one year would drain away the following. Joyce again pointed to Unipol as the key to resolving this dilemma. Unlike high-pressure technology, Unipol, as a brand new technology, could be licensed out to other producers for fees and royalties. While Carbide traditionally did not license its new and most important processes—there were in fact those in the company extremely antagonistic to the idea—Joyce could mute the dissenting voices by calling attention to three simple facts: first, by pursuing this tactic, Carbide would no longer have to suffer the problem of cyclicality; it could receive a constant income stream from its innovation from its licensees: "When we put the licensing business into our portfolio, we had a source of more steady income. Then we could ride the cycles, using licensing income."[34] Second, Carbide would not have to pay to put up these plants; the costs would be borne by companies who license the process. Third, Carbide faces little risk when demand for polyethylene slackens since its licensees will have sunk their—and not Carbide's—money in the plants; they, not Carbide, will face the brunt of the financial problems that come with underutilized capital. Joyce further pointed out that the fact that the Unipol process is so flexible—different products can be made within a single plant—will entice more companies to become licensees at terms highly favorable to Carbide. As Joyce explained it:

> One of things we learned when we were selling polyethylene licenses [was that] if our customers thought we could make polypropylene in the same equipment, anybody who was sitting on the fence would make a decision [to license] right away: "I better go this way because I could make whatever I want."[35]

Essentially, Joyce presented Carbide with a way to extricate itself from its current polyethylene problem and obtain a (more or less) steady—and robust—cash flow in the process. Furthermore, he stressed the minimal risks involved in this approach: "There are ways to take the risk out of things ... – we figured out how to keep it manageable—so if it [the Unipol licensing strategy] went bust we [Union Carbide] didn't die."[36] Joyce painted a very appealing scenario to Carbide's top executives and, although it involved significantly changing the way Carbide had historically done business, they took the bait. In fact, his arguments actually "changed the perspective [within Carbide] of the polyolefin business."[37] By offering a convincing answer to an immediate problem, Joyce actually effected a major cultural shift and along the way got management to welcome a long-term strategy for the company. In the final analysis, Joyce persuaded Carbide to revamp its traditional culture over two dimensions: from a company that was static (noncannibalistic) to one that accepted a certain level of dynamism (cannibalistic) and from one that was internally oriented (no licensing) to one more externally positioned (licensing).

Carbide's first major Unipol licensees were Exxon and Mobil (when they were still separate companies), who hoped to get into the polyolefin business. As Joyce predicted, the number of licensees continued to grow so that by 1997 there were nearly 70 Unipol reactors operating around the world, with this number reaching over 90 reactors by the turn of the 21st century and 150 reactors in 2012. In 2012, Unipol reactors worldwide turned out over 35 million tons of polyethylene annually and accounted for roughly one-third of all HD and LLDPE in the world (with license fees and royalties now collected by Carbide's parent organization, Dow Chemical). As of this writing, the technology continues to expand its presence globally, and by 2025 it is expected that the Unipol process will account for around 60% of the world's polyethylene.[38]

In our chapter on Nucor and advanced steels, we saw how an advanced process (specifically, thin strip casting) led to the production of a whole new group of advanced materials known as microalloys. Similarly, the power of Joyce's and Karol's Unipol process has made possible the creation of an important family of polymers known as the metallocenes. These are discussed in the following section.

6.3   THE UNIPOL REVOLUTION AND THE METALLOCENE POLYMERS

The importance of the Unipol process extends well beyond the polyolefins. It also ushered in what some consider one of the most significant technological revolutions in advanced materials in the late 20th and 21st centuries, namely, the metallocenes, or more precisely, man-made materials produced with metallocene catalysts. These catalysts, which first operated commercially within the fluidized reactors of the Unipol process in making LLDPE, have permitted for the first time different types of synthetic polymers to be created with an especially high degree of preplanned accuracy. As a result, the metallocenes have been at the center of a new age of custom-made commodity synthetics. The demand for metallocene catalysts is expected to continue its rapid growth during the first half of the 21st century. Intensification of competition resulting from expanded synthetics capacity within China and the Middle East is forcing producers in the United States, Europe, South Korea, and Japan to find new polymers with special market applications. This need, in turn, has put the spotlight on metallocene technology with its ability to create diverse types of materials with great precision and so differentiate the product lines offered by advanced materials companies.[39]

6.3.1   Science and Technology of the Metallocenes

The metallocene catalysts are a subset of the organometallic family of compounds, so called because they are composite materials made of metal particles encased in some organic substance.[40] The chemists Karl Ziegler and Giulio Natta—both sharing the prize in 1963—discovered the first industrially important organometallic catalysts in the 1950s. As previously noted, these catalysts became extremely important in polyethylene production as they could make LDPE without recourse to high pressures. Until the arrival of the Unipol process, the Ziegler–Natta catalysts were the most important innovations in polymer production in the postwar era. As impressive as they are, they have their problems. The most important drawback of these catalytic materials is that they cannot be made in such a way as to predict the exact nature of the polymer they produce. In fact, they tend to turn out complex mixtures of polyethylene molecules from one batch to another that tend to dilute—even destroy—specific properties needed for targeted applications. The problem is that the catalyst's lattice structure contains many metal atoms or ions, each providing a different reaction site that synthesizes different types and weights of polyethylene material. The final product that comes out of the reactor is actually an unpredictable amalgam. This means that manufacturers cannot make a pure form of the plastic nor can they know for certain what physical properties a particular shipment of polymer will have.

The metallocene catalysts revolutionized polymer materials by introducing a much greater precision in the synthesis of man-made materials. Metallocenes are organometallics with a difference: they have only one metal atom or ion sandwiched in by two layers of "bread" in the form of cyclic, five-sided organic molecules. This means that there is only a single reaction site where reaction actually occurs and so only one type of polymer produced. Further, by varying the shape of the organic molecule surrounding the metal atom, and changing the type of metal atom, chemists can "dial in" the precise structure and properties they want the final polymer to have for particular applications. For instance, metallocenes can produce thinner polyethylene film with tighter sealing ability for packaging, clearer polymer for use in electronics and fiber optics, and stronger and moldable plastic for advanced engineering applications. The metallocenes also enable the combination of previously incompatible polymers—or comonomers—to make entirely new materials. One such "interpolymer" results from the polymerization of ethylene and styrene resulting in an ethylene–styrene material exhibiting superior flexibility, processability, and formability.

6.3.2   The Metallocene Era and Advanced Materials

Chemists actually have been studying the metallocenes for most of the 20th century but did not consider them candidates as practical catalysts until the 1980s. At that time, two developments came together to propel the metallocenes as major players in the advanced materials field. Firstly, two German chemists—Walter Kaminsky (University of Hamburg) and Hans H. Brintzinger (University of Konstanz)—demonstrated experimentally that the addition of a cocatalyst (methylaluminoxane or MAO) to a metallocene greatly increases the latter's chemical activity, that is, it allowed greater throughput under lower temperatures and pressures. These results alerted the advanced materials community of the industrial potential of the modified metallocene catalyst system. This possibility set off a flurry of research activity in the 1980s by the likes of Dow, Exxon, Fina, Mitsui, and other major petrochemical producers, including Carbide itself, all of whom were in search of tailored design and highly engineered polymers.[41] The second important development was the commercialization of Carbide's Unipol process. In the 1980s, Carbide and its global licensing strategy showed the world the effectiveness of a fluidized plant in the synthesizing of the most important polymers known to man. Since the catalysts first used in these systems were related to the metallocenes, the next step was to test actual metallocene catalysts in these reactors. In 1991, Exxon, Carbide's first major licensee of its Unipol technology, took the lead with introduction of its own proprietary metallocene catalyst called "Exxpol" (a combination of "Exxon" and "Unipol"). This catalyst in conjunction with the gas-phase process was the first to produce volume amounts of high-performance polyethylenes, including LLDPE polymer. The cow was now out of the barn. Within the next few years, other companies, including Dow, Chevron, Mitsubishi, and Hoechst came out with their own metallocene systems. By 1994, total world production of metallocene polymers reached nearly two billion pounds with the industry's leaders, Carbide and Dow, together accounting for over half of this amount.[42]

Despite these early successes, the metallocene industry faced its share of problems that has slowed down its expansion in the mid-to-late 1990s and early 2000s. These included patent battles fought between various companies, including Exxon,

Mobil, and Phillips and difficulties that customers found in processing the metallocene polymers into different forms and shapes on their existing equipment. But, by the early 2000s, these difficulties had been significantly reduced. Extensive merging and joint venturing activity promoted patent agreements between companies and so reduced the number of major combatants in the patent wars. An important example of this trend was Exxon and Union Carbide forming the joint venture Univation Technologies in the late 1990s to develop and market advanced metallocene systems by combining Exxon's Exxpol metallocene catalyst with Carbide's gas-phase process.[43]

The merging of the new catalyst with gas-phase reactor technology had another very important benefit: these flexible systems could turn out all sorts of different types of polyethylenes—with different degrees of branching—that were more readily processed on different clients' own equipment. Thus, customers could purchase the metallocene materials without having to scrap existing production lines. This was of enormous importance in winning over customers and expanding the market for metallocene polymers. Thousands of mom-and-pop plastics processors could now buy the polymer for their operations. This is not only true in the United States and Europe but increasingly in Asia where fabrication equipment suppliers tend to sell older vintage equipment. Their customers, once they install this machinery, do not have the money to replace it with the latest equipment technology from the west. In this region especially, fabricators need the capability to run widely different polymers (e.g., LDPE, LLDPE, HDPE, and even PP) on the same processing equipment. Gas-phase metallocene technology satisfies this demand.

With the intellectual property tensions dying down and the housing of these advanced catalysts in fluid bed plants, the metallocene industry has grown rapidly since 2001. Between 2000 and 2014, the portion of the important LLDPE market taken over by metallocene polymer grew from less than 15% to more than 20% (with the rest dominated by polyethylenes made from Ziegler–Natta-type catalysts). By 2025, metallocene technology will have extended and (many believe) solidified its domination of the global polyolefin market.[44]

REFERENCES

1. The Economist. (2001), "Designer Plastics," *Technology Quarterly*, Q4. http://www .economist.com/node/885143. Accessed November 25, 2015.
2. The following discussion on DuPont's Polychemicals Department and its experiences with Delrin and polyethylene comes from Hounshell, D. and Smith, J. (1988), *Science and Corporate Strategy: DuPont R&D, 1902–1980*, Cambridge, UK: Cambridge University Press, pp. 474–497.
3. This narrative of DuPont's struggle with polyethylene comes from a number of sources including Hounshell and Smith, pp. 491–500, and Spitz, P. (1988), *Petrochemicals: The Rise of an Industry*, New York, New York: John Wiley & Sons, pp. 257–268; 331–338.
4. Hagley Library, DuPont Company Records, and Chilton, T.H., "Plans for Chemical Engineering Work at the Experimental Station," Accession 1784, Box 22, FN 27.

5. Starkey, J. (June 12, 2011), "Dupont Pays No Tax on $3B Profit, and It's Legal," New Castle, Delaware: The News Journal. http://www.webcitation.org/5zOsjuvA5. Accessed November 28, 2015. See also Dreibus, T. (December 22, 2015), "Dow-DuPont Merger Could Create Opportunities for Investors," *Yahoo Finance*. http://finance.yahoo.com/news/dow-dupont-merger-could-create-145923802.html. Accessed December 24, 2015.

6. See for example December 1, 1968, "Giant with a (Giant) Headache," *Forbes*, pp. 24–26.

7. Frederick J. Karol, interview by James J. Bohning in Bound Brook, New Jersey on January 10, 1995 (Philadelphia, Pennsylvania: Chemical Heritage Foundation, Oral History Transcript #0125), p. 31.

8. For a history of Union Carbide Corp., see Spitz (1988), pp. 65–82; 101–107; 251–264; 398–409; 468–474; 507–510, and Trescott, M. (1981) *The Rise of the American Electrochemicals Industry, 1880-1910: Studies in the American Technological Environment*, Westport, Connecticut: Greenwood Press, pp. 173–206. Aftalion (1991) discusses Union Carbide at various points in his narration of the US chemical industry. See also "Union Carbide Corporation History" at: http://www.fundinguniverse.com/company-histories/union-carbide-corporation-history/. Accessed March 29, 2016.

9. The following narrative of the Unipol process is based on interviews conducted by the Chemical Heritage Foundation with the two major players in the development and commercialization of the Unipol process. One is a 2004 interview of William Joyce (Joyce, interview by Arnold Thackray at Nalco Company, Naperville, Illinois, August 17, 2004, Philadelphia, Pennsylvania: Chemical Heritage Foundation, Oral History Transcript #0288). The other is an interview of Frederick Karol (Frederick J. Karol, interview by James J. Bohning in Bound Brook, New Jersey on January 10, 1995, Philadelphia, Pennsylvania: Chemical Heritage Foundation, Oral History Transcript #0125). See also Spitz (1988), pp. 507–510 and Karol, F. and Jacobson, F. (1986), "*Catalysis and the Unipol Process*" *Studies in Surface Science and Catalysis*, 25: 323–337.

10. CHF interview of William H. Joyce, p. 17.

11. Ibid., pp. 20–21.

12. Ibid., p. 22.

13. Ibid., p. 31.

14. CHF interview of Frederick J. Karol, p. 4.

15. Ibid., p. 18.

16. CHF interview of William H. Joyce, p. 33.

17. Ibid., p. 31.

18. Ibid., pp. 31–32.

19. CHF interview of Frederick J. Karol, p. 6–10.

20. Ibid., p. 5.

21. Ibid.

22. CHF interview of William H. Joyce, p. 32.

23. Ibid., p. 32.

24. Ibid., p. 33.

25. Ibid., p. 36.

26. Ibid., p. 31.

27. Ibid., p. 32.

28. Ibid., pp. 33–34.

29. CHF interview of Frederick J. Karol, p. 7.

30. Spitz (1988), p. 527.

31. Ibid., pp. 317–331. Spitz also credits forced information sharing between companies during World War II and a growing tendency of chemical companies to license out new technologies for new petrochemical-based technologies becoming more broadly available to smaller start-up companies.

32. CHF interview of Frederick J. Karol, pp. 9–10. In his interview, Karol gives Joyce full credit for conceiving and championing the strategy of licensing out the Unipol process.

33. Ibid., pp. 9–10.

34. Ibid., p. 27.

35. CHF interview of William H. Joyce, p. 34.

36. Ibid.

37. CHF interview of Frederick J. Karol, p. 27.

38. Fraser, W., et al. (1997), *"Manufacturing Efficiencies from Metallocene Catalysis in Gas-Phase Polyethylene Production,"* Bound Brook, New Jersey: Univation Technologies (Whitepaper), p. 3; Ferenz, P. (2013), *"UNIPOL PE Process Technology: Advances for the Dynamic Indian Market,"* Bound Brook, New Jersey: Univation Technologies, LLV (Powerpoint), pp. 3–7.

39. Tullo, A. (October 18, 2010), "Metallocenes Rise Again" *Chemical and Engineering News, 88*(42): 10–16. See also, Fraser et al. (1997), pp. 10–16.

40. The following discussion on metallocenes is based on a number of sources including Thayer, A. (September 11, 1995), "Metallocene Catalysts Initiate New Era in Polymer Synthesis" *Chemical & Engineering News, 73*(37): 15–20; Tullo, A. (October 18, 2010), *"Metallocenes Rise Again"*; Seymour, R. and Cheng, T. (1986), *History of Polyolefins,* Dordrecht: D. Riedel Publishing Company, See also The Economist (2001); and Flank, W, Abraham, M., and Matthews, M. (eds.) (2009), *Innovations in Industrial and Engineering Chemistry: A Century of Achievements and Prospects for the New Millennium,* Washington, DC: American Chemical Society.

41. Thayer (September 11, 1995), pp. 15–18.

42. Ibid.

43. Fraser et al. (1997), pp. 1–3.

44. 2011, "Prospectus: Global Metallocene LLDPE Markets, Technologies & Trends: 2011–2020," Houston, Texas: Chemical Market Resources, Inc. and Report Abstract. (2009), *"Linear Low Density Polyethylene (LLDPE),"* San Francisco, California: Nexant, Inc.

PART III

FUNCTIONAL MATERIALS

Semiconductors

7

ADVANCED MATERIALS AND THE INTEGRATED CIRCUIT I

The Metal-on-Silicon (MOS) Process

It has today occurred to me that an amplifier using semiconductors rather than vacuum [tubes] is in principle possible

William Shockley, 1939

In the last part of our story of advanced materials, we focused on the so-called structural materials. These are the metals, fibers, plastics, and composites that are made in metallurgical plants and chemical and petrochemical complexes to make components and assemblies for automotive, aerospace, industrial, infrastructural, and biotechnical applications. The upstream advanced materials firm makes the actual metal, polymer, or composite and sells it to its immediate customer, the fabricator, who turns it into a useable part for an original equipment manufacturer (OEM).

We now turn to a group of products known as the functional materials more generally known as semiconductors. The primary purpose of the semiconductor firm is to design and make electronic components, devices, and (sometimes) systems for a wide variety of customers. At the heart of these components, devices, and systems—transistors, memory chips, microprocessors, and solid-state lasers—are advanced materials in the form of semiconductor compounds and composites.[1] If the structural materials dominate new developments in construction and transportation, the functional materials do so for 21st century information and communications. In this chapter, we explore the rise of the transistor and integrated circuit, focusing in on the role of advanced materials technology. We particularly note the importance of

Advanced Materials Innovation: Managing Global Technology in the 21st century, First Edition. Sanford L. Moskowitz.

a company—Fairchild Semiconductor—in the early development of these devices and discuss how and why it faltered so badly in taking semiconductor technology to the next level. As we shall see, failure to perfect the "metal-on-oxide" (MOS) transistor and integrated circuit permanently marginalized the company within the semiconductor industry.

7.1   BACKGROUND

Through the 19th century, scientists knew from their research that materials could be classified in one of two ways: conductors—usually metals—and insulators, where the former allowed electric charge to move through them while the latter did not. But by the later 19th century, physicists and chemists were intrigued with a third type of material, one that appeared to possess both properties in the sense that under some conditions they readily conducted electricity but under other conditions they hardly did so at all. The semiconductors occupy groups III–V of the periodic table and include such elements as silicon, germanium, gallium, indium, arsenic, boron, and antimony.[2] But why these strange materials behave as they do remained a scientific mystery until the 1930s and the rise of quantum physics. Research into their possible application in electronics did not begin in earnest until the end of World War II. Before the rise of the transistor, however, the vacuum tube reigned supreme as the king of electronic components in the early years of the information and communications revolution.

7.1.1   The Vacuum Tube and Advanced Materials

Before there were semiconductors and transistors, there was the vacuum tube, which evolved from the work done on Thomas Edison's electric light bulb.[3] Invented by the American physicist Lee de Forest prior to the First World War, the vacuum tube is composed of three distinct components—filament, wire mesh, and metal plate—all enclosed within a vacuum by a glass bulb. The vacuum tube revolutionized the radio industry in two ways: it was the first practical device that could convert the alternating current signals sent out by radio stations and picked up by antennas of listeners into direct current needed to operate the radio's speakers and earphones, and it could amplify long-distance signals.

The coming of the vacuum tube meant that wireless radio could finally reach a mass audience. In the years leading up to World War II, it also played an important role in the telephone industry. American Telephone and Telegraph (AT&T), looking to remain competitive following expiration of its original Bell patents, purchased the rights of the invention to use as amplifier in its telephone network by which it periodically strengthened voice signals sent by users over long distances. This technology meant that AT&T could monopolize transcontinental—coast to coast—and eventually international telephone service. Soon, other electrical equipment manufacturers on both the East and West Coasts entered into vacuum tube production. In addition to the large eastern companies—RCA, Motorola, and General Electric—manufacturers around the San Francisco Bay area began to emerge as important vacuum tube makers. In addition to well-known outfits such as Federal Telegraph Company and Heintz & Kaufman, a number of start-ups came on the scene during the decade.

Engineers began pushing the vacuum tube to its limits by designing it to send and receive short-wave signals with frequencies many times greater than radio waves. This set the stage for the device's future applications in radar for the military and as the central component in such post-World War II inventions as television and the early computers.

The rate and direction of vacuum tube development depended mightily on the application of advanced materials. The filament material, for example, evolved from oxide-coated fibers to tungsten filaments modified with the radioactive alloy thorium. This improvement in materials technology increased the tube's power output and efficiency. Advanced heat-resistant sealants were developed that could withstand high thermal shocks and tightly joined the metal electrodes to the glass envelope. New types of "getters" made of the relatively rare metal tantalum replaced the magnesium pellets patented earlier by RCA and helped established better vacuum conditions within the tube by absorbing any fugitive gases and thus heightened the device's operating performance. By the 1940s, tantalum would be replaced by a totally new material developed by the firm Eitel and McCullough called "Pyrovac." This zirconium–carbon composite material was superior to tantalum as a getter and thus sped up the manufacturing process and increased the operational lifetime of the vacuum tube. The tube grid or mesh—the component De Forest added to the electric light bulb to create the vacuum tube—also came in for some radical changes. Originally made of platinum, the wire grids did not do well under high-temperature conditions. Their structure weakened often resulting in tube failure. A new grid-making process, involving the coating of molybdenum wires with carbon, platinum, and zirconium and the sintering of the metal composite in a high-temperature furnace, produced grids that were stronger structurally and possessed superior electrical characteristics (including reduced instances of electrical malfunction).

The vacuum tube remained the essential component in electronic systems for both military and commercial use through the decade. It was employed in radar systems during the Korean War and in radio and television sets for over a decade. It also found itself as the central device in the first computers. But, by the late 1950s, scientists and engineers began to understand the severe limits that vacuum tubes placed on future electronics devices and systems. They were large and bulky. They took time to warm up and had to remain heated during downtime. This meant that they consumed massive amounts of power. They were fragile and easily broke in transport, during installation, and in service. They generated an immense amount of heat that had to be removed by some means; fires posed constant danger. Degradation of surrounding components and wiring because of the heat that the tubes generated was a continual problem, especially in computers. The early computer ENIAC, which the Army used to plot antiartillery trajectories, weighed 30 tons and required the service of 18,000 vacuum tubes.[4] With some tube breaking every few minutes, generally through overheating, an actual basketful of tubes had to be on hand and a number of men stationed at different locations to replace the ones that burned out. In addition to all this, vacuum tubes were expensive because they were not easy to mass produce, and quality control was always an issue. To a large extent, cold war strategy and the space program rendered the vacuum tube obsolete: the intercontinental ballistic missile's electronic systems and NASA's rocket booster and space craft designs called for the use of highly compact, extremely sturdy, and ultrareliable electronic devices. Even the most sophisticated and powerful vacuum tubes featuring the most advanced materials could not come close to meeting these rigorous requirements. Bell Laboratories found a solution to this and other problems posed by early electronics components by inventing a handy little device that, quite simply, changed everything.

7.2 BELL LABS AND THE POINT-CONTACT TRANSISTOR

The transistor is often cited as one of the most important inventions of the 20th century. The wide range of innovations that, in aggregate, makes up our 21st century information technology infrastructure, derives in some way from the transistor. Even before the existence of the PC and Internet, the transistor was a revelation. All of the problems of the vacuum tube—bulky glass tube, tricky vacuum requirements, heat generation and "burnout," downtime for tube replacement, time-consuming warm-up—disappeared in an instant. And beyond all this, the transistor did everything the vacuum tube could—rectification, amplification, and "on/off" switching—and did them far faster, efficiently, and reliably. By the mid-1950s, the military began using the transistor in its missiles and communications systems and businesses in their computer technology. But it wasn't just for the specialist. The public welcomed it as well, first as a component in hearing aids and portable radios and then, by the 1960s and 1970s, in the form of the integrated chip (IC), pocket calculators, watches, and other consumer products. By the 1990s and 2000s, smaller and more powerful ICs became inextricably linked to the rise of the PC, portable computers, and mobile communications.

More than is true with the vacuum tube, the story of the transistor is the story of advanced materials in the late 20th and 21st centuries. As we have seen, the vacuum tube relied extensively on new materials in order to operate at higher powers and frequencies. But this dependence on materials had its limits. As Christophe Lecuyer reminds us in his history of Silicon Valley, the advances in vacuum tube technology depended as much, if not more, on tube design and advanced manufacturing processes as on the application of new materials.[5] In contrast, the transistor was born out of and is really inseparable from the story of new materials and the most advanced materials processes.

7.2.1 Bell Labs: The Early Years

The invention of the transistor took place within a corporate research setting and specifically within AT&T and the legendary Bell Labs.[6] The first transistor was no more than a simple demonstration prototype. It proved the principle of transistor action but by itself was fairly useless. It was as amazingly complex and unwieldy piece of jerry-rigging that ever announced a technological revolution. The inventors, although steeped in the relatively new theories of quantum mechanics, never really could explain why it worked. They came upon it only after a series of missteps led them to the Rube Goldberg-like contraption that achieved something never pulled off before: the amplification of an electrical signal using electron currents within a solid material. This invention placed solid-state physics firmly on the map of American—and world—technology. It also raised Bell Laboratory, the place where this technological upheaval began, to the very pinnacle of American post-World War II research and development activity.

At the end of WWII, Bell Labs, which began life in the late 1930s as a simple joint venture between AT&T and its manufacturing arm, Western Electric, made the conscious decision to restructure itself around solid-state physics and in particular semiconductor research. Its goal was to take the results from quantum mechanical theory developed by the great European physicists and transferred to their American students and harness these to the creation of a modern communications technology. Bell Labs took this step because it was well aware of how important corporate R&D had already been to the success of such American companies as General Electric (GE), DuPont, and Standard Oil of New Jersey (Exxon) earlier in the century. It too hoped to generate a stream of new technologies to upgrade its products and services and so be well-positioned to beat back a growing postwar competitive threat.

By 1946, Bell Labs employed around 11,000 people finding ways in which to apply Bell's wartime experiences to the AT&T communications systems. One-third of these employees were professional scientists and engineers, one-third technical assistants, and the remainder clerical and support staff. AT&T soon moved its Bell facilities from its Manhattan space along the Hudson River to a spacious research campus in Murray Hill, New Jersey. This layout became the model for other postwar R&D facilities around the country including IBM. The US government funded many of Bell's projects in the postyears. In the decade 1949–1959, it backed more than $600 million of research, which accounted for more than half of Bell's R&D budget. In addition to research money from the military, AT&T funded its own R&D through a built-in "R&D tax" on every customer within the AT&T system. Bell spent most of this money on smaller projects. While Bell is often thought of as dedicated to pure "blue sky" research, it must be remembered that during the 1940s and 1950s, the vast majority of Bell personnel—over 85%—spent their time on incremental innovation for the day-by-day improvement of AT&T's national and international phone systems. Nevertheless, it is certainly true that AT&T wanted a small hub of its work to be dedicated to fundamental long-range projects, with the important caveat that the research team show potential—if not immediate—application at the end of the day.

By the 1950s, this elite group of scientific researchers was the cream of the crop, the most influential of the organization. Their reputation for important scientific research then shaped the predominant culture, one that embraced the academy and its priorities: fundamental scientific research, publications in leading journals, and presentation of papers at major scientific conferences. Bell Labs came to value recognition and reputation within the international scientific community as much, if not more than, innovative output. It naturally embraced the new physics and conducting research in it as a way to gaining a reputation for scientific excellence as well as a road map to major innovation. Quantum physics diffused into Bell by two main routes: new hires from the major universities who studied under European scientists and professional conferences and summer school programs put on by university departments of physics at which the "greats" of quantum mechanics, mostly Europeans, came to impart their knowledge and to inspire a new generation of American scientists and engineers.

7.2.2 Bell Semiconductor Research: The Leading Players

The guiding spirit of Bell Labs and its pursuit of science-based electronics was its legendary postwar director. Mervin J. Kelly, a Ph.D. in physics from the University of Chicago, has been described as a visionary.[7] And indeed it appears that he was. His main goal was to modernize AT&T's archaic telephone switching systems and get it ready for the projected increased demand that was expected to take place starting in the 1960s. Specifically, he wanted to replace the old electromechanical relays in telephone exchanges with the more efficient solid-state devices. He strongly believed that semiconductor research had to be based on the most advanced science, and this meant quantum physics. He expected quantum mechanics to provide a "unified approach to all of [Bell's] solid-state problems." Kelly also wanted the multidisciplinary and practically oriented approach that proved so powerfully successful during the Manhattan Project. The men he hired for Bell would therefore need not only be top ranked in their field—and therefore mentored by the great European physicists—but also very un-European in the sense of being less interested in grand theories and more intrigued with practical application.[5]

Happily for Bell and AT&T, Kelly recognized talent when he saw it and took pains to nurture and direct it. Even prior to the war, he latched on to the three future Nobel laureates who would play the foundational role in transistor technology: John Bardeen, Walter Brattain, and William Shockley. These three scientists had much in common. All three came from the mid- or far-western part of the United States. All were born in the first decade of the 20th century and showed interest at young ages in the inventions of the new age, especially electrical devices and radio technology. They all were deeply steeped in quantum theory through contacts—in academia and at international conferences—with the European pioneers in the field. They were all committed to exploring the potential applications that they believed could be coaxed out of new scientific developments. All three, who died in the last quarter of the last century, lived long enough to witness their invention, the transistor, stand at the center of the greatest transformation of society since the industrial revolution. These similarities, notwithstanding, these three scientists possessed very different personalities. Bardeen, who would go on to win two Nobel Prizes in his career—the only physicist to ever do so—was the grand theorist uncomfortable with hands-on experimentation; he could explain laboratory results but not be the one to actually carry out the experiments. Brattain, in stark contrast, was a brilliant experimenter who had an uncommon instinct for laboratory work but was not particularly interested in constructing theoretical models. It comes as no surprise that these two scientists perfectly complimented one another. They worked closely together in their pursuit of the transistor. Each provided vital input in the search; neither could have succeeded alone.

And then there is William Shockley. Shockley was in many ways the outsider of the three. To the general public, of course, Shockley is famous (or infamous) for his views on intelligence and race and as the unflinching champion of eugenics. This is unfortunate indeed for the man was a truly exceptional scientist in the field he knew best. There is little question Shockley was an immensely creative spirit and one of the most important physicists of the 20th century. As Robert Noyce, the father of the integrated chip, recalls:

> Shockley was certainly one of the most creative men that I've ever known in my life.... He had a marvelous way of simplifying a problem and getting at the fundamental part of it, cutting away all of the extraneous information and getting a model simple enough to be handled mathematically or experimentally.[8]

A close colleague at Bell, Morgan Sparks, reminisces:

> There is no doubt that [Brattain and Bardeen] were brilliant physicists, but they were conventional, Shockley was unconventional ... [Shockley was a catalyst and was] highly competitive and hard-driving leader [who] knew how to challenge people and how [to] put things together."[9]

He was also highly cultured in language, literature, and music. Having been raised in Los Angeles in the post-World War I period, he readily absorbed and identified with Hollywood culture, which is to say, he had a sense of the dramatic and a narcissistic streak utterly lacking in his two colleagues. He was indeed extremely charismatic and so could galvanize a team of creative researchers around him and his research agenda. The other side of this coin shows a polarizing personality who divided loyalties so that those not on the "in" with Shockley and his research program were, "the other," not to be trusted and therefore shunned.

By many accounts, Shockley was the motivating force for Bell's work on the transistor. He was always ahead of the group in his thinking. He could work with many individuals. Even his penchant for excluding those who did not agree with his research approach could prove useful as he was able to ban from his circle individuals who he felt championed research paths that would lead to dead ends. This had the salutary effect of his research agenda pushing ahead efficiently toward useful results; on the other hand, it kept out potentially valuable information and insights, a problem that plagued Shockley when he left Bell to go out on his own.

Kelly certainly recognized Shockley's particular genius—whether he was blind to his deficiencies of personality is an open question. Even the often difficult and self-promoting Shockley came to appreciate Kelly's contribution to his own successful career leading to the Nobel Prize. When Shockley first came to Bell in the late 1930s, Kelly put him to work on improvement of vacuum tubes. But both Kelly and Shockley understood the limits of the tube. Both latched onto the idea that solid-state physics would lead to a new technology to replace Lee de Forest's now aging invention. More than any of his other scientists, Kelly took pains to indoctrinate Shockley in his vision for Bell and its role in leading AT&T into the modern world. Shockley proved a staunch ally of Kelly and his future plans for the company. In 1938, Kelly reorganized Bell's Physics Research Department. He put Shockley at the head of a newly formed group, concentrating on the discovery and understanding of new solid-state materials and devices. For a few years, Shockley and his team investigated the potential in electronics of different semiconductor materials such as copper oxide. They rejected them and looked to other elements and compounds. Shockley began to extend Bell's work on silicon and germanium that it had undertaken prior to the war. He soon learned that when a battery was connected to silicon, three sections formed on the inside: one region having a predominately negative charge (an abundance of electrons), one region with a predominately positive charge (a deficiency of electrons), and one very narrow band—a junction—separating these two larger regions. He learned further that this junction appeared to be a very active place where "things happened." Specifically, electrons from the negative region and strange entities (which Shockley would call "holes") from the positive area flowed together and met up at the junction. He was especially intrigued with the strange behavior of this boundary—which soon came to be known as the "p–n" junction. He believed it held the key to a new world of solid-state technology. He began to think of the possibility of amplifying electrical currents in either silicon or germanium by creating a device that manipulated what went on at this junction. Such an invention would be a revolution indeed.

Shockley turned to the theories of his beloved quantum mechanics to see how he might apply them to exploit the possibilities offered by the p–n junction. He conceived of a horde of electrons and holes, induced by an electric field applied outside of the semiconductor, to leave their positions deep within the semiconductor and move toward and congregate at the p–n junction. Shockley predicted that this subatomic traffic jam of charge carriers would enter the circuit causing the current to be augmented or amplified. A whisper of sound in a microphone would come out as a loud scream through the speaker. This describes the famous—or more appropriately infamous—"field-effect" phenomenon. But the theories proved to be useless in the real world. Applying his theory, developing his concepts, and completing his calculations, Shockley tested out experimental prototypes. The results were severely disappointing. Shockley found that his external electric field had virtually no effect on current in the semiconductor; even worse, he could not figure out why. With this failure, Shockley's ever-shifting attention moved on to other pursuits at Bell. He would not come back into the center ring of semiconductor research until the late 1940s.

7.2.3 The Point-Contact Transistor

Enter Bardeen and Brattain.[10] When faced with a device that would not behave as predicted, Shockley went to Bardeen to ask for help in explaining why the field-effect amplified current couldn't be observed. Bardeen, a theoretical physicist, concurred with Shockley's assumptions and calculations; he too believed that the external electric fields used by Shockley should have been more than sufficient to produce amplification. He and Walt Brattain got together to find an answer and come up with a working field-effect solid-state amplifier. He instinctively felt that the problem was occurring on the surface of the semiconductor. We have seen in past case stories of advanced materials innovation the importance of certain critical pieces of information obtained outside of the current research effort that find their way into the research phases of technology creation and that help push a current project along. In this case, we can identify academic work previously done by Bardeen as particularly relevant to the problem he was now facing. Bardeen recalled his doctoral dissertation, which examined the behavior of electrons on the surface of metals and was struck by the similarity in the type of issues he struggled with then and the kind of problem he was facing now. In applying his dissertation results to the issue at hand, he came upon the intriguing idea of electrons being "trapped" on the surface forming a barrier or shield preventing any external electric fields from entering into the semiconductor material. This would explain why Shockley's field effect was not observed. Now at least the researchers had a plausible model with which to work.

Bardeen and his team, now reinvigorated with direction and purpose and with renewed hope, could begin to construct experiments to test out Bardeen's model. Bardeen called upon Brattain to work with him on the experimental side. Of enormous importance now was Brattain's own piece of outside knowledge he obtained when working with Bell Labs' scientist Russell Ohl shortly before the war. Ohl, interested in studying the semiconductor behavior of silicon, had made silicon rods in a furnace, cut out sections, and then measured their electrical conductivity profiles. Brattain was extremely impressed with Ohl's early work and particularly how much signal amplification Ohl showed could occur using silicon. He began applying Ohl's methods in attempting to prove the existence of Bardeen's surface states and how they might be minimized so as to permit an outside electric field to enter the semiconductor in order to be amplified through the electronic activity occurring at the p–n boundary. This insight fueled a series of experiments that led the scientists to the chance discovery that if a drop of electrolyte (such as water) contacts the surface of the silicon, the troublesome protective electron barrier seemed to disappear, thus letting the electric field penetrate into the semiconductor material and, in turn, control and amplify the internal current. In other words, this simple solution allowed the field-effect device to actually work at least in a very limited way. This discovery led naturally enough to the next insight in designing a new device: jab a sharp metal point into the silicon and then surround that part of the surface with an electrolyte. By varying the voltage—which could be accomplished with a simple battery—an electric field would be created and, thanks to the presence of the electrolyte, there would be no protective film of trapped surface electrons to block that field. A drop of electrolyte (distilled water) was placed on the slab. The researchers pushed the tungsten wire down through the drop and into the piece of silicon. Over the next few weeks in the fall of 1947, they performed numerous experiments on this contraption to see if they could create an amplified current by increasing the strength of the electric field. However, while able to boost current, they could do so only minimally and the device did not seem to work with high-frequency signals, a requirement if it was to find practical application. They continued to tinker with the apparatus. They used empirical methods informed by a soupcon of theory tempered by their scientific instincts and Brattain's hands-on knowledge of p–n junctions. They replaced silicon with germanium, which was a "faster" semiconductor material; they also substituted electrolyte with an oxide layer on top of which they placed a gold dot to act as the outside field source. Their results were promising. But the greatest improvement came when Brattain had unknowingly removed the oxide layer leaving the gold dot physically touching the surface of the germanium. At that instant, the gold spot stopped working as an external electric field source and had become a simple electrode. This meant that they no longer had a field-effect device, but one stimulated by internal voltages that pumped electrons and holes directly into the semiconductor. In fact, it wasn't clear to either scientist how or why this device worked at all. For some unknown reason the gold contact was inserting into the germanium "charged carriers" that appeared to travel within a narrow channel just under the surface. Given the direction of the observed current, these had to be positively charged carriers.

The design of this first working solid-state device—the so-called "point-contact" transistor—was less than impressive. It was nothing more than an inverted polystyrene wedge about an inch high, its edges covered in aluminum foil, and one of its points imbedded into a slab of germanium. And neither Bardeen nor Brattain really understood the details of how it worked. But it worked and that was all that mattered. It took a high-frequency signal, converted it into direct current and then amplified that current 400–500%. In engineering terms, an AC signal in the input circuit induced a greatly enhanced signal response in the output circuit. This was, in short, a solid-state technology that could potentially replace the vacuum tube.

Scientists and managers realized almost immediately that they had a revolution on their hands. The transistor was first demonstrated to the public via a well-publicized press conference on June 30, 1948. But much work still needed to be done to get the transistor out of the laboratory. For Bardeen and Brattain, their task was to find the theoretical mechanism behind their invention. Until they did, they felt they could not improve upon it. But pressure from the outside forced Bell to accelerate development of the transistor. Soon after announcing its invention, demand for samples of the so-called "type A" point-contact transistor began to pour in from government laboratories, major electronics firms, and university researchers who wanted to see, test, and use the device. But Bell was still far from being able to satisfy these requests, and the world was beginning to think that Bell had oversold its new invention. Many problems delayed progress. Because of trace contaminants in the air or even the electrical wires that got into the transistor, no two of them behaved the same, and any one of them could stop working at any time. Further, the transistor had to be completely redesigned to make it manufacturable as a standardized device. The ability to turn the transistor into a practical electronic component was important to AT&T. In the late 1940s, the Truman administration targeted the monopoly as in violation of antitrust laws. Specifically, the Department of Justice was considering divesting it of its manufacturing arm, Western Electric. This would have been devastating for AT&T and, even worse, set the stage for even more action by the government down the line. This was a risk that had to be deflected. The transistor offered a way to do this. Because of its potential use by the military—in communications systems, radar technology and navigation—successful production of a working line of transistors that could be delivered to the Pentagon upon request would go far in proving to the government the critical importance to national security of the Bell Labs–Western Electric relationship. Moreover, if Bell working with Western Electric could produce effective transistors and then offer the device—along with extensive information on how to use it—at low costs to any and all who wished to license it, AT&T could claim that, far from restraining competition, it openly encourages it by basically giving away its most important technology. Very much like DuPont three decades earlier, Bell Labs had to develop—and quickly—an external culture. It could no longer just think about internally generating new ideas just to add to its own reputation or only for use at some point within AT&T's phone system.

Ultimately, the buck—and all the risks attached to it—stopped at Kelly's desk. It was his laboratory that invented the transistor that had its reputation on the chopping block and that had the responsibility of making sure the Pentagon—or any company that asked—got the transistors they wanted. It would seem that Kelly was taking quite a chance in publically announcing the transistor unless of course he felt very certain in his ability to produce a (more or less) commercially valuable device. He had, in fact, three reasons to believe this. First, he had the facilities of Western Electric at his disposal. In the previous decades, Western Electric had excelled in designing and mass producing some of Bell's most important innovations, including advanced phone equipment (e.g., handset containing both the transmitter and receiver in the same unit) , innovative electric components (e.g., repeaters, rectifiers), and radar systems used during World War II. In the same way its Ammonia Department gave DuPont the courage to proceed to the development of nylon, so Western Electric permitted Kelly the luxury of feeling fairly certain of success in the development and production of the transistor. Kelly had another card up his sleeve, namely, the electrical engineer Jack Morton. Morton had a superb reputation at Bell and, more critically, with Kelly who saw him as a "hard-driving, aggressive, and visionary man" who could get the job done. He already had an excellent track record at Bell: he led a team of engineers in developing a new type of amplifier that permitted AT&T "to transmit TV signals from coast to coast." Morton's colleagues looked up to him as they did few others. Kelly knew that Morton was just the man who could act as intermediary between Bell scientists and the production team at Western Electric.

In addition to these two assets, Kelly had a two-step plan to speed the creation and production of a practical transistor. Soon after the press conference announcing Bardeen and Brattain's remarkable gadget, Kelly had Bell Labs build an experimental production line to produce a limited number of type A transistors to satisfy immediate demands of the military (and to make transistors for in-house research). This facility also provided key information on the limitations of the current transistor design and on production problems that had to be faced. He also created a new group called "Fundamental Development," headed by Morton, to take the transistor as it now existed and redesign it so that it was a more practical component that could be mass produced and reliably used in real-world equipment and systems. By forming this group, Kelly was making sure that the scientists in Solid-State Physics did not get diverted from their basic mission of pure research and, at the same time, did not have the opportunity to stick their noses into and interfere with an area they knew little about, that is to say, making a practical machine.

Kelly's confidence paid off. By the fall of 1951, Bell could begin offering seminars and symposia on the nature and applications of the great invention to companies, government bureaucrats, and military officials from across the country. By then

Morton could claim that his group " … had overcome most of the knotty problems experienced with the point-contact transistor so that mass production could finally begin."[11] Western Electric of course would be the epicenter of that production. Kelly now had a weapon he could wield when arguing the case against divestiture to his contacts in Washington. By 1954, the government backed down from its attack against AT&T (a development in part due to the transistor and in part to an Eisenhower administration's reluctance to pursue antitrust actions). The reputation of both Kelly and his labs not only survived the transistor but, for a time at least, thrived because of it.

7.3 SHOCKLEY SEMICONDUCTOR AND THE JUNCTION TRANSISTOR

While a landmark achievement, even in the early 1950s, Kelly and Morton's point-contact devices still had serious problems, and the market for these transistors remained restricted. Standardization of the production process was difficult to achieve, and transistors coming off the line continued to differ significantly in their electrical characteristics. This variability in performance from batch to batch and even between transistors within batches seriously threatened market acceptance because of the uncertainty of performance and risks of failure of networks and systems that depended on the long-term reliability of its critical components. Even when working properly, the germanium transistor fell short in a number of respects. It was electrically "noisy," even more than the vacuum tube, coupled with severe limitations in its power output and frequency range. Because of these issues, the military, which required reliability under very intense conditions, continued to depend on vacuum tubes far more than it did on transistors. AT&T also hesitated to place its own technology within its network. It would not do so for many years and when it did, it did not use the point-contact transistor. Soon another type of transistor challenged the point contact for market supremacy.

7.3.1 The Junction (Bipolar) Transistor

Not surprisingly, Shockley was severely disappointed that he had little to do with discovering the first transistor.[11] He meant to regain the spotlight by inventing his own superior device. He did remain bothered by the fact that no theory existed to explain how the point-contact transistor worked and that it was so unpredictable and therefore unreliable. He had not forgotten what Kelly had wanted from the beginning: the possibility of an electronic switch technology for AT&Ts network. He knew that the point-contact transistor was too delicate and erratic to do this job. He wanted to invent a whole new type of transistor that would sidestep these problems. Critical information that he obtained early in his research came from a demonstration by and discussion with Russell Ohl on a visit to the AT&T Holmdel labs (where Ohl had set up his shop) in April 1945. Ohl's work on the p–n junction greatly stimulated his thinking about the possibilities of using a block of semiconductor to amplify a signal. Soon after his visit with Ohl, Shockley came up with the idea of the field-effect transistor (FET). The failure of this transistor to work as expected compelled him to understand the science behind the transistor effect. So he set to work developing a comprehensive theory of transistor action, especially that of the critical p–n junction. This effort culminated in his 1950 magnum opus *Electrons and Holes in Semiconductors.*[12] A fundamental discovery that came out in the book and that Shockley derived using quantum theory was the nature of those positive charge carriers—actually, semiconductor regions containing too few electrons—first observed by Bardeen and Brattain and which he called "holes."

From these insights Shockley and his team continued to push ahead fundamental understanding of how electrons and holes flow between the two electrodes—called the emitter and collector—in a transistor. The series of historic experiments showed that, contrary to what Bardeen believed, holes not only traveled in a channel just under the surface but actually moved through the bulk of the semiconductor, even in the face of a sea of electrons that try to "gobble" them up (since negative and positive charges tend to cancel one another). This proved Shockley's idea of the ability of these so-called "minority carriers"—the holes—to survive in a rather hostile electron-rich environment and to congregate and build up at the crucial "p–n" junction within the body of the transistor. Thinking about and synthesizing these ideas in a New York hotel room on a New Year's Eve 1948, Shockley came up with his design for a new type of transistor—or as he put it "improved solid-state amplifier"—very different from the one Brattain and Bardeen conceived. Shockley called it a junction transistor. This was not by any means his elusive field-effect device, for he had given up on this because of the surface effects problem. His new device, like the point-contact transistor, made electrons and holes move by directly wiring the semiconductor to a battery. Still, he had every reason to be satisfied for he believed it would soon knock Bardeen and Brattain's point-contact transistor off its pedestal and take its place as the solid-state invention of the day.

The so-called junction transistor that he designed was quite compact and even elegant compared to the clunky point contact. It was made up of a chunk of silicon divided into a sandwich-like structure. The two larger outer sections—the "bread"—possessed an excess of negatively charged electrons and were said to be made up of "n-type" silicon. In between these regions was a thin layer of silicon with an excess of positively charged holes. This region was electron deficient and was called "p-type" silicon. This three-layered structure would then be known as an "NPN" junction device. Since this thin middle zone bordered "N silicon" on two sides, there were two p–n junctions in the structure that would attract hordes of electrons and holes, doubly bolstering an applied current. Once hooked up to a battery and a small voltage applied, electrons and holes would pour across both junctions in opposite directions from one electrode to another. In this way, a miniscule initial voltage (incoming signal) would create a very large response by the mobile electrons and holes within the junction device, in other words, amplification of the outgoing signal took place. Shockley believed his junction (otherwise known as bipolar) design would outperform the point contact in many ways: greater amplification, less noise, more reliability, more easily standardized, and manufactured.

While Shockley's design for the junction transistor appeared flawless on paper, it actually could not work in practice without major advances in materials processing.

Shockley was still the leader of the Solid-State Physics Group at Bell, and it was at Bell Labs that many of these new technologies arose. One important requirement was purification of the semiconductor to a degree never before attempted. The purity and uniformity of the semiconductor crystal made a crucial difference in the effectiveness of the transistor since defects and unwanted contaminants in the crystal structure interfered with the internal movement of electrons and holes, even short-circuiting the system and ruining the transistor. Bell researcher William Pfann invented an innovative process called zone refining, a technique adapted from the metallurgical industry that for the first time produced extremely pure grade silicon and germanium. Crystal growth also had to be reconsidered. Up to this point, transistors used slabs of silicon and germanium composed of many crystal units. But the grain boundaries separating these crystals, just as contaminants, tended to slow down electrons and interfere with device performance. Bell physicist Gordon Teal (who later would help transform transistor research at Texas Instruments) came up with a way to make a single semiconductor crystal, thus removing the barriers to electron mobility (the microchip of today uses single-crystal silicon).

Beyond the question of purity and crystal uniformity, there was also the critical issue of how to form clearly defined "N" and "P" regions in the silicon in order to form the NPN sandwich that defined the junction transistor. In doing this, it was very important to create sharply outlined junctions (less clearly formed junctions caused havoc with the flow of charge carriers). These regions were formed by inserting a specified number of atoms of certain elements into those regions of the silicon (or germanium) that needed to be either positive or negative. If a P region was desired, then it was known since the early 1940s that you had to insert atoms of an element located in Group III (the third column) of the periodic table, such as boron or aluminum; if you needed to construct an "N" region in the silicon, then you had to inject atoms from an element in Group V (the fifth column), such as phosphorous or arsenic. This process is called doping. It was accomplished by introducing these atoms or "dopants" into the silicon at specific predetermined places through a procedure called thermal diffusion. In this process, round wafers of silicon—with a diameter of a 45 rpm record and thickness of five pages of paper—were placed in a furnace, where two types of dopants entered and diffused into the wafer at slightly different times. To create an NPN junction transistor, for instance, phosphorous would first be introduced into the furnace over a desired area of the silicon wafer until the phosphorous atoms diffused into the wafer a certain distance to create the first N region. Then boron would be added to create the middle P section and then phosphorous reintroduced so that it diffused into the silicon to make the second N region. The slab was continuously reoriented as required to create an NPN layer cake-like structure (N layers on top and bottom and a thin P layer in the middle). Researchers then cut the wafer into about a hundred tiny three-layered NPN transistors.

If made correctly, these devices performed impressively. Laboratory tests showed that the transistor could amplify signals with frequencies that were 10 times higher than could be handled by even the improved point-contact transistor. In short, when it worked at its full potential, the junction transistor was "smaller, lighter, faster, more sensitive, more reliable, and far more power efficient" than either the vacuum tube or point-contact device. This meant that the junction transistor would eventually become the central component in advanced radar, communications systems, and early computers and justly deemed the first "workhorse of the semiconductor industry."

7.3.2 The Creation and Fall of Shockley Semiconductor

By the mid-1950s, Shockley's patience with Bell management had reached its breaking point.[13] Troubled by the slow pace of development of his junction transistor and even more upset that Bell would not promote him past midmanagement level, his ambition and sense of his own worth overcame any remaining loyalty he might have had for Kelly and his labs. He left Bell and went west to California, taking the best and brightest technical talent with him to start Shockley Semiconductor of Palo Alto, the first semiconductor company to set up shop in what would in a few years' time be known as Silicon Valley. Shockley certainly had every reason to believe that his new company would control the country's junction transistor business. He was after all the inventor, and he had with him a superb group of engineers and scientists who knew and understood the transistor as well if not better than anyone alive. But expectation is not by any means reality, and Shockley's company failed to produce any new product before eight of his best people deserted him to start their own company.

Certainly, Shockley's outrageous, divisive, and abusive management style, evident even when he was at Bell, has often been cited as the cause of this fiasco. But we should not lose sight of what the fundamental bone of contention was in this case. Shockley saw himself as the premier physicist of the age, a Nobel Prize-winning scientist who had a reputation to uphold. Thus he felt he needed to continue to probe fundamental transistor phenomena by creating new and advanced types of semiconductor devices. He knew of course that he was running a business but felt great science will lead to great (and profitable) products. So it was to science that Shockley paid homage. He was then very excited about an advanced device known as the four-layer diode. Something like it had never been made before, and that was just fine with Shockley, for it would surely lead to a greater understanding of how electrons and holes operate in silicon and of course to a batch of excellent articles for the leading physics journals. The one problem is that it would take many years before this transistor could ever be made commercially. There was no certainly in fact that it could be produced at all on a large scale or that there would be a market waiting for it even if it could.

None of this sat well with a number of Shockley's team. They did not join up with Shockley to spend their best creative years bolstering up the scientific legacy of their boss. They had their careers to think about. Their standing in the semiconductor industry depended on being part of making something important and as soon as possible so that they could be earning real money. And they felt they had just the thing: Shockley's old three-layer junction device. Much work had been done on it already and a prototype created at Bell. They now felt they needed to go to the next step and quickly: prototype improvement, semiscaled production, and then full throttle manufacture for customers anxious to use them. They wanted to be the first ones in the ring and didn't want to dither around doing laboratory experiments on Shockley's four-layered will-o'-the-wisp.

In essence there was a great gap in what each party believed to be risky. Shockley couldn't afford to sully his reputation as a world-class scientist by wasting time on commercially doable but scientifically unexciting projects. After all, he didn't want to be thought of as just a mere producer of transistors. For their part, his antagonists could only bolster their reputations and open up opportunities for themselves in the expanding semiconductor industry by showing the world they could make something commercially important. They wanted Shockley to build and integrate into the company pilot, semicommercial, and full-scale production facilities for the three-layer transistor. At most, Shockley was willing to build a small pilot plant but only to make his four-layered contraption for scientific exploration and limited sales to research organizations. This complete focus on science over product and the resultant lack of a vertically integrated facility that could speed the development and production of a commercial technology proved to be the real undoing of Shockley Labs.

Without the support needed from Shockley, the group of malcontents, led by the physicist Robert Noyes, mutinied. Only a year and a half after they began working for Shockley, they resigned en masse and, on the basis of Noyes' reputation, obtained venture funding to start a new company. The "Traitorous Eight," as Shockley purportedly called them, founded Fairchild Semiconductor, a subsidiary of Fairchild Camera and Instruments in September 1957. Now it was Fairchild, not Bell or Shockley Semiconductor, that contained within its walls the best brains in the business. The eight founders of the start-up, operating out of a facility only a mile from Shockley's Palo Alto's shop, planned to make the best and most advanced transistors on the market.

7.4   FAIRCHILD SEMICONDUCTOR: THE BIPOLAR COMPANY

Not surprisingly, Fairchild soon became the most important semiconductor outfit in the world. It was a great innovator with many firsts in junction transistor technology. It far outpaced both Bell and Shockley as a technological leader in the field. It established once and for all the primacy of the junction transistor. Junction or bipolar technology (so-called because in a junction transistor both electrons and holes flow simultaneously in opposite directions) came to define Fairchild as a semiconductor company. Its two major achievements in the evolution of the junction transistor were the silicon transistor and the planar process.

7.4.1   The Silicon Transistor

Given the impatience of the traitorous eight to build an integrated transistor production facility, it is no surprise that they "moved quickly to put out a product" as soon as they set up a shop on East Charleston Road.[14] They managed this by creating what they could not have with Shockley: closely linked research, development, and production operations. Noyce took on the role as head of research while his close colleague Gordon Moore, a physical chemist, led the production engineering group.

Information flowed freely between them as they designed the prototype and set up the production line for their three-layer junction transistor. At this early stage, Fairchild was still considering using germanium in its transistors. Fairchild's first major decision was to abandon germanium and dedicate its efforts to using silicon. While germanium transistors are much faster than silicon, silicon was fast enough and moreover was a far more practical material in terms of manufacturing. Silicon was both cheaper and more durable than germanium at elevated temperatures. Because of its atomic composition, silicon, unlike germanium, did not produce unwanted reverse currents—resulting in "leaky" switches—that drain the energy from, and compromise the electrical properties of, the circuit system. Because of the importance of transistors as tightly controlled switches in computers, the choice of silicon at this point was an absolutely crucial decision for the company (Texas Instruments was the first company to actually make silicon junction transistors and it had been doing so since 1954).

Noyce and his Fairchild crew, seamlessly linking research with process development, experimented with different diffusion techniques and with improving furnace technology. At this time they made their own equipment for their own use. Their attention to designing and building advanced diffusion furnaces made all the difference in Fairchild's success in semiconductors. As one of the original eight, Jay Last firmly believed, "The diffusion furnace ... was the key item."[15] They then targeted the largest market for transistors at the time, the government. The military and the newly created space program looked to the transistor for application in aircraft, satellites, missile systems (such as *Explorer* and *Vanguard*), and planned manned space programs. This meant a need to achieve miniaturization of the transistor and to satisfy increasingly severe restrictions on operating parameters, including reliability, durability, and uniformity.

To meet this demand, Fairchild set out to make the most advanced silicon junction transistor that existed up to that time. It was called a silicon "mesa" transistor because it resembled the elevated flat-topped mountain formations of the southwest. Fairchild's first major contract, arranged by Fairchild Camera Executives, was to design and make junction transistors for use in computers being designed and built by IBM for military aircraft. Moore and his team made a number of engineering improvements to the diffusion furnace, including incorporating an innovative heating alloy ("Kanthal") first developed in Sweden, replacing inductive-type heating with the more effective and simpler resistive element and applying an adjustable transformer to control the temperature range within the heating zone. They soon were operating seven such furnaces in their facility. The men pushed the wafers through manually and placed the "diffused" silicon into a cooling area. In the early days of the company, the furnaces still resembled laboratory apparatus rather than industrial manufacturing equipment. Within a year, Fairchild placed the diffusion process on a large-scale manufacturing basis. The company pioneered a new type of industrial manufacturing process—a hybrid of batch and continuous flow transistor manufacture carried out within a "clean room" environment and controlled through an "unprecedented level of industrial discipline"—that would become standard in the semiconductor industry in the years to come. The focus of the group on the underlying production process paid off beyond even what they were expecting. Jay Last, for instance, reminisces at the close cooperation that existed within this essentially flat organization and the impressive technology that resulted:

> It was remarkable to me ... I was just amazed to realize that 10 months after we went into this empty building we had a commercial product....which showed the way we were working and cooperating ... each one of us depending on the rest of the group to do their part ... We worked together really without too much of an overall leader telling us what to do ... we just solved whatever had to be done.[16]

The start-up delivered its first product, the NPN junction transistor, to its client, IBM, in the summer of 1958. A few weeks later, it introduced it to the larger market. Fairchild learned that Western Electric also manufactured junction transistors but their product did not equal in quality with Fairchild's transistor and, in any case, was designed exclusively for use in Bell's system. Texas Instrument's silicon transistors also fell short technically compared to the mesas of Fairchild. Robert Noyes, feeling particularly pleased that no other company could compete with them at this time, told a colleague, "We scooped the industry [on the NPN transistor]," and he happily noted to colleagues, "Nobody [is] ready to put something like this on the market."[17]

Noyes' elation over this success clearly reflects an early feeling by the founders that the new company would be able to quickly turn out improved and even more commercially sophisticated products at a rapid pace. This growing confidence soon began to transform itself into a highly dynamic culture but one which was deeply rooted in another cultural dimension: short-term thinking. This may sound surprising to some who closely link technological advance with blue sky, that is, long-term research, but in fact, after their experience working with Shockley and fighting his infatuation with such an impractical device as the four-layer diode, Noyes and his colleagues were more than ready to quickly introduce to the market a whole battalion of new, superior, and commercially friendly semiconductor products. As part of this rapidly developing dynamic mind-set, they were more than prepared to jettison an existing product to make way for a better one. In fact, the quicker they could do this, the more competitive they could be. No other semiconductor company at this time had a better sense of this possibility than this one. Within a few years, Gordon Moore would publish his now-famous article the gist of which came to be known as Moore's Law, the very foundation and "bible" of short-term dynamic thinking. This attitude began to take shape at Fairchild, but, as we will see, it would only reach full fruition a decade later—and 5 years after Moore wrote his article—when Noyes and Moore settled in at Intel.

7.4.2 The Planar Process

But problems with Fairchild's transistors soon surfaced, and the company's government clients began complaining.[18] Too many transistors were malfunctioning. This failure rate was unacceptable in missiles and computer systems. The military—and IBM—demanded that Fairchild increase the reliability of their transistors. Contamination of the semiconductor material appeared to be the problem. Dust particles and other impurities in the air entered into the silicon and interfered with the delicate operation of the region around the transistor's p–n junctions. Growing competition from new entrants into the transistor field, notably Motorola and Texas Instruments, provided further incentive for the Fairchild group to improve transistor performance.

Bell's work on silicon oxide films, known to Fairchild through personal contacts and conferences, inspired the Swiss-born physicist and former Shockley employee, Jean Hoerni, to develop a radically new manufacturing process that solved the reliability problem. Why not, he reasoned, design a transistor device by growing a flat, protective oxide layer on top of the silicon wafer and introduce dopants through small carefully positioned holes etched into the thin shell? These dopants would then diffuse into and make contact with the underlying silicon to produce the required "NPN" regions. Furthermore, the oxide film would become an integral part of the device and serve as a protective shell keeping outside contaminants from entering into and disturbing the delicate electrical operations of the safely enclosed NPN cake. As Hoerni graphically described his idea, "It's like setting up a jungle operating room. You put the patient inside a plastic bag and you operate inside of that [plastic bag], and you don't have all the flies of the jungle sitting on the wound."[19] In fact, both Bell Labs and RCA were independently working on the same idea, and at the same time, but did not proceed beyond the concept stage due to what they perceived as great risks in attempting to manufacture such novel and complex devices. But Hoerni was a driven man who would not be thwarted from pursuing his process. He initially worked on his design by himself and often at night, as he found no funds from Fairchild's management to support his effort. Eventually, Hoerni's work reached the point when it was clear he was onto something practical, important, and quite timely. Pressure from one of Fairchild's main subcontractors, the Autonetics Division of North American Aviation, to solve the reliability problem for the guidance and control systems of the Minuteman Missile compelled Fairchild to give Hoerni's planar process a trial. Ultimately, the company began devoting significant engineering and financial resources to understanding and perfecting the process. In time, the company succeeded in obtaining a reliable and efficient manufacturing process. Introduced publicly at the 1960 Institute of Radio Engineers show in New York, the "planar" junction transistor was a revelation. It was a very different transistor than the mesa. When looked at under a microscope, the planar device was clearly flat compared to the mountain-shaped mesa device. It was the most dependable and efficient transistor in existence. It was made by an extremely flexible process, capable of economically making different types of superior silicon transistors for various applications. Virtually overnight, planar technology relegated all other transistor designs obsolete in terms of performance and long-term reliability. The importance of the planar process did not end with the transistor.

7.4.3 The Integrated Circuit

The success of the planar transistor led soon thereafter to an even more important innovation: the integrated circuit.[20] The person behind this groundbreaking technology was Hoerni's colleague and boss, Robert Noyce. As many physicists and computer engineers in the late 1950s, Noyce wanted desperately to solve the difficult "tyranny of numbers" problem, that is, the inability to increase the performance of complex electronic systems due to the rapidly growing number of components (transistors, diodes, resisters, and capacitors) that needed to be soldered together by hand—which he considered *the* major issue looming for the semiconductor industry in the 1960s and beyond. He also saw it as an opportunity for Fairchild to extend and truly solidify its lead over its competitors. At the heart of Noyce's conception was an advanced materials process.

As so often happens in the early phases of a research project, the integrated circuit of Noyce depended on a critical piece of information or technology, in this case the planar process itself. From here, Noyce moved to the next step from transistor to a slab of silicon containing many integrated transistors (and other electrical components) that would solve the tyranny of numbers problem. As Noyce describes it, in thinking of how he first conceived of the integrated circuit, Noyce thought that it " ... would be desirable to make multiple devices on a single piece of silicon ... in order to be able to make interconnections between them as part of the manufacturing process, and thus reduce size, weight, etc. as well as cost per active element.[21]

How to achieve this occupied Noyce at this time. In a three-step process, Noyes worked his way to the ultimate answer. The first problem that bothered him involved the bulky wires that had to be handled and connected into circuits. Especially troubling to Noyce was the fact that, with the human hand thicker than the wires, mistakes proliferated in assembly. This meant that Fairchild had to test each transistor, one at a time, which greatly increased man-hours needed and costs expended in production. His first thought was to push connecting wires through the planar process' oxide layer at the exact location needed until they touched the underlying silicon surface. He reasoned that this approach would ease the attachment process because the tough oxide layer would support the wires where inserted, thus reducing the need of physical contact between wire and hand.

From this first concept, he moved on to a better one that completely eliminated the need for the wires at all. Thinking of how photographic systems work, he imagined using photolithographic technique to "print" tiny lines of a conducting metal, such as copper or aluminum, on top of the oxide layer (instead of poking wires through it). This new process created these circuits on a piece of silicon by using light to transfer circuit patterns from a set of reusable templates known as "masks" onto the wafer surface. At this point, doping technology came into play to insert the appropriate atoms and ions to create the many junction (bipolar) transistors and other elements that conformed to the dictates of the circuit design. Through an etching procedure, portions of the wafer were then cut away to create the circuit pattern on the silicon. These actions would then be repeated many times employing "hundreds of different steps using dozens of different types of equipment" to create a complex chips with 30 or more layers of interconnected circuits. Noyes' photolithographic method—and its use of reusable masks—was a true flash of genius. It meant that the bipolar transistors and other components—resistors and capacitors—could all be made from silicon and interconnected into many miniature circuits and that these tiny circuits could be linked together on the same chip by the printed grid defined by the set of photolithographic masks. Noyes' concept of the integrated chip opened the gates for bipolar integrated circuits to be replicated precisely and quickly using mass manufacturing technology. Wielding the planar process and the integrated circuit, Noyce and his company soon pulled ahead of their other competitors so that, by 1962, "Fairchild Semiconductor had grown from a small semiconductor start-up concerned with quickly developing a product, to a large semiconductor company increasingly concerned with the long term."[22]

In only about 2 years, thanks in large measure to the young company's fluid organizational structure, the technical competence of its eight founders, the power and elasticity of the planar process, and its evolving short-term dynamic culture, Fairchild had become one of the major players in bipolar semiconductor technology. But another and competing advanced materials device was beginning to make itself known. Called the "metal-on-silicon" (MOS) process, the transistors made by this technology and the integrated circuits composed of these MOS transistors were designed very differently than Fairchild's bipolar transistors and ICs. Destined to eventually supplant junction bipolar technology in computers and many other applications, MOS did not easily find a place in companies like Fairchild where it had to unseat a trusted, tested, and familiar technology. The remainder of this chapter will trace the early research undertaken on MOS at Bell Labs and discuss how and why it failed to take off at Fairchild despite persistent attempts by talented individuals to champion this new way to tame silicon.

7.5 THE MOS TECHNOLOGY AT BELL AND FAIRCHILD

Despite its success, the planar junction transistor had its own problems with which to contend.[23] Most importantly, it was a fairly bulky device and difficult to manufacture on a mass production basis, which limited it to a number of specialized applications. Scientists and engineers believed that only a field effect transistor (FET), the type that Shockley first conceived of in the late 1940s but never could get to work properly, held out the hope of a compact, truly mass produced transistor that could be miniaturized for a wide range of uses. But the only way for the FET to work was to find a way to eliminate the troublesome surface state barrier that prevented the external electric field from penetrating into the material. A major step in this direction was the invention of the "MOS" process. Early MOS research began at Bell Labs but never evolved very far there. Fairchild then moved in and pushed MOS technology further along but in the end never managed make much of a business out of it. We will first discuss the origins of the MOS process at Bell and then consider Fairchild's heroic but failed efforts at making it into a working technology.

7.5.1 MOS Research at Bell Labs

It is no surprise that the initial research into MOS technology occurred at Bell Labs. By 1956, it had been less than a decade since Bardeen's and Brattain's first observation of the transistor effect through their point-contact device and only 5 years since Shockley came out with his diffused junction transistor. In the face of persistent problems with both the point-contact and junction transistors, Bell management decided to renew its efforts to come up with a working FET. No one knew what this new component would ultimately look like, but what was known was the importance of solving once and for all the surface state issue. An Egyptian with a doctorate in mechanical engineering from Purdue and who had just joined Bell's development group, John Atalla, immediately began working on this problem. This was certainly an area important to Bell and which would be a potential career maker for Atalla. As a new recruit in the organization, he took it upon himself to talk to other scientific researchers within the various disciplines who had worked on surface states. Such easy communication between research groups was common at Bell. Doing so, of course, provided important information and insights into the issue. It also afforded Atalla a foot in the door of surface states research.

Atalla's inquiry uncovered an important line of research conducted at Bell just the year before. In 1955, a Bell chemist, Carl Frosch, working on better ways to make diffused transistors, quite by accident created a layer of silicon dioxide over the surface of a piece (or wafer) of silicon. A few years before Jean Hoerni's insight into planar transistors, Frosch became keenly aware of the advantages offered by such a layer: it protected the wafer surface from outside contamination that could short-circuit or otherwise interfere with the device's operations; it *passivated*—that is, eliminated—part of the unwanted surface effects, thus allowing an external electric field to break through the outer shell and enter the inside realm of the silicon wafer; and it made an excellent "mask" or template that, through holes cut into the layer at predetermined locations, resulted in more precise placement of dopants (phosphorous and arsenic atoms) into the silicon wafer than was possible by using a diffusion furnace only. This meant the creation of much more sharply defined "p–n" junctions within the silicon material than was heretofore possible and the creation of higher quality, more reliable transistors.

Frosch's discovery supplied the critical knowledge Atalla needed to pursue his research. It also galvanized his team and made it so his superiors could justify his continuing this research. As he experimented with the oxide films formed over the silicon surface, he began to construct devices that closely resembled the FETs first conceived of by Shockley back in 1945. These were the first devices made with what came to be known as the MOS process. The MOS transistor looked and operated very differently from the bipolar device. Rather than a three-layered NPN sandwich, it had a metal film on top of a layer of silicon dioxide. In contrast to a junction transistor, which required a battery to get the electrons and holes to move, a MOS device was triggered by an electric field—such as provided by and incoming electrical signal—applied to the top metal layer called the gate. Because of the oxide film, the surface effects that had acted as a barrier were canceled, and the electric field emanating from the gate could now penetrate deep into the silicon wafer causing electrons and holes to flow to the "p–n" junction as required for amplification. Atalla first developed a theory explaining how the device worked and discovered that its most important application would be as transistors in the newly invented integrated circuits. He also identified his transistor's major selling point: its ease of fabrication.

Here we have a common situation where an individual working more or less on his (or her) own comes across an important invention and recognizes it as something that could "change the world" but needs to convince the rest of the organization that this project, as potentially important as it might be, does not pose a serious risk for those who stick their necks out for it and for the company as a whole. But the cards were stacked against Atalla from the start. Most importantly, Bell's senior management could not identify a need for such a technology. One of the potentially important applications for Attala's device was in integrated circuits since ICs that incorporated MOS rather than bipolar transistors would be a lot easier to manufacture. If Bell had decided to pursue integrated circuits for its systems, then its management may have taken MOS transistors more seriously. But, in 1960, AT&T and Bell had no interest in IC technology. "Not invented here" syndrome no doubt played a part in this attitude (it was independently invented in 1959 at Fairchild Semiconductor and Texas Instruments). An even more compelling reason for its resistance to the integrated circuit was the fact that, the great inventions of Bell Labs notwithstanding, AT&T was a "profoundly conservative organization" with $20 billion already sunk into their existing system.[24] Attempting to inject such new and unproven technology into AT&T's telephone network—which worked well enough already—meant considerable risk. AT&T had to maintain steady and reliable service in order to hold onto their favored monopoly status with the government. While management within AT&T and Bell was willing enough to cautiously experiment with transistors as components in their switching systems, it rejected integrated circuits as too complex and uncertain. Bell scientists and engineers believed (incorrectly it turns out) that the chances of failure of an integrated circuit—and thus of the telephone system within which it operated—rose rapidly as the number of transistors placed on it increased. As far as Bell was concerned, the integrated circuit was too unreliable to be useful to the AT&T.

Of course, if Atalla had been a forceful champion of his work, he may have been able to continue pounding his superiors with the importance of his MOS transistor and eventually turned the tide in his favor. We have seen—and will see again—examples of how individuals can alter the outlook, strategy, and even culture of organizations. But Atalla was not in much of a position to be an agent of change within Bell. He was a fairly new member of Bell Labs who had not yet had a chance to prove himself. While he did have a team behind him, they were not particularly dedicated to him as he and his colleagues had not been working very long together. Besides, the group consisted of new recruits like their leader and thus no one who could defend the project (or Atalla) to top managers. The fact that Atalla was a mechanical engineer did not particularly help bolster his reputation within Bell, that palace of science where pride of place went to physicists, physical chemists, and mathematicians. Atalla actually proved himself to be a quick study in physical chemistry and semiconductor physics, judged by his superb efforts in making the first operating FETs. He was clearly a generalist at heart and a potentially effective one. If he had had the chance to pursue MOS devices, his mechanical engineering background, with its emphasis on manufacturing technology and production economics, would have been vitally important as he moved his technology through development and scale-up. But if Bell Labs did not follow up on Atalla's research, the possibilities of MOS intrigued others outside of Bell. Some of the technology's most forceful champions found their way to Fairchild.

7.5.2   MOS Research and Development at Fairchild

An integrated circuit "integrates" many components onto a single piece of silicon, the most important of these are the transistors. When Noyes invented his integrated chip at Fairchild, he did so thinking that those transistors designed into the slab of silicon would be the bipolar junction type. With respect then to both transistors and ICs, "bipolarism" was the company's founding religion. Yet, certain members of that community who championed the very different MOS approach to semi-conductors desperately tried to bring about, what they fervently believed to be, a much needed reformation toward focusing on MOS transistors and ICs. Their inability to bring Fairchild to their way of thinking and the difficulties the company thereby experienced sadly proved them right. The following section narrates the early work Fairchild undertook on MOS. This is followed by a discussion of the many problems faced by the company that prevented it from capitalizing on this initial effort.

7.5.2.1   The Fairchild MOS Project: Initiation, Research, and Early Development

That Fairchild did initially pursue the MOS transistor had everything to do with the persistence and persuasiveness of a creative midlevel inventor. It also depended on Fairchild doing well and feeling flush with the success of its existing product line to allow its employees to pursue research into new and different areas. The person at the center of Fairchild's MOS research was Frank Wanlass who joined the company in early 1962. Wanlass, who obtained his Ph.D. in physics from the University of Utah, was a true individual who liked to go his own and not be confined by organiza-tional structures. While a bona fide scientist, he was also an inventor who understood the practicalities of circuit design as applied to device development. Thus he did not see much difference between research and development and in fact firmly believed in a fluid two-way communication between these two areas. Happily for Wanlass, Fairchild, in contrast to Bell Labs, did not have strict barriers between its research arm (the Physics Section) and its development area: scientists and engineers worked closely together in developing new devices. For Wanlass, nothing was of greater inter-est than MOS transistors. He appears to have known about MOS independently of Atalla's work at Bell.

While Fairchild's senior executives did not actively push the project—the com-pany had its hands full with improving its planar transistors and integrated circuit technology—it was willing to give Wanlass a fairly wide berth in exploring this field. While every engineer knew that MOS transistors were slow, they also knew that they required fewer process steps to place on a chip than the bipolar variety. Wanlass' ideas particularly intrigued Gordon Moore who soon became an advocate for MOS technology. Moore did not throw caution to the wind however. As far as he was con-cerned, Wanlass' MOS research would have very short-term practical results: a better understanding of and ability to produce better bipolar transistors, Fairchild's bread and butter. But he also felt that low-cost manufacture of integrated circuits with MOS transistors would be an appealing prospect for the company's future. The big question was whether Wanlass could produce stable and reproducible MOS devices.

Wanlass' research therefore had relevance to Fairchild and its existing business. Wanlass also found ways to further entice management to expand its support of his MOS project through a "carrot and stick" approach. The stick he wielded was to generate interest in his work beyond Fairchild and particularly with other semiconductor competitors by presenting his findings at international conferences on solid-state technology. His carrot was in finding new and promising applications for MOS transistors that would resonate with Fairchild. He particularly pushed Fairchild to leverage his work and its expertise in integrated circuits to pioneer semiconductor memories. By early 1963, Wanlass convinced his superiors to hire three young Ph.D. researchers to help him with his work. One of these bright new hires was a young chemical engineer from UC Berkeley, a Hungarian and future Intel legend, Andy Grove.

Grove and his colleagues worked in the Physics Section on the scientific issues facing MOS transistors. These devices had stability problems: transistors performed differently—sometimes better, sometimes worse—over time and under different voltages and temperatures. Moore, in keeping with the Fairchild's fluid organizational structure, made sure that the three researchers met weekly with personnel from Device Development and the Materials and Processes Departments. Grove and his fellow researchers continued to probe the nature of MOS transistors and factors that might stabilize their performance. They discovered that contamination of the silicon dioxide with sodium was a major cause of transistor instability, leading them to find ways to minimize this element in the device. The group also studied the nature of surface states and ways to minimize these destructive influences. Over a course of 4 years, they published dozens of scientific papers on the physics and chemistry of the MOS transistor through which they—and Fairchild—earned recognition as leaders in the field. They also attempted to make working MOS transistors. Their scientific work actually had little bearing on the quality of these transistors, the design of which was based on empirical "hit-or-miss" experimentation. The operations of these devices continued to vary wildly and in ways that could not be adequately explained by increased scientific understanding.

Ultimately, Fairchild lost momentum in its MOS effort. It was the Fairchild spin-off, Intel, which ended up exploiting the technology to its fullest extent. Fairchild suffered competitively as MOS turned out to be the technology of choice for integrated chips powering the personal computer revolution. In retrospect, Fairchild had the opportunity to dominate MOS and thus the microchip industry in the late 20th and 21st centuries but let it slip through its fingers.

7.5.2.2 Development and Early Attempts at Scale-Up: Risk Analysis   Despite

Fairchild's failure in MOS technology, it must be emphasized that no one in the company wanted this outcome. In fact, once Wanlass convinced the company's founders of the potential production advantages of MOS over bipolar, they did everything in their power to keep the project alive. Gordon Moore, in particular, became a very

strong champion for the technology to the point of pushing its development forward
even when it failed to satisfy prior benchmarks.

Here we have a situation that highlights the fact that the background and experience of top managers are extremely important in the decisions made on whether
a company will or will not continue to support a new venture. When the executives
at the top are trained in finance and marketing and so are distant from, even alien to,
technical matters, they tend to be very loyal to the product that they know and that has
been bringing in revenue and see any alternative technology as potentially dangerous
to the bottom line. Their requirements for continuing to fund a new venture are quite
strict. They need much convincing to ease their fears of new approaches engendered
by their high perception of risk. Like the executive committee of DuPont, before they
will continue to throw money at a project, they need to see successful completion of
each step in the innovation process; they also need to see that the next phase to be
undertaken is likely to succeed. But top managers in an organization who are themselves creative scientists and engineers look upon a new venture in a different way.
Even though a new product or process might conflict with the current strategy, they
are more willing to consider different possibilities as long as they can be convinced
on technical grounds that the novel approach is worth exploring. Then they begin
understanding what the innovator—in this case Wanlass—is up to. To these technically savvy managers who understand the economies of scope possibilities of a new
process, the potential of an exciting new device or component outweighs the risks in
pursuing it.

This is not to say that technically sophisticated and alert managers, like Nucor's
Iverson or Fairchild's Moore, are willing to take more chances than are other types of
executives with the power to grant or withhold resources. Iverson, for example, was
very cautious in planning and carrying out his Crawfordsville plants, and, as we have
just seen, Moore's initial interest in MOS technology was to be able to improve his
company's current products. However, they are also more knowledgeable of what is
possible and what isn't when it comes to carrying out innovation. They know from
their own experience that setbacks will occur in any technically challenging project.
They also know if and when these problems are likely to be overcome and so are more
likely to push forward in the face of an immediate difficulty. Moore, Noyce, and Grove
faced serious problems with MOS to be sure but, particularly at the urging of Moore,
kept the enterprise moving ahead. It was only when these three were swept from
their leadership positions by corporate headquarters and replaced by more traditional,
bottom-line managers did the MOS project experience serious push back from above.
The new crew saw greater risks than did the more technically savvy founders and was
not willing to take the project any further. We conclude this chapter by discussing the
various risks that the MOS effort entailed and how they finally forced the termination
of serious work at Fairchild on this vital technology.

Human Resource Risk: Defection of Initial Champion   Around this time—late
1963—Fairchild lost its great champion of MOS technology. Wanlass, feeling
Fairchild was not moving fast enough in commercially pursuing MOS products,
pulled up stakes and went to General Microelectronics, which promised him a more
central role for MOS in the company. This defection was a serious blow to Fairchild.
Wanlass practically singlehandedly introduced MOS research into Fairchild and,

through various strategies, kept the project going into the development stage. While Fairchild's top managers, including Noyce and Moore, did not originally think the MOS transistor was something to pursue, Wanlass convinced them that it was very relevant to the company. When no one at the company knew how to proceed into MOS, Wanlass, with his wide-ranging knowledge encompassing science, engineering, and circuit design, showed the way forward: he took the information he needed from each discipline and synthesized this intellectual input to design and make Fairchild's first MOS devices while at the same time identifying possible applications. As he proceeded, Wanlass brought an increasing number of scientists, engineers, and managers at Fairchild into the fold and convinced them of MOS' potential. Before he came to Fairchild, the company was totally committed to junction (bipolar) devices; by the time he left, top management had at least begun to put some of its resources into the MOS route to the transistor.

It is a good bet that, had he remained, Fairchild would have become a pioneer in the manufacture and marketing of early MOS technology. In less than a year, Wanlass led General Microelectronics on to make and sell the first commercial MOS integrated circuits. As for Fairchild, now without the Wanlass' passion and guiding hand, the MOS project languished. In the post-Wanlass era, Fairchild stumbled so badly when it came to MOS that the company, like Shockley Semiconductor before it, ended up losing its most valued scientists and engineers, including Noyes, Moore, and Grove. And, also like Shockley, it never really recovered from the defection.

Structural Risk—Horizontal: The Problem of Dispersed Efforts   One problem that continually plagued the project, especially during development stage, was the extreme difficulty in designing a reliable device and in predicting how it would perform under various conditions. Scientific understanding seemed to have little bearing on how well a transistor or integrated circuit worked. While MOS technology may have required fewer processing steps, those that had to be undertaken—preparing the surface, forming the silicon dioxide film (oxidation), thermally implanting the dopants, and evaporating and depositing the metal gate—were unique to MOS and very different from what Fairchild's scientists and engineers understood about bipolar electronics. Not only that but the processes were very difficult to coordinate with one another. Just as we found with high-pressure catalysis, one could not predict how a change to improve upon one process variable (say, oxidation) would influence the others. This meant development involved an extended—and seemingly unending—iterative, hunt-and-peck method: fixing one problem resulting in another popping up somewhere else, attending to that causing unforeseen issues in another part of the fabrication process, and so on. For example, with a possible method of layering metal on top of the oxide film in a MOS device came another problem: drifting of sodium ions from the metal into the silicon oxide layer below causing transistor instability. This drifting issue led to extensive research into other ways to evaporate and deposit metals and to finding appropriate metals to use (preferably free of sodium). Soon, Fairchild's Materials and Processes Department initiated an extensive study on analyzing the sodium content of possible metals that could be used. In another instance, attempts to minimize the negative effects of surface states in MOS transistors through improved thermal processes in turn affected the movement of dopants within the silicon thus weakening the sharp delineation of the "p–n junction" and further destabilizing the device. And attempts to resolve this issue—such as by using very high temperatures to accelerate diffusion of dopants into the silicon—tended to make another problem more likely to happen: that other unwanted contaminants would diffuse as well causing the device to behave erratically.

The fact that the solution to one problem often led to other difficulties underscores a structural deficiency at Fairchild. Consider that these various challenges could have been managed in a way to reduce uncertainty if the MOS effort at Fairchild had been localized within a single department and, even better, within one person who could see how all the parts of the MOS' complex jigsaw puzzle fit together. Wanlass could have been that person but he was no longer at Fairchild. As it was, Fairchild dispersed device development throughout the company so that its various R&D departments—Solid-State Physics, Chemistry, High Speed Memory Engineering, Device Development, and Digital Circuits—worked on MOS technology virtually independently and without close coordination with one another. This scattering of the research effort—and the absence of a champion who could span the different department and absorb and synthesize the different pieces of information—caused confusion and even conflict throughout the company in its pursuit of MOS. The result, however, was very predictable: escalating costs without discernible progress toward a workable device.

Technology–Market Interaction Risk: The Challenge of the "Killer App" Market
Even as Fairchild attempted to come up with a working MOS transistor, the question remained of how to make that transistor do what was needed for various applications:

> After Fairchild introduced its ... MOS transistor, it still faced the challenge of conceiving an integrated circuit product. Although one could imagine dozens of applications, no one knew which one would be successful.[25]

Development involved continuously adjusting the MOS process as target markets shifted, but this proved to be a terribly difficult and expensive task. In other words the process was not flexible. This increased the costs and uncertainty of the endeavor. If, like in the case of nylon, MOS technology could have found one major ("killer") market that did not require changes made to the underlying process *and* if a feasible process perfected that could meet this demand, an early success may have been achieved. One example of where MOS would have found its "killer app" market was in memories. Indeed, Fairchild did pursue this possibility under the direction of a new hire from General Electric, John Schmidt. Originally brought into Fairchild to build thin-film magnetic memory, he examined what Fairchild had already done with MOS and saw potential for much better and cheaper memories through MOS integrated circuits. Gordon Moore was enthusiastic about this possibility, especially because of the low-power requirements. He thus gave Schmidt the go ahead to come up with a family of MOS memory chips. But making such devices for different applications proved to be very difficult. Schmidt could not build uniform, reliable MOS transistors over a chip's surface. Processing problems soon arose due to the complexity and size of the chip. Ultimately, Schmidt could not tame the MOS process. The reality was success in memory " ... required great strides forward in many areas and ultimately proved to be beyond Fairchild's capabilities at the time."[26] While, as discussed in the chapter on nylon, high-pressure technology was no easy process to conquer and did not readily adjust to different products, at least, once time and money had been spent on it, it could lead to great rewards as long as there was one big market to tap. But the MOS process couldn't even do this, and that one killer market for semiconductor technology had to wait for a later, better process to come along.

Structural Risk—Vertical: The Problem of the Divisive Champion Another very important risk that hindered success of MOS at Fairchild was structural in nature: the lack of creative communication between research and development on the one hand and manufacturing on the other. To a large extent, this "disconnect" between research and production came about due to the divisive personality of another MOS champion who came to set up his own sphere of influence within Fairchild.

In 1966, Fairchild's MOS program got a shot in the arm with the coming of Les Boysel, a disciple of Wanlass. Like his mentor, he had the clearest vision for the future of MOS, particularly its importance and its role in computers. Like Wanlass, he was a true individual and champion. Armed with a Masters in electrical engineering from the University of Michigan, he initially worked at Douglas Aircraft, working on Missile Systems. When Wanlass, then employed by General Microelectronics, visited Douglas to discuss the possibilities of MOS circuits in missile systems, he met Boysel. This was the beginning of "years of apprenticeship" under Wanlass. Feeling that Douglas was a dead end for MOS, and after a stint at IBM (during which time he continued to go to Wanlass for help), he came to Fairchild, the leader in MOS—at least in theory—up to that time.

Fairchild gave Boysel a great deal of freedom to pursue his vision. Unlike Wanlass, however, he was a "divider" and "disconnector" rather than uniter and connector. He avoided working or even interacting with the R&D group in Palo Alto feeling that group was too bureaucratic and more concerned with developing theory rather than turning out working devices. He set up his shop at Fairchild's main manufacturing facility at the Mountain View location where he could pursue his work on developing integrated circuits populated by MOS transistors. While Boysel's success meant antagonizing Fairchild's existing customers who were dedicated to integrated circuit composed of bipolar transistors (Bipolar IC), Fairchild's senior managers let him proceed because they wanted their customers to know that the firm continued to explore cutting-edge semiconductor technology. Besides, they really didn't expect him to succeed anytime in the near future and, by the time he did, they felt they would be able to convince their customers to accept the new devices for their products. So they allowed Boysel to add people to his project as he saw fit. He hired designers and technicians—people who worked with their hands, not theoreticians with advanced degrees. This worked well for Fairchild since his team didn't cost as much as would one inhabited by Ph.D.s. In short order a sharp divide existed between the Palo Alto scientists led by Gordon Moore and the Mountain View "inventors" and production crew led by Boysel. These two sites now had two very different cultures, values, and even goals: the Palo Alto team wanted to learn more about what makes the MOS transistor tick and place production on a sound, scientific basis; their focus was on the behavior of advanced materials. Boysel's men at Mountain View, on the other hand, simply wanted to make commercial MOS transistors anyway they could; understanding the reasons why the transistor worked or didn't work was of no consequence to them. They approached their work through cutting-edge circuit design. Clearly, Boysel championed the technology but only on his terms. Employing his talents and undeniable charisma, he ended up dividing Fairchild into two opposing and increasingly fractious groups.

By 1967, the conflict reached a head with disagreement over what process to use to make MOS transistors. Mountain View favored the so-called "Vapox" process for creating the silicon dioxide layer. Rather than using a furnace to form that oxide shell, Vapox worked by making the oxide first and then physically depositing it on the silicon surface without the use of high temperatures. Boysel claimed several advantages over the furnace including the ability to create much thicker oxide layers—thus achieving greater protection to the semiconductor underneath—and to do so much more quickly than possible with thermally grown oxide layers. Moore and his scientists over at Palo Alto opposed this route believing that too many contaminants would invade the silicon layer resulting in catastrophic device failures. Boysel, with no higher authority telling him otherwise, ignored this advice and pushed on with attempting to produce his Vapox MOS transistors. He believed he could design his MOS circuits around this problem. He was wrong. The instabilities and failures mounted on the production line and Boysel was forced to abandon this avenue. Moore's team at Palo Alto, sticking with thermally growing its oxide layers, also failed to come up with a commercial MOS device. Had the Palo Alto group been able to merge its expertise in materials processing with Boysel's particular genius for design and Mountain View's expertise in production, it is possible that far more progress would have been made on developing and producing commercial MOS technology.

Cultural and Strategic Risks: A Regime Change at Fairchild   As long as MOS fit into Moore's long-term strategy, many things were forgiven and kept alive at Fairchild, even in the face of the numerous problems. A promising new line of development called the silicon gate process looked to be a way to finally nudge MOS transistors to stabilize. But once that strategic environment changed, this no longer applied. Up until 1967, Fairchild Semiconductor—Noyce, Moore, and the others—had control over their company. But by the late 1960s, the competitive landscape began to change. The inability of Fairchild to excel in the MOS technology exposed them to the competitive pressures of other firms. While Fairchild remained the leader in bipolar devices, especially silicon transistors, this was now a maturing technology. Established companies like Texas Instruments jostled for position in transistors and integrated chips. There was also now greater competition as former Fairchild scientists and engineers defected to form their own companies—outfits like Signetics and General Microelectronics—to enter the market for advanced semiconductor technologies. Without leading edge products to offer, Fairchild had to compete the old fashioned way by cutting prices on its bipolar line to extend—and as time went on just to hold on to—its market share. In this destructive "war of attrition," Fairchild's earnings plummeted and the parent company, Fairchild Camera and Instruments, felt compelled to move in and do something to staunch the flow of red ink. Fairchild Camera's CEO, Sherman Fairchild, called in Lester Hogan, head of Motorola's Semiconductor operation, to be the new president of the parent company and to place the semiconductor operations on a tighter leash controlled from the center. To this end, Hogan brought in a team of managers he knew well from Motorola to run the Semiconductor division.

In this radically new culture, top executives at Fairchild were businessmen with little in common with the scientists and engineers working on MOS. One former employee referred to the new order as "Mahogany Row" meaning "well-appointed executive offices isolated from engineers and workers." Along with this greater degree of bureaucracy with more organizational layers and less informal communication taking place between them came a significant cultural shift. To be sure, this new crew rejected long-term research in favor of short-term thinking and quick results. In this, they and Noyes' team saw eye to eye. But Mahogany Row and Noyce's group split on what exactly should be accomplished so quickly. The former wanted the existing and proven products—such as the silicon bipolar transistor—to be the focus of attention. If there was going to be any innovation, it was to be incremental in nature to improve the production process or slightly modify the performance of tried and true products in order to gain additional markets and improve the bottom line. In other words, they pushed a "short-term stagnant" culture on Fairchild. Noyce's culture was willing to support a difficult project—such as MOS—as long as at the end of the day it was believed such a process would result in a number of new and important products in relatively quick succession. Moore's Law said that such should happen with MOS. But the horde from the northeast would have none of this.[27]

Now, in this new, less forgiving cultural climate, all of the tensions and conflicts between Palo Alto and Mountain View came to a full reckoning. With profits on the decline, Fairchild's new culture could not tolerate dedicating any more significant resources to what they now considered no more than a faraway possibility. For the new managerial elite at Fairchild, the risks appeared far too great, and attention and resources went to improving its existing bipolar products and processes. During this period, Gordon Moore and Robert Noyce left to found their start-up, Intel. As for Fairchild, it became a has-been in semiconductors as it "essentially sat on the sidelines while Intel developed the first generation of memory products."[28] The company never could recover from Sherman Fairchild's "Mahogany Row" strategy. By 1979, the once powerful semiconductor company was purchased by Schlumberger, suppliers to the oil and gas industry. While Fairchild Semiconductor would be reconstituted in the 1990s, and exists to this day, it has never regained the influence it once had in the semiconductor industry when Noyce, Moore, and company were at the helm. As for these two pioneers and Intel, armed with a new and truly revolutionary advanced materials process, they dominated the semiconductor industry over the next four decades up to the present day.

REFERENCES

1. This identification of functional materials with the semiconductor revolution is emphasized in Cahn, R. (2001), Functional Materials, in *The Coming of Materials Science*, Oxford, UK: Elsevier Science Ltd., pp. 253–306.

2. Lecuyer, C. (2007), *Making Silicon Valley: Innovation and the Growth of High Tech, 1930–1970*, Cambridge, Massachusetts: The MIT Press, pp. 13–128; Sturgeon, T. (2000), "How Silicon Valley Came to Be," in Kenney, M. (ed), *Understanding Silicon Valley: The Anatomy of an Entrepreneurial Region*, Stanford, California: Stanford University Press, pp. 15–47; Reid, T. (2001), *The Chip: How Two Americans Invented the Microchip and Launched a Revolution*, New York, New York: Random House, pp. 10–11; 24–61; 86–87.

3. Ibid.

4. Riordan, M. and Hoddeson, L. (1997), *Crystal Fire: The Invention of the Transistor and the Birth of the Information Age*, New York, New York: W.W. Norton & Company, Inc., p. 200.

5. Lecuyer (2007), pp. 13–89; 129–167.

6. Gertner, J. (2012), *The Idea Factory: Bell Labs and the Great Age of American Innovation*, New York, New York: The Penguin Press, pp. 14–91; Riordan and Hoddeson (1997), pp. 56–63; 108–110; 116.

7. Riordan and Hoddeson (1997), pp. 11–27; Gertner (2012), pp. 75–91.

8. Lojek, B. (2007), *History of Semiconductor Engineering*, Berlin, Germany: Springer, p. 67.

9. Ibid., p. 32.

10. This section on the point contact transistor is taken from Riordan and Hoddeson (1997), pp. 115–141; Gertner (2012), pp. 84–97; Lojek, B. (2007), *History of Semiconductor Engineering*, Berlin, Germany: Springer, pp. 13–23. See also Ross, I. (December 1997), "The Foundation of the Silicon Age" *Physics Today*, 50(12): 34–37.

11. This section on the junction (bipolar) transistor is taken from Riordan and Hoddeson (1997), pp. 142–194; Gertner (2012), pp. 100–114; Lujek (2007), pp. 26–49. See also Ross (December 1997), pp. 37–39. For information on advanced crystal materials and transistors at Bell see Teal, G. (July 1976), "Single Crystals of Germanium and Silicon—Basic to the Transistor and Integrated Circuit" *IEEE Transactions on Electron Devices*, ED-23(7): 621–639 and Pfann, W. (July 1952), Principles of Zone Melting" *Transactions of the American Institute of Mining and Metallurgical Engineers*, 194, 747–753.

12. Shockley, W. (1950), *Electrons and Holes in Semiconductors: With Applications to Transistor Electronics*, New York, New York: D. Van Nostrand Company, Inc.

13. There are numerous accounts of the rise and fall of Shockley Semiconductor, including Riordan and Hoddeson (1997) pp. 225–253; Reid (2001), pp. 87–91; Lecuyer (2007), pp. 129–139. Particularly interesting is a 2006 discussion between four major actors in the rise and fall of Shockley Semiconductor conducted by the Computer History Museum ("Oral History of Shockley Semiconductor Laboratory," Participants: James Gibbons, Jay Last, Hans Queisser and Harry Sello; Mountain View, California: Computer History Museum, pp. 1–14.)

14. Bassett, R. (2002), *To the Digital Age: Research Labs, Start-Up Companies, and the Rise of MOS Technology*, Baltimore, Maryland: Johns Hopkins University Press, pp. 45–47; Riordan and Hoddeson (1997), pp. 262–263; Lecuyer (2007), pp. 139–150; Reid (2001),

pp. 91–95. See also Laws, D. (January 2010), "A Company of Legend: The Legacy of Fairchild Semiconductor" *IEEE Annals of the History of Computing*, 32(1): 60–64 and Moore, G. (January 1998), "The Role of Fairchild in Silicon Technology in the Early Days of 'Silicon Valley'" *Proceedings of the IEEE*, 86(1): 53–62.

15. Jay Last, interview by Craig Addison in Beverly Hills, California on September 15, 2007 (Mountain View, California: Computer History Museum, Oral History, Reference Number: X4158.2008), p. 11. http://archive.computerhistory.org/resources/text/Oral_History/Last_Jay/Last_Jay_1.oral_history.2007.102658211.pdf. Accessed November 25, 2015.

16. Ibid., pp. 5–6.

17. Lecuyer (2007), p. 148.

18. Bassett, R. (2002), *To the Digital Age: Research Labs, Start-Up Companies, and the Rise of MOS Technology*, Baltimore, Maryland: Johns Hopkins University Press, pp. 45–47; Riordan and Hoddeson (1997), pp. 262–263; Lecuyer (2007), pp. 150–154; Reid (2001), pp. 91–93. See also Laws (January 2010), pp. 64–67 and Riordan, M. (December 2007), "The Silicon Dioxide Solution: How Physicist Jean Hoerni Built the Bridge from the Transistor to the Integrated Circuit" *IEEE Spectrum*, pp. 51–56.

19. Berlin, L. (2005), *The Man Behind the Microchip: Robert Noyes and the Invention of Silicon Valley*, Oxford, UK: Oxford University Press, p. 108.

20. The major sources for the following section come from Reid (2001) and Berlin (2005).

21. Berlin (2005), p. 104.

22. Bassett (2002), p. 50.

23. This account of MOS at Bell and Fairchild comes mostly from Bassett, Ross, and Knox, pp. 22–33; 45–53; 107–138. An important article also is Sah, C. (October 1988), "Evolution of the MOS Transistor—From Conception to VLSI" *Proceedings of the IEEE*, 76(10): 1284–1300. See also Laws (January 2010), pp. 67–70.

24. Bassett (2002) p. 19.

25. Ibid., p. 117.

26. Ibid., p. 120.

27. Ibid., pp. 168–173.

28. Ibid., p. 206.

8

ADVANCED MATERIALS AND THE INTEGRATED CIRCUIT II

The Silicon Gate Process—The Memory Chip and the Microprocessor

... the basic difficulty in those days [in order to make] microprocessors was having a process technology that could actually do it ... Well, in 1970, the technology to make microprocessors was really available only at Intel because Intel had developed the silicon gate [process, and this] was the only way to do it ...

Federico Faggin, 1995

By 1968, scientists and engineers knew that the integrated circuit's only hope of ever becoming a mass-produced consumer device was if it ran on field-effect transistors (FETs). As useful as Shockley's bipolar junction—"sandwich-structured"—transistor was in many respects, it was no FET; it operated very differently. Its very structure prevented it from being able to be reduced in size beyond a certain point, which meant that ICs made from these components could not contain very many of them. This, in turn, limited the size and power potential of IC technology. In contrast, MOS transistors, which are FETs, have a very different structure which does allow virtually unlimited miniaturization. They promised a tantalizing future of extremely small and powerful microchips for a whole host of possible products, including small computers for home use. By the time Intel came along, MOS transistors and ICs based on them were certainly a very tantalizing prospect for champions of that technology. Both Bell Labs and Fairchild had their chances to dominate the technology, but they failed for reasons explained in the previous chapter. Intel was now next up at bat, and it hit a home run that shook—and continues to

Advanced Materials Innovation: Managing Global Technology in the 21st century, First Edition.
Sanford L. Moskowitz.
© 2016 John Wiley & Sons, Inc. Published 2016 by John Wiley & Sons, Inc.

shake—the world. Its success depended not only on getting MOS right but, at least as importantly, on harnessing it to another advanced material innovation called the silicon gate process. And so we come to a rather unique chapter in our story of advanced materials. In contrast to other cases we consider, this one demonstrates the additive nature of technological change, particularly how two processes come together to create a single technology more potent than either its component parts. For the silicon gate process—the subject of this chapter—combined with and greatly augmented by MOS integrated circuits, what seemed impossible to do with MOS alone could be done, and in a powerful way, when it embraced the silicon gate technology. The combined process made semiconductor memory possible, but, even more importantly, it created the microprocessor, a device absolutely vital for 21st-century computers, mobile technology, smart TVs, and all such devices that inhabit and even define our digital age. The company that managed to link the two processes and pioneer both the memory and the microprocessor—and continue to dominate the microchip industry today—is the Fairchild spin-off Intel. That company began for the same reason Shockley Semiconductor, Fairchild, and many other spin-offs arose, rebellion by creative employees against what they feel to be a conservative culture, namely within the parent company that stifles innovation. They want to break free to build their own business in their own way around their own ideas. In this case, this meant transferring and expanding the "short-term dynamic" culture they began to build up at Fairchild but was so rudely destroyed by Sherman Fairchild and his bureaucratic minions. This time around, they had no intension of being interfered with; this time around they would continue what they had started and nourish and develop and fully realize their unique strategic vision. The silicon gate process allied with the MOS platform would be the tool that would give them to leverage to do this.

8.1   BACKGROUND: CREATING INTEL

In 1968, Noyce and Moore sold their interests in Fairchild and established their own start-up operation.[1] They set up a shop in an abandoned semiconductor plant once owned by Union Carbide and located in Mountain View, California, not far from Fairchild's manufacturing facilities. Arthur Rock, Silicon Valley's leading venture capitalist, helped secure initial funding for the enterprise. They chose the name "Intel" as a contraction of integrated electronics, a clear sign that their venture was about the development and exploitation of the integrated circuit.

   Noyce and Moore had just come from Fairchild and its failed MOS program. They took away from that experience the importance of rethinking organizational, decision-making, and human resource strategies. They were especially concerned that they create commercially viable products and do so quickly. Learning from the troubles that Fairchild had with MOS, they believed their new firm had to have the type of flat organization, such as existed when they first formed Fairchild, before it became large and segmented. Accordingly, they concluded research, development, and manufacturing had to work extremely closely. They in fact didn't even create a separate R&D department. All research and development took place alongside manufacturing and only as the on-site need arose. Then too, they wanted decisions on where to place scarce resources to be economically rational and self-regulating; they arranged it so that money flowed automatically—and without the need for explicit orders from above—to those products that actually performed the best in the marketplace. By the same token, resources did not go into products that performed poorly. Further, they initially avoided hiring pure researchers. They wanted practical scientists and experienced engineers who would immediately understand how to carry out—and improve upon—the critical materials processes necessary to make MOS integrated circuits bankable. Two early hires were Andrew Grove, assistant director of the R&D lab at Fairchild, and Les Vadasz, head of Fairchild's MOS development. Both men played a major role in the future of Intel.

Once they determined how Intel was to be organized, resources allocated, and people hired, Noyce and Moore had to decide what technology and products they would pursue. From their work while at Fairchild, they had intimate knowledge of a number of closely linked advanced materials processes. These included crystal growing and purification, surface treatment, doping (using thermal methods), oxidation (also using furnace technology to create the silicon oxide surface layer), metal evaporation, photolithography, masking (adapted from the photographic industry), and chemical etching. They not only understood these processes separately but also how they overlapped and intimately fit together to create planarized systems, or integrated circuits. Using these processes, Noyce and Moore concluded they could succeed where Fairchild had failed: creating and selling semiconductor memories. They went ahead and began delving into fashioning such devices from bipolar integrated circuits (i.e., ICs made of bipolar or junction transistors). But Moore particularly believed that the future of mass-produced, low-cost, and high-capacity semiconductor memories was in MOS integrated chips, that is, integrated circuits composed of MOS transistors. Here he thought Intel could really make its mark on a truly breakthrough innovation. But he and Noyce also knew very well the hurdles the MOS approach presented to anyone who attempted to take it on to make actual, working products. However, they had a card up their sleeve that might give them a winning hand in the MOS game. Moore was well aware that Fairchild had begun looking into an interesting and potentially game-changing technology known as the silicon gate process that, if linked up with MOS technology, could lead to a very effective approach for making semiconductor memory products. With Fairchild's R&D program in shambles and with its most talented personnel leaving (or wanting to leave) to work in—or start—other companies, it would not be too difficult to entice the leaders of that company's silicon gate project to defect over to Intel. Noyce's and Moore's reputation in the industry, combined with financial incentives, would make this a fairly easy thing to do. This is exactly what happened and it didn't take long for Intel to have their silicon gate process up and running. This made all the difference for the company and its future growth:

The silicon gate process was the foundation of both Intel's success in MOS technology and its success as a company, for it proved to be extremely manufacturable and applicable to a wide range of products.[2]

Once research on the silicon gate (which I'll call from now on the "MOS-SG") process had been completed, development, scale-up, and commercialization went in different directions, depending on the product Intel wished to turn out. Silicon gate technology became the main technological driver of the semiconductor industry beginning in the 1970s.

8.2 THE MOS-SG PROCESS: RESEARCH AND EARLY DEVELOPMENT

The idea of the silicon gate process came from an electrical engineer working on Fairchild's MOS project.[3] Thomas Klein helped develop MOS technology at the Mullard Research Labs in the United Kingdom before joining Fairchild 1966. At about this time, a vital piece of information came to Klein's attention that started him along the path leading the silicon gate approach. Klein, always on the lookout for any insight that could improve the performance of MOS devices, learned something very intriguing from a physicist with whom he shared a cubicle. The scientist, who had nothing to do with MOS research, showed Klein an interesting type of capacitor he had been working on. Its distinctive feature was that it incorporated a small amount of evaporated polysilicon, a material composed of many silicon crystal units in contrast to the single crystal silicon used in making transistors and integrated circuits. This one change appeared to significantly improve the performance of the capacitor. Klein then began to wonder if such a material might work its magic on MOS devices. So he began experimenting with replacing the metal gate (the "M" in MOS) with evaporated polysilicon. He found that indeed MOS transistors with polysilicon gates were faster and more stable. He attributed this improvement to the lower work function of the polysilicon compared to metal. This meant that it didn't take as much energy—that is, it was easier—to remove electrons from the former. This lower work function meant chips that were faster and more stable and had greater reproducibility. Klein then took the next significant step: he created laboratory devices using a special photolithographic technique employing his new evaporated polysilicon gate process.

A problem then surfaced that threatened to derail progress on the technology. The evaporation of polysilicon proved to be difficult to control resulting in polysilicon gates that cracked causing device malfunction. Attempts to use different techniques such as sputtering did little to help, in large part because they injected destructive impurities into the gate site. Fairchild's "Materials and Processes" group also worked on the problem and came up with a promising technique called chemical vapor deposition. This process used an electrically heated horizontal tube hot enough to decompose a silicon compound known as silane when forced through it. This method produced a uniform and smooth layer of polysilicon at the gate. This approach proved to be a major breakthrough in semiconductor manufacturing.

8.3 THE MOS-SG PROCESS: DEVELOPMENT PHASE—PERFECTING THE PROCESS

Soon after Intel began its life, Noyce, Moore, and Grove decided to become a semiconductor memory company.[4] But the question remained, would that semiconductor memory be bipolar or MOS? They decided to conduct both research efforts simultaneously using two separate teams in a sort of competition to see which made the greater progress. If MOS were ever to become a possibility as the way to make memory chips, the silicon gate process would have to be perfected. After a slow and fitful start, MOS-silicon gate memory started looking very promising in terms of reliability and ease of manufacturing. It soon took the prize and Intel jettisoned its bipolar research. During this crucial MOS-SG development phase, Intel avoided the decentralized approach of Fairchild, by placing responsibility for process development on one man, Tom Rowe, who had a Master's degree in metallurgy from MIT and had worked at Fairchild at their Mountain View location. Rowe had a "deep intellectual curiosity" about science and technology but also a practical manufacturing bent, which he could exercise while at Fairchild. He understood the vital interactions that must occur between research, development, and manufacturing. At Intel the human and structural barriers between these activities had been largely—if not totally—eliminated. It was Rowe—as both scientist and engineer—who absorbed and dealt with the various issues that came up with the MOS-SG process. In doing so, he understood how the different problems linked together and, with research, development, and manufacturing carried out simultaneously and in one place, how to quickly apply these solutions to actual working circuits and devices. This concentration of problem solving in one department and in one person had enormous advantages that accelerated the company's march to the market. A case in point was producing stable MOS-SG devices. Fairchild had earlier discovered that traces of sodium infiltrated into its MOS devices causing device instability. This was a very serious problem. But because of the lack of communication between Fairchild's R&D at Palo Alto and its manufacturing at Mountain View and because there was not one person with the authority and capability to translate this scientific observation into practical terms, Fairchild fumbled badly dealing with this issue. Specifically, the R&D department dealt with the problem in terms they understood, attempting to completely remove all traces of sodium in the production process. While this approach might be reasonable at the laboratory level, real-world manufacturing could never totally eliminate every vestige of sodium. It is no surprise then that Fairchild's devices never really worked satisfactorily. Intel took a different approach. It accepted the fact that some sodium would get into the production process but searched for a way to render it harmless. The solution was to design the MOS transistor in a different way than had Fairchild so that it was more rugged and capable of functioning even if contaminated with small amounts of sodium.

The concentration of the MOS-SG process in one person and one department also worked well when Rowe and his team were faced with a series of interrelated problems. Rowe understood how one bottleneck fed into a second and how the solution for the former could be used to help solve the latter. Difficulty in depositing the silicon gate onto the silicon dioxide layer continued to plague the project. Fairchild's approach (using the electron beam evaporator) turned out not to work so well outside the laboratory as the evaporated silicon had trouble "sticking" to the oxide surface. The evaporator was also slow and dangerous; it tended to explode when in the hands of unskilled operators. Rowe's answer was to chuck the electron beam evaporator and see what was immediately available in manufacturing that might be used. He settled on something called an epitaxial evaporator, which Intel was already using in the production some of its bipolar junction transistors. But use of an epitaxial reactor caused another problem involving masking and photolithography. These used a chemical called a photoresist that had to adhere very tightly to the polysilicon gate. But the resist did not cling as well to a silicon gate that had been deposited by epitaxial reactor than when an electron beam evaporator had been used. Rowe, understanding the benefits and costs of the epitaxial reactor, came up with a solution. Keep the technology but now grow an extra oxide layer on top of the polysilicon gate and then put a film of photoresist on top of this. But this answer had its own costs to deal with. The extra film layer created a very steep steplike structure at the gate that was very difficult to cover with metal, as required in making MOS transistors. Rowe handled this situation by adapting a type of machine tool used in bipolar manufacture, a special dome-shaped planetary jig (such as used in metal grinding operations), that would hold wafers in place while being covered by evaporated metal.

The fact that Rowe and his team solved one problem after another and kept the project moving forward continued to energize the effort and push Intel's management to keep backing Rowe and the silicon gate technology. But by late 1969 Rowe still did not have a working process. He had produced an actual silicon gate memory chip—called the "1101"—but the yields (only 2%) were a disaster. At the same time, the much better known junction bipolar effort at Intel were far more successful and getting ready to ramp up production. The question for Intel's management now was whether to abandon the field-effect approach to transistors and the integrated circuit, as embodied by MOS and silicon gate and dedicate its resources totally to junction bipolar chips or to keep MOS going and hope for improvements to come, and come soon. Moore and Noyce decided to stick with it. They were after all extremely savvy about the promise and possibilities of MOS. Moore had been keen on it since his Fairchild days. He saw it as the most likely way his great law of improved chip performance with lower costs could continue on course. Grove of course had worked on MOS since he first got to Fairchild. He did not want to abandon it now. The fact that Rowe continued to make progress sealed the deal, and the hunt for a workable MOS transistor using the silicon gate process continued. Grove, with Moore's urging, told Rowe to soldier on until he knew beyond any doubt that the process was unworkable—by Intel or any other place.

It is a very good thing that Intel allowed Rowe to continue. For he soon came upon the breakthrough insight that solved the yield problem. In thinking about the various difficulties he had confronted and how he solved them, he realized that all the improvements made up to that point involved the process of layering, that is, setting down layers of material through such methods as evaporation and epitaxy. He recognized that what had not been addressed was another important part—the most important part—of the process: etching, or removing materials to form the components within the silicon. Then he recalled his Fairchild experience working on the bipolar transistor: if a material that had to be removed was not taken away cleanly and completely, layers did not adhere well to one another. The device could then fail and even come apart physically. Since chemical rinses were used to remove materials, finding more powerful rinses was the key to better devices and improved yields. He soon developed his own rinse—which he called "super dip"—that increased tenfold the performance and reliability of his chips and greatly accelerated chip yields.

All of these innovations and improvements to the MOS-SG process took place in a relatively short amount of time. By September 1969, just a little over a year after Intel's founding, the silicon gate process was in place, and it had made MOS transistors and chips commercially viable. The next step was to apply this process to making actual products.

8.4   THE MOS-SG PROCESS: PRODUCT DEVELOPMENT

With the success of the silicon gate process, Intel had themselves an extremely pliable tool—its basic template could be used for making different revolutionary products. Noyce and Moore, of course, knew how they wanted to cash in on silicon gate. They knew that now Intel's competitiveness could be demonstrated in semiconductor memories. Intel, in fact, bet its future on developing and commercializing memory chips for the then dominant mainframe computers.

Existing memory components in mainframe computers operated according to 19th-century electromagnetic principles. They consisted of magnetic cores in which coils of magnetic material were interwoven with copper metal. These coils could move in only one of two dimensions—up or down—at any one time depending on the current flowing through it, thus allowing binary operations to be performed. But MOS-SG technology offered a much better way to go in terms of denser circuitry, greater storage capacity, and more rapid movement of data between memory circuit and memory storage area than ever before. In addition, MOS-based memory chips required fewer manufacturing steps—and thus were less expensive to make—compared to magnetic core technology.

8.4.1   MOS-SG and Memory I: The "DRAM"

Because the important immediate goal for Intel was to get a "killer" product on the market as soon as possible, and in keeping with the closeness between research, development, and manufacturing, Intel's test chips—those chips used to develop and prove the effectiveness of the silicon gate process—actually were also the very first memory chips.[5] The most important of these was a 256-bit random access memory (RAM) chip that Intel designated as the "1101." This was a relatively simple device known as a "static" random access memory (SRAM) chip; it held memory only as long as the power to the chip was on; when the power was turned off all stored data was lost. The 1101 was an ideal chip to test the MOS-SG process. Intel wanted to see if this one worked before moving on to a more complicated family of chips. It went on the market in September 1969. Despite its simplicity, it was ahead of its time and Intel charged a premium price for it. Feeling very confident in its MOS-SG process, the fledgling company adopted a "technology first" strategy and sold their new chips at top prices. Noyce and Moore assumed that customers would be willing to pay for these chips because they outperformed the older, magnetic memories. The risk of course was the customers would be wary of replacing what they knew with new components. In fact, Intel did have trouble finding customers for its 1101. Noyce, Moore, and Grove understood this risk and had no intension of high-wiring it without a net. As a fairly secure source of income, they knew they could design and build some of the most advanced shift registers using MOS-SG and be competitive. Shift registers are circuits that can be used either in parallel or in serial, a capability in great demand by computer makers. This market brought in revenue until the memory chip business took off. It didn't take long before memories did just that. The demand for semiconductor memories by computer makers, learning of Intel's technology at conferences, at the trade press, and through the company's sales efforts, exploded.

Intel soon applied its MOS-SG process to create its second-generation memory chip, a "dynamic" random access memory (DRAM) product, which continued to hold on to its information even after power had been shut down. Called the "1103," the impetus for this 1024-bit chip came from a request from Honeywell that wanted semiconductor memories for its line of minicomputers. Intel had to work very closely with Honeywell to make a chip that was not only technologically superior but also that would match up with the computer maker's exacting requirements. It came to appreciate the great flexibility of its MOS-SG process on this project. Silicon gate could not only adjust chip structure to fit into Honeywell's computers, but the process could modify the chip and its architecture to fit into many other computers. Intel started shipping out the 1103 chips in June 1971. It soon became the standard memory chip in the computer industry; Motorola, Fairchild, National Semiconductor, Signetics, and other major semiconductor companies all wanted to make it. They couldn't compete directly with Intel, which was the only company that controlled the critical MOS-SG process. But they could second source the chip, with help from Intel. Intel allowed them to do this since computer makers, not wanting to be totally dependent on just one supplier for such an important component, demanded that Intel find other companies willing to make it. Intel knew of course that no other competitor could be a second source without its help. In fact, without the capability that Intel had in MOS-SG, many of them faced numerous problems in attempting to manufacture such an intricate chip: "The 1103 was a very difficult part to make and ... [many gave up when] they saw what a large head start Intel had."[6] Intel clearly was in the driver's seat. By end of 1972, the 1103 was being used by nearly 80% of mainframe computer manufacturers in the United States, Europe, and Japan, and it was the top-selling semiconductor part with the world's largest dollar volume. In the early 1970s Intel controlled up to 85% of the 1103 market.

But this success was only just the start of Intel's domination of semiconductor memory market. The great elasticity of the MOS-SG process became a blueprint and active force for Intel's strategy of "infinite expansion" and the foundation for its closely related short-term dynamic culture. Through exploitation of the MOS-SG process, and its initial insistence of not being diverted from this path by venturing into tangential technologies, Intel was able to place more and more transistors on a smaller real estate of silicon, coming out with new, denser, and more powerful chips every 18 months or so, in close adherence to Moore's law. The computer industry came to expect—and demand—these chips and was willing to pay dearly for them. Intel profited greatly soon from these new high-end chips, while competitors were always a generation behind, making older chips that Intel had introduced a couple of years earlier in volume and selling them as best they could at commodity prices for lower profit margins. Intel's culture not only accepted the cannibalization of the old by the new, but counted on it to occur every year and a half. It didn't take long before Intel began dictating to their customers their own memory requirements as computer manufacturers, hungry for more and more memory for their machines, came to Intel hat in hand waiting for the company's newest and most advanced chips. As one historian of the industry observes:

> … it was as if the semiconductor industry was a virus that had successfully inserted its genetic code into the computer industry host so that henceforth the host would follow its dictates.[7]

The range of Intel's processes extended beyond new and more powerful versions of the DRAM, and it soon created entirely novel products that catapulted the company into the undisputed leadership position of the semiconductor industry. The first of these was a breakthrough in memory technology, which a few years earlier even the founders of Intel never dreamed possible.

8.4.2   MOS-SG and Memory II: The "EPROM"

One of the most significant of Intel's memory products that came out of the use of the MOS-SG process in the 1970s was the EPROM, which stands for "erasable programmable read-only memory."[8] The EPROM, which many consider to be a radical leap in memory technology, did not have any obvious market. As its inventor tells us: "It resulted from a creative engineer's attempts at improving the performance of the DRAM." In particular, low yields and reliability problems with the 1103 chip led one of Intel's physicists, Dov Frohman—a Ph.D. electrical engineer from Berkeley who defected from Fairchild to come work at Intel—into making a remarkable discovery. In closely examining the yield problem, he noted that some of the MOS transistors created by the MOS-SG process seemed to become dislodged from their proper positions. Frohman understood that this "gate instability" caused the problems with the DRAMs. But he also began to understand that it might also be the source of a whole new type of memory product. In a flash of insight, he showed how this phenomenon of the "floating" gates could, by adjusting the parameters and operating conditions of the MOS-SG process, actually be turned into great advantage for Intel engineers. He conceived of a radically new and important memory device, one that could actually use floating gates to speed up the production cycle and, as importantly, allow each device to be programmed "with incredible ease."

But this was not all. As Frohman continued to explore the floating gate effect, he discovered something even more remarkable: if floating gate transistors were arrayed properly on the memory integrated circuit, programs imprinted on the memory chip could actually be erased if the chip was exposed to ultraviolet (UV) light, and a new program could be written by the user in a matter of minutes. This meant that, for the first time, engineers could virtually on the spot correct design errors or simply improve program content, and they could do so as often as they liked. This ability to instantly change and upgrade memory devices meant that the MOS-SG process as embedded in the EPROM had reached a new level of sophistication.

While Intel's ruling executives did not plan for this innovation, once Frohman showed them what it could do, they gave their blessing to it. To them, the EPROM clearly was another and very promising species of fruit ready to be picked from the MOS-SG process tree. As Stanford's Robert Burgelman relates: "Gordon Moore committed Intel to the production of the EPROM even though no one could tell where the device would have applications....."[9] From a technical point of view, Intel felt very comfortable with the EPROM since it derived from the company's mainstream silicon gate technology. Nevertheless, despite Moore's acceptance of the concept of the EPROM, converting that idea into a commercial product was no easy matter. Frohman worked alone on devising a prototype and with only a very limited staff to take it and turn it into a manufacturable device. Problems presented themselves almost every day. Intel was willing to let Frohman continue to work on the project, but only if he did so invisibly, or as he tells us in a 2006 interview:

> ... there were two phases. There was the phase of the concept, and there was a phase of developing the product. I didn't get any support. I did not ask for it either. It was up to me to convince people it was worthwhile to produce. In effect, I had to do a lot of the early marketing negotiations internally to try to convince people because the first responses were not very amicable. In the concept phase I was alone ... [10]

Frohman continued to chip away at the various roadblocks that came his way. Once he created a workable prototype, Intel at least allowed him to work "above-ground" on company time. But the question of markets still loomed. Moore and Grove felt that the EPROM could start generating income fairly quickly as a specialized component for computer manufacturers in designing and modifying their prototypes. But they were also concerned that, once a dominant design for a computer model had been established—which is what the prototype does—and its microcode frozen, there would be little need for the EPROM. This posed a serious risk for the future of the device. But test marketing soon indicated that the EPROM itself was a far more fluid device than initially realized. In fact, Intel's management soon realized that the EPROM could be adapted for many market applications and that the demand for EPROMs once available would take off. Buyers of computers wanting the ability to program and rewrite code for their machines on a regular basis demanded EPROMs, and they were willing to pay dearly for them, up to ten times more than for other types of memory chips. Sales—and prices—continued to climb as Intel worked the MOS-SG process to continually make the EPROM denser and more powerful. Given that the EPROM was a large chip, considerably larger than Intel's DRAM memory, there were clearly going to be challenges in manufacturing the device:

> We had to really take new steps with almost every step we took. We were considered crazy. The mask making itself was a problem because the masks were very big and of course there were problems in the process.[11]

However, a number of factors came together to ease Intel's fears about proceeding. Most importantly, the EPROM embodied Intel's core technology (the MOS-SG process). Then, too, Gordon Moore particularly had great confidence in Frohman's ability to solve any remaining problems that might come along. And Frohman showed Moore and his colleagues that there would be great "synergies" between the EPROM and the newly invented microprocessor, that in time they would be able to be produced together on the same production line, and that the EPROM would enable " ... microprocessors to do things that they couldn't do with [the older] mass programmable ROM [read only memory chip]."[12]

The EPROM turned out to be way ahead of its time, and as a result, Intel faced absolutely no competition in this technology for many years. No one at Intel could predict that the company would for years come to depend on the EPROM financially and that it would be its most profitable product through the 1970s.

8.4.3   MOS-SG and the Microprocessor

The microprocessor, of course, is the most important product that came out of the MOS-SG process.[13] While the memory chip is undeniably significant, the microprocessor is something else entirely. It is a totally unique invention that opened up entirely new applications. It has also proven to be the key innovation that has allowed Moore's law to continue on its impressive trajectory. The microprocessor "proved to be the key engine of growth, with new applications appearing with each of the endless rounds of price cuts and performance improvements."[14] Because of the microprocessor, the PC revolution burst on the scene in the 1980s and thrust the computer, until then the creature of government and business, into virtually all aspects of everyday life.

The microprocessor is a solid-state version of a mainframe's (or minicomputer's) central processing unit (CPU); it contains all of the essential parts of a CPU but on a single chip. Together with memory chips and peripheral equipment components, it makes up the essentials of a fully functioning computer. The idea of putting a computer—or the major functions of a computer—on a single chip was not a totally new concept in the early 1970s. As early as 1964, companies large and small—Westinghouse, IBM, General Instrument, Viatron, and Four-Phase Systems—were writing about it and thinking about it and making some attempts to do it. The one problem they all faced was the limitations in how much information could be placed on a silicon chip. Gordon Moore also liked the idea of it, but in the late 1960s, even he did not believe it possible for the simple reason that such a chip would cost too much to make. He and Noyce thus rejected the idea as too unrealistic and scotched the notion that their new company ought to pursue such a technology. Even though Intel was still a young, small start-up, the push for such a radical innovation would have to come from below. The two men most responsible for Intel's pursuit of the microprocessor—its two champions—were Ted Hoff and Federico Faggin.

8.4.3.1   Ted Hoff, Circuit Design, and Inventing the Microprocessor   Hoff is generally credited with inventing the microprocessor. With a Ph.D. in electrical engineering from Stanford, Hoff brought to Intel extensive computer skills, something that the company had in short supply in the early 1970s. Because of this much needed expertise, Hoff was the perfect person to be the technical liaison between Intel and the computer industry. In the summer of 1969, he held discussions with a Japanese company called Busicom for Intel to develop a group of seven chips for Busicom's new electronic calculator. Intel was just getting started and was glad for the business. But it also had a problem. A circuit design capability now was what was needed to make the chip set for the new Busicom calculator. But Intel had only a few people there who knew circuit design, and they were otherwise engaged working on what was becoming Intel's bread and butter, memory chips.

By default, the job fell on the newcomer Hoff. Luckily so since he was something of a genius in the field. Working with another engineer Stan Mazor, he found a way to greatly simplify chip architecture so that all of the tasks involving logic (the performing of the calculations) could be carried out on one chip (the CPU or microprocessor). This discovery reduced the total number of chips needed for the Busicom calculator from seven to three: the processor and two memory chips. This simplification greatly appealed to Hoff's passion for finding the most economical way to design an integrated circuit. He also was quite pleased that he found a way to design a computer on a chip that could actually work. It also made it much easier for one man (Hoff)—Intel's only available circuit designer—to proceed with Busicom's assignment. While the Busicom engineers in Japan initially opposed any change from their original concept, Hoff convinced them to go along with his new design.

8.4.3.2   Federico Faggin, the MOS-SG Process, and Making the Microprocessor

Designing the microprocessor was one thing, and turning that design into a working device was something else entirely. Doing this required advanced materials processing, and this was something Hoff and Mazor knew very little about. The project stalled until Intel found someone who could take over from Hoff and get the job done. Federico Faggin, an Italian scientist, got his doctorate in physics from Padua University. He had done work for Fairchild on both the MOS and silicon gate processes in the late 1960s. Faggin was an extremely talented scientist who greatly impressed Moore while both were at Fairchild. Moore was the one who brought him to Intel. With the Busicom project significantly delayed and the Japanese breathing down Intel's neck to make progress, Moore put Faggin immediately on the project. What Hoff and Mazor had done was the chip architecture, basically a diagram indicating what the main components were and how they related and communicated with one another. In other words, it was a paper tiger at this point. Faggin's job was to convert this map into a working chip. This required figuring out how the circuit elements—mainly transistors—needed to be organized to perform logic functions and then make the chip that could do this. Faggin soon understood that the only way Hoff's microprocessor could be realized was by making MOS transistors using the still unproven silicon gate process. He would need to adapt the process to "build on a chip a series of layers" of interconnected transistors in highly intricate patterns. A deep understanding of how the process worked was here absolutely essential. It would be quite an easy matter to miss implanting one of the layers or to incorrectly interconnect one layer with another. Faggin knew what he was up against as this would be a far more complicated device than a memory chip; in fact it would be the most complex chip ever created. Faggin worked at breakneck speed over the next few months, and by March 1971, he and his team had made their first microprocessor called the "4004." A year later, at the request of another customer (A San Antonio producer of computer terminals who ultimately backed out of the project), Hoff, Mazor, and Faggin worked together to make a far more sophisticated, second-generation microprocessor, which Intel designated as the "8008." The ultimate success of these two microprocessors and particularly the 8008 depended on the power of the Intel's MOS-SG process:

> A crucial factor in the successful realization of the microcomputer was Intel's use of silicon gate MOS technology, which made a large complex chip like the 4004 or the 8008 manufacturable. The silicon gate process was denser than the standard MOS processes in use at many other firms so Intel could build chips that had more components on them than other firms could.[15]

8.4.3.3   The Competitive Advantage of Intel's Microprocessor   To illustrate the absolute dominance of Intel over its competitors because of its expertise in the MOS-SG process, we need only consider that Texas Instruments had also been approached by that same San Antonio firm with the same request; TI failed to deliver despite earnest attempts at making the type of chip needed. Nor could Intel's nearest threat in the memory business, Mostek (the TI spin-off), succeed when contacted by Busicom to come up with its own chip set for the Japanese company's calculator. Other companies—including Rockwell and National Semiconductor—also had developed their own version of MOS technology. But what all these companies did not have was the silicon gate process, and this made all the difference. MOS transistors and chips without silicon gate were perfectly fine for making certain memory products, which are much simpler chips than are logic devices, and for custom designing and building a number of integrated circuits to carry out logic for very specific applications, such as electronic calculators. But mass-producing standardized microprocessors—with all the logic functions on one chip—useful over a broad range of applications, required a far more involved chip-making capability that only silicon gate could provide.

At first, Intel's competitors made the most of what they could do. They first made less sophisticated chips for consumer products and then, like Texas Instruments, integrated forward into electronic calculators and watches. They initially enjoyed high profit margins. But soon other companies in the United States and Japan learned how to make these relatively simple chips and also pushed forward into consumer products. The Japanese in particular excelled in these markets and could make their chips, calculators, and watches very cheaply. While sales rose, prices—and profit margins—plummeted. Through the 1970s, prices of electronic calculators fell from thousands of dollars to 20 dollars or less. The smaller companies, like Mostek, went out of business, while the larger ones suffered such severe losses that it took them years to recover.[16]

By the mid-1980s, Intel itself faced the biggest crisis of its fifteen-year existence. The Japanese had decimated its memory business, Intel's bread and butter. But unlike Mostek, it had its golden lifeboat—the microprocessor—to save it. Intel not only survived; it thrived as prime mover of the personal computer revolution.

8.4.3.4   Championing the Microprocessor at Intel   As early as 1971, through an announcement in the trade journals, Intel told the world about its new product.[17] While Intel was nowhere near the point of wrapping corporate strategy around the microprocessor—this would not happen until the mid-1980s—even this limited acceptance of Hoff's and Faggin's innovation by Noyce, Moore, and Grove did not arise automatically. It took some convincing by the technology's creators to bring top management to their point of view that the microprocessor was worth pursuing as a part—perhaps important part—of Intel's future's growth. To understand this reluctance from above, it is important to remember that, even though in the early 1970s Intel was still a relatively small start-up and even though the company adopted a more or less flat organizational structure, it still had a clear hierarchy. There were the top managers—the two founders and Andy Grove—and there were the middle managers working on R&D and manufacturing. At this time, Intel's power structure was most concerned with building up their silicon gate process and applying it to memories. Recall that Noyce and Moore had been refugees of two companies, Shockley Semiconductor and Fairchild, in which they tried and failed to get an important technology that they believed in to market. In no way would they allow Intel to be sidetracked by future possibilities, as intriguing as they might be; they were laser focused on here-and-now profits. These profits must come from semiconductor memories, and they meant to control this increasingly competitive market. They understood the laws of "first-mover advantage," especially in the semiconductor business, a winner-take-all world, and that they had a superior process (MOS-GATE) that could make them winners in the field. They also appreciated the power of economies of specialization by which market supremacy comes from marshalling your resources toward excellence in one thing. By the same token, attempting to spread your forces over too many projects weakens the firm on all fronts resulting in losing all advantage in any one. From its very beginning, it had become dogma that Intel's "one thing" was memory. For Grove, Intel's primary order of business was clear: "Our Priorities [at Intel] ... were formed by our identity; after all, memories were us" (italics in Groves').[18]

As far as Intel's ruling triumvirate was concerned, Hoff's computer on a chip had one application: the processor for Busicom's electronic calculator. It might also be individually designed for other customers but only as specifically requested. But Hoff, Mazor, and Faggin, being much closer to how a microprocessor was made and to its power and elasticity of application, knew that it offered Intel an impressive tool with virtually unlimited possibilities as a standardized logic component. But Hoff and Faggin had a struggle on their hands. It took active campaigning by both of them working together to convince their bosses to consider the microprocessor as something that Intel should pursue over the long term.

> Ted Hoff and Federico Faggin recognized the strategic importance of the micro-processor ... It took intensive championing on their part to convince top management of the strategic importance of microprocessors for Intel's future.[19]

In a telling interview years after his great success, Faggin has a lot to say about his difficulties with senior management and its acceptance of the microprocessor. The major hurdle that he faced was in convincing Intel that the microprocessor was not just a one-trick pony and that it had general application beyond Busicom. He did it by demonstrating that his microprocessor could do other things beyond the electronic calculator:

I really wanted that product [4004] to be on the market and so I really pushed Intel management. The first thing that I did was: I developed a tester. A tester for the 4004 as a matter of fact, and I used the 4004 as the controller of the tester and so I could show, "Look, you know, this is not a calculator right? I'm using the 4004 to do a control function for the tester and it's doing the job, and it's doing the job well." And that certainly got into the ears of Noyce and certainly that was an important event because it turned their minds toward the potential of the 4000 [family of microprocessors].[20]

Showing Intel's top managers that the 4004 had general application was critical in getting the company to secure its rights to the invention by buying them back from Busicom. If they didn't, Intel could not have proceeded with further development of the microprocessor. But this negotiation depended on important strategizing on the part of Faggin. Faggin, realizing that only Busicom could legally use the 4004, meant to push Intel to buying its release from the exclusive contract it had signed with the Japanese company. Noyce and Moore, now seeing that their 4004 had potentially wide applicability, were very much interested in finding a way to be released from its existing contract but did not see how this might be done. Faggin came up with a possible avenue based on information provided by one of Busicom's former employees with whom he was acquainted:

I knew [from my contact] that they [Busicom] were really hurting because of the cost they were paying for the chipset to Intel and the company was not doing very well. And I realized that if Intel was to give a price break to Busicom, they would have a chance to get released ... from the [exclusive contract].[21]

Faggin relayed this information to Noyce who, when he took his next trip to Japan, proceeded to renegotiate the contract in such a way as to secure Intel rights to the 4004. But Hoff and Faggin continued to have problems with management as they persisted in pushing the envelope of the MOS-SG process to produce ever more complex and powerful microprocessors. The problems in pursuing Faggin's beloved "8080" microprocessor is a potent example of continued resistance from above to new and pioneering technology. The 8080 was the third-generation microprocessor, the successor to the 8008. To Faggin and other engineers there was little question that, at the time of its invention, the 8080 chip was going to be the game changer in integrated circuits. Indeed, it did become "the breakthrough part [that] ... made the industry ... and ... was ... the beginning of the microprocessor revolution."[22] Yet,

Intel's leadership was far from eager to chase the 8080 as it would take revenues away from its main business for a still commercially untested technology. As Faggin lets us know in no uncertain terms:

> ... top brass [at Intel] felt [the 8080] was too risky to start a new microprocessor when still they had not seen how the 4004 and 8008 were doing in the marketplace ... [23]

Faggin pushed back. He managed to convince his superiors that the microprocessor fit very comfortably within the company's know-how and business strategy. He emphasized that the microprocessor and memory chip were closely linked and that sales of the former increases those of the latter. He also argued that the new microprocessor was simply an extension of the process technology already used in making the previous two microprocessors. Consequently, Intel already had the core competency to see the 8080 through and without having to spend much money or manpower to do it. And if Intel did not proceed with it, other companies would and, if they succeeded, steal Intel's thunder as the most technologically advanced player in the industry, something in which the company always took great pride and that made Intel attractive to customers. Using these and other arguments, Faggin persuaded Noyce and Moore to agree to develop the new chip. This success came only after a long, difficult struggle with his superiors: "And so it took me a long time, it took me about nine months of really pushing and lobbying, to finally get permission to do the 8080."[24]

So Faggin and Hoff did succeed in getting Intel to stick with the microprocessor, at least on a limited basis. The fact that Intel did not kill the microprocessor but let it evolve was a big win for these middle managers. In truth, it is likely they would not have been so fortunate if Moore and Noyce had not themselves been very close to the very process—that is, they would not have been able to convince Intel leadership to support their device even on a limited basis—made the microprocessor possible and could be made to appreciate its technical kinship with the memory chip. In other words, the technical experience and sophistication of Intel's leadership kept the microprocessor alive within the company. Still, Intel accepted the microprocessor only as a "back burner" technology, one that would be tolerated but not given significant resources that were better spent on what Intel knew and accepted as its main business line. Memory chips continued to dominate the company long after they had ceased to be a winning product. For all this time, Intel refused to fully scale up production of the microprocessor or to include it in its central strategic planning.[25]

8.5 MOS-SG: SCALE-UP AND COMMERCIALIZATION

From the early 1970s, when Hoff and Faggin came out with their first microprocessors, until around 1985, Intel treated the device as a second-class citizen. It was given only a small area in Intel's factory for its limited production in order not to take too much manufacturing space from memory. Under such circumstances, the microprocessor could not achieve anything close to its economies of scale nor benefit from focused specialization. Intel had a lot riding on memory. It built its entire culture on it. It hired people—technical, marketing, and sales—committed to it. While competition from the Japanese had seized a growing slice of its market share, there was reason to believe Intel could recapture memory customers with its superior technology:

> The DRAM group led the company in linewidth reduction [in the photolithographic process]. They were already developing a 1-micron process while the logic [microprocessor] group was still developing a 1.5 micron process.[26]

Abandoning memory then would mean squandering the technological momentum Intel had built up. It also meant losing talented people who had spent their professional lives dedicated to the DRAM. Noyce and Moore had only to recall what a disaster the loss of gifted employees had been to Shockley Semiconductor and to Fairchild. They certainly did not want their firm to suffer in the same way. And there were powerful voices within Intel against making the microprocessor a more central part of the company. The marketing and sales people in particular strongly urged Intel's management to stay the course on DRAMs and EPROMs and not divert limited resources to the microprocessor. Not being process people, they could not see beyond what the current markets told them. They reasoned that—with the total global demand in 1971 for mainframe computers at 20,000 units and with at best a 10% share of that market—even if it continued to improve and gained acceptance in the computer industry, the microprocessor could never become a major profit center for Intel.[27]

Upper management, with its own loyalty to memory products at stake, willingly went along with the arguments of its marketing crew, at least for a while. And it is true that Intel did not accept the microprocessor as its central product for a number of years and only well after it had lost control of the memory chip market. Indeed, Federico Faggin himself became so frustrated at Intel's apparent inability to see the true market potential of the device that he left the company to start his own firm, Zilog. Faggin's departure not only slowed Intel's efforts in microprocessor development but by 1976 when Zilog first introduced its microprocessor chips, it also brought a new and powerful competitor to Intel in Silicon Valley.

Finally, at long last, Noyce, Moore, and Grove took the plunge. They abandoned memories and wrapped themselves in the microprocessor as the company's premier product. In making this decision, they took up the challenge of scaling microprocessor production and creating a business strategy to fully commercialize the technology. In the final analysis, this was not a particularly risky path to take. Intense competition, especially from Japan, and Intel's own internal resource allocation mechanism had already pushed the company in this direction. Even more importantly, Intel's sole control of the most flexible and powerful process in the field meant it, and it alone, could predict fairly precisely using Moore's law the future technical and economic trajectory of its microprocessor line of products. The MOS-SG process allowed Intel to peer into the future and see that it was likely to control the market for this up-and-coming product. Mistakes and errors of judgment might be made from time to time by Intel's management, but the overarching trend for the company was one progressing toward smaller, faster, and more powerful devices more rapidly than any known competitor. Noyce, Moore, and Grove jettisoned memories and took on microprocessors only when they knew they had lost the former but was positioned to dominate the latter. The following sections considers the important ways in which perceived risks toward the microprocessor were lowered to the point where Noyce, Moore, and Grove felt comfortable enough to replace memories with microprocessors as Intel's main product.

8.5.1 Competition and Resource Allocation

By the early 1980s, serious competition at home and abroad eroded Intel's lead in memory. Intel's expertise in making memory chips remained embedded in the superior MOS-silicon gate process. However, over time, other companies began to catch onto how Intel did what they did. Diffusion of Intel process know-how occurred in a number of ways, including the licensing of the process technology to second-source companies, transfer of knowledge through the movement of people from Intel to other companies within and outside of Silicon Valley, and skillful reverse engineering by competitors, who worked backward in analyzing how Intel made its memory chip. In some cases, a company might develop its own and advanced process technology that, for a time, gave it a competitive edge. Mostek is one important example. It used its proprietary ion implantation process to gain market lead in memory technology for a number of years.[28]

Most critically, Intel and the US semiconductor industry as a whole faced serious competition from abroad and in particular Japan. Companies like Fujitsu, Hitachi, Toshiba, and NEC learned how to make memory chips in a number of ways, not least of which through reverse engineering and pirating Intel technology. But Japan did more than simply parrot US technology; it found ways to beat American companies—and Intel in particular—at the mass production game. They did so in ways that completely undercut American ability to compete. Japanese companies that entered into semiconductors were large, diversified, and growing. A company could afford to spend massive amounts of cash on upgrading equipment and production lines by siphoning profits from its other and often unrelated businesses. For example, while over three-quarters of Intel's revenue came from semiconductor sales, only between one-fifth and one-ninth did so within Intel's Japanese competitors. Then too, Japan's militaristic educational system—highly regimented, disciplined, and inflexible—also fed into this push for superiority in production quality and efficiency.[29] Even more than in Silicon Valley, the Japanese semiconductor companies established and maintained extremely close relationships with their suppliers. This meant that producer and supplier worked together intimately and in harmony to make even small but crucial adjustments to the fabrication process. In contrast, US chip companies, and Intel in particular, viewed their processes as proprietary and deeply embedded in the firm. Thus, they often made changes to equipment and process without conferring with suppliers, often leading to delays and production problems when the equipment makers tried to replicate the changes. Finally, and not a little ironically, the Japanese embraced the ideas of the American W. Edwards Deming much earlier and more completely than US companies. Deming's system, which centered on the importance of cutting costs by improving product quality, was particularly relevant to chip making where yields often fell short due to defects on the silicon surface. By the early 1980s, the leading Japanese semiconductor companies achieved a very high yield rate compared to even the best American producers. Startlingly, American chips from the best firms had six times the error rate as compared to the devices from the worst of the Japanese companies.[30]

By the early 1980s, Japanese semiconductor firms were far better prepared than their American counterparts to engage in a long and savage price war; and they were also willing to win this war at virtually any cost. They succeeded in cornering the world market in high-quality large-volume memory chips. The integrated chip in the form of the DRAM had become a commodity, and Japanese companies managed to become the most competitive in memory components in the world. By 1980 Intel controlled less than 3% of the global memory market. But Intel's resource allocation mechanisms, established during the early days of the company, had continued operating. It assured that resources shifted automatically from the less profitable memories to the up-and-coming microprocessor. Even if Intel outwardly continued to claim itself to be a memory company, it had already been shifting its resources to microprocessors. Its elastic internal selection mechanisms, allowed—even mandated—that middle managers put more men, money, and materials to what contributed to the most to corporate profits, and increasingly that was microprocessors. By the time the top managers made the decision to switch from memories to microprocessors, Intel was essentially a microprocessor company. As Grove himself admits, there was not much risk involved in the decision as it had already been made de facto, if not officially, by middle managers continuing over the years to adhere to the resource allocation rules of the company:

> While [top] management was kept from responding [to the need to switch from memories to microprocessors] by beliefs that were shaped by our earliest success, our production planners and financial analysts dealt with allocations and numbers in an objective world.[31]

8.5.2   The MOS-SG Process, Moore's Law, and Intel's "Internalized Short-Term Dynamic" Culture

Even with Intel's resource allocation mechanism, Noyce, Moore, and Grove could have continued to resist turning to the microprocessor as the company's main product. If they felt that the Intel memory chip still had life in it, they could have easily overrode internal resource transfers and undertook a Manhattan Project-type effort to get back in the memory game. This would require really abandoning economies of scope strategy and focusing all effort on improving product quality and reducing production costs. This is how the Japanese gained their advantage, and this is how Intel could have fought back.

Intel however was never this sort of company. Its competitive edge in the early memories' market never really depended on its exceptional manufacturing prowess but rather on its ability to create new and more powerful chips at premium prices. Intel's supremacy in the combined MOS and silicon gate processes was the key enabler of this strategy. But by the mid-1980s, it became very clear that there were no more exciting new memory products to be had. Staying in memories meant putting more and more resources into incrementally improving the same old products and improving quality control and production efficiencies. The founders of Intel, although they learned from their battle with the Japanese the importance of heeding their production capability, never really had much interest in making this sort of thing their main competitive weapon. They had imparted to their company a more vibrant culture, one that valued producing new, exciting, and highly valued semiconductor products incorporating the latest technology.

In 1985, the question Noyce, Moore, and Grove had to ask themselves was whether the microprocessor was that sort of promethean technology. Thanks to the persuasive power of Hoff and, even more, Faggin—and to the string of new microprocessors they created through the 1970s and early 1980s—they realized that it indeed was, that it derived from the same MOS-SG process that gave them the DRAM and EPROM, and even better, that, because of the extreme flexibility of the process, an exciting period of new and powerful microprocessor products stood before them, if, that is, they didn't blow it by stubbornly sticking with memories. In short, Intel's top managers finally came to realize that the microprocessor was perfectly suited to the deeply rooted dynamic culture—enabled by their superior MOS-SG process—that they themselves had created in the very early years of the company.

By the mid-1980s, then, Intel's management could feel very comfortable with the microprocessor once they came to terms with the notion that it was closely related to memories through a common process and that it conformed very closely to Intel's deeply rooted dynamic culture. For this reason, jumping to the microprocessor no longer seems like such a leap into the unknown. Furthermore, once working within this world, Intel's leaders knew very well they could actually predict the trajectory microprocessor development would take in the months and years to come. They could because the MOS-SG process was the only semiconductor technology existing that could closely shadow the trends dictated by Moore's famous law, which I have mentioned in earlier pages. Moore's law first surfaced in a 1965 in an article Gordon Moore wrote for *Electronics*, while he and Noyce still helmed Fairchild Semiconductor and before the MOS transistor met the silicon gate process.[32] But even in 1965, Moore understood the potential of the technology he and Noyce had already created from their experience at Fairchild. Moore believed that continual progress could and would be made in the essential process leading to more product possibilities. Purer silicon crystals, greater precision in doping, more sophisticated and exact photolithographic processes, and improved furnace design would lead inexorably to progressively smaller and more powerful circuits for a broadening range of applications. As a result, engineers should be able to double the number of transistors—and thus the speed, power, and ability to perform complex operations—on a silicon chip every 12–18 months and do so at a minimal economic cost (per transistor) that would continue to decline over time.

Moore's conclusion was thus a statement mapping out technological progress year by year into the foreseeable future. It was also a prescription for how a company like Fairchild (and later Intel) could successfully compete in an increasingly crowded field. As long as Intel controlled the best process, it could expect to regularly turn out better and smaller chips for a broadening range of applications. It could charge a premium on these chips because they will be the most advanced on the market and so in great demand. And as production costs go down, company profits will grow. As one generation of chip matures, the company can jettison it because a brand new and much better one is just around the corner. The technology is so dynamic that patenting no longer becomes an important competitive tool, for by the time patent protection takes hold, the company has come out with the next-generation technology that makes the older chip virtually obsolete.

Fairchild, as we know, blew its chance to build up its semiconductor process to the point where it could realize Moore's predictions. Intel picked up the pieces, grabbed hold of silicon gate technology, and applied it to MOS to create the "killer" process that turned Moore's law into a practical, working reality. But the memory chip was a limited device technically; MOS-SG could do so much more. The microprocessor is the product that one really tests its very limits. By the mid-1980s, Intel's management understood that MOS-SG and the microprocessor were made for each other and that Moore's law would help them to envisage the future course for this revolutionary product. Moore's law served as the guiding strategic force at Intel and also as a practical way to forecast what the most advanced semiconductor chip would look and act. Intel, in fact, placed bet on it; Noyce, Moore, and Grove had simply to extrapolate on an almost perfect straight line to know where they needed to go to keep on top of its competitors.[33]

So by the time Intel made the decision to concentrate on microprocessors, it could be reasonably certain of the future direction that technology would take the company and that even the large and established firms would not be able to keep up with them. While it is true a company like Motorola could come up with extremely sophisticated chips, Intel (helped enormously by Grove's aggressive marketing) continued to beat the competition eventually winning that historic IBM contract that pushed the company into the front ranks of semiconductor firms.[34] Intel's nearest competitor, Mostek, did not have the technology to maintain such a policy of aggressive, internal creative destruction and as a result fell by the wayside. In a revealing 1995 interview conducted by the Institute of Electrical and Electronics Engineers' (IEEE) Center for the History of Electrical Engineering, the first president of Mostek, Richard Petritz, explains why Intel succeeded and implicitly why Mostek faltered:

> Intel's management ... recognized that ... [it is important for the growth of the company to] kill your own children, to obsolete your own products by the next generation. And you do it more timely so that other people really never catch up to you, and this comes back to this learning curve ... I think Intel did a marvelous job of managing to obsolete their own products at a rate that made it extremely difficult for anyone to compete ... I think one of the smarter things Intel did was they got into this mode of 'Let's just obsolete our own products and we don't care that somebody copies us. They copy us two years too late anyway'.[35]

The late 1980s and 1990s witnessed the development and introduction of Intel's most formidable chips yet, the Pentium series of microprocessor. While marketing strategies played a most important role in the success of the Pentium chip, ultimately, its triumph rested on its technological content, and that, in turn, hinged on Intel's supremacy in materials processing. By 1995, Intel could reduce gate length, and thus the size of the transistor, to 0.35 μm (where 1 μm equates to a millionth of a meter). This landmark accomplishment, in turn, opened the way for the most sophisticated chips ever designed. Intel increased the complexity of their microprocessors from 270,000 transistors per wafer (the "386") in the late 1980s to three million transistors on a chip (Pentium Pro) by 1995. Through the 1990s, Intel continued to come out with newer Pentium designs possessing higher speeds and greater production yields (which translated into lower prices) and lower failure rates. Soon, the computer industry, more precisely personal computers, became dependent on Intel's chip and its ability to follow Moore's law. So for Intel, the way ahead became clear, for "... here was a growth path ... into more and more complex ... integrated circuits that had not existed [previously]."[36] Between 1995 and 2015, the silicon chip has continued to evolve as the number of transistors placed on the surface of a piece of silicon has rapidly multiplied from a few million to over a billion. This has resulted in extremely small, dense, and powerful devices known as ultra large-scale integrated (ULSI) chips. In 2015, IBM reported that it actually created transistors at the nano- or atomic level.[37] Some believe that, as the integrated chip continues to shrink in size, silicon will reach the limit of its ability to keep up with the demands of Moore's law. They argue that new materials will have to be found. Ongoing research into "strained" materials, nanotubes, quantum dots, and graphene continues to explore how these new forms of matter might be used to extend the useful life of silicon or even replace it altogether in future microchips in order to keep Moore's implacable rule in play.

REFERENCES

1. Tedlow, R. (2006), *Andy Grove: The Life and Times of an American*, New York, New York: Penguin Group, pp. 113–187; Bassett, R. (2002), *To the Digital Age: Research Labs, Start-up Companies, and the rise of MOS Technology*, Baltimore, Maryland: Johns Hopkins University Press, pp. 167–180; Reid, T. (2001), *The Chip: How Two Americans Invented the Microchip and Launched a Revolution*, New York, New York: Random House, pp. 160–163; 244–248.

2. Bassett (2002), p. 181.

3. This account of the research and early development of the silicon gate (MOS-SG) process comes from Bassett (2002), pp. 181–183. See also Lojek, B. (Fall 2009), "Early Development of Polysilicon-Gate MOS Technology at Fairchild Semiconductor" *IEEE Solid-State Circuits Magazine, 1*, pp. 18–24.

4. This section on the MOS-SG development phase is taken from Bassett (2002), pp. 183–190.

5. This account of the DRAM technology is taken from Bassett (2002), pp. 190–202; Tedlow (2006), pp. 132–176; Burgelman, R. (2002), *Strategy Is Destiny: How Strategy-Making Shapes a Company's Future*, New York, New York: The Free Press, pp. 29–76.

6. Bassett (2002), p. 196.

7. Ibid., p. 197.

8. The secondary sources used in constructing this account of the EPROM technology come from Burgelman (2002), pp. 78–90; Bassett (2002), pp. 199–200; Tedlow (2006), pp. 143–144.

9. Burgelman (2002), p. 79.

10. Dov Frohman, interview by David C. Brock, May 10, June 6, and June 12, 2006 (Philadelphia, Pennsylvania: Chemical Heritage Foundation, Oral History Transcript #0341), p. 30.

11. Ibid., p. 35.

12. Ibid., p. 37.

13. The secondary sources used in constructing this account of the microprocessor come from Burgelman (2002), pp. 55–59; 103–111; Bassett (2002), pp. 251–281; Reid (2001), pp. 174–180. See also Malone, M. (1995), *The Microprocessor: A Biography*, New York, New York: Springer-Verlag and Noyce, R. and Hoff, M. (1981), "A History of Microprocessor Development at Intel" *IEEE Micro*, *1*(1): 8–21.

14. Bassett (2002), p. 251.

15. Ibid., p. 269.

16. Ibid., pp. 277–278.

17. At this time, Intel took out a now-famous advertisement in the trade press. See "Announcing a New Era in Integrated Electronics" (Advertisement), *Electronic News*, November 15, 1971.

18. Grove, A. (1999), *Only the Paranoid Survive: How to Exploit the Crisis Points That Challenge Every Company*, New York, New York: Doubleday, p. 88.

19. Burgelman (2002), p. 112.

20. Federico Faggin, interview by Robert Walker, March 3, 1995 (Silicon Genesis Project/Stanford University), Los Altos Hills, California: Oral Histories of Semiconductor Industry Pioneers, p. 8. http://silicongenesis.stanford.edu/transcripts/faggin.htm. Accessed November 30, 2015.

21. Ibid.

22. Ibid., p. 11.

23. Ibid.

24. Ibid.

25. Andy Grove refers to Intel's decision to abandon memories in favor of microprocessors as the company's main product as the time the firm was facing its most important "inflection point" requiring it to travel through the dangerous "Valley of Death." See Grove (1999), Only the Paranoid Survive, pp. 139–172.

26. Tedlow (2006), p. 210.

27. Marcian (Ted) Hoff, interview by Robert Walker, March 3, 1995 (Silicon Genesis Project/Stanford University), Los Altos Hills, California: Oral Histories of Semiconductor Industry Pioneers, pp. 5–6. http://landley.net/history/mirror/intel/Hoff.html. Accessed November 25, 2015.

28. Lecuyer, C. and Brock, D. (September 2009), "From Nuclear Physics to Semiconductor Manufacturing: The Making of Ion Implantation" *History and Technology*, *25*(3): 193–217.

29. Reid (2001), pp. 210–236.

30. Ibid., pp. 227–236.

31. Tedlow (2006), p 209.

32. Moore, G. (1965), "Cramming More Components onto Integrated Circuits" *Electronics*, *38*(8): 114–117.

33. Reid (2001), pp. 153–161; Bassett (2002), pp. 168–170; 196–198; 277–278.

34. Burgelman (2002), pp. 103–112.

35. Richard Petritz, interview by David Morton, June 21, 1996, IEEE History Center, New Brunswick, New Jersey: Oral History Program, p. 8. http://ethw.org/Oral-History: Richard_Petritz. Accessed November 15, 2015.

36. Ibid., p. 279.

37. Markoff, J. (October 1, 2015), "IBM Scientists Find New Way to Shrink Transistors (Measuring in Atoms)," *New York Times*, p. B3.

9

THE EPITAXIAL PROCESS I

Bell Labs and the Semiconductor Laser

[Bell Labs'] so-called "million hour [research] paper" … demonstrated to us and to the world's laser community that it was possible to construct semiconductor laser devices with very long lifetimes.
Richard W. Dixon (Bell Labs Solid-State Laser Pioneer), 2012

In the second decade of the 21st century, silicon remains the most important semiconductor material in large part due to its ease of fabrication through the MOS-SG process. Continued improvements manage to keep silicon in the forefront of the digital revolution. But silicon is not the only important functional material in play today. Another group of semiconductor materials very different in composition and structure from those based on silicon command attention. These "superlattices" or "heterostructured" composites began their rise to prominence in the 1970s. One of the most important technologies based on these materials is the semiconductor laser, used today in many applications including barcode systems, medical surgery, and fiber optics networks. Silicon does not play any significant role in these semiconductor technologies. A second major category of superlattice materials does employ silicon but in conjunction with its close cousin germanium. The silicon–germanium microchip combines the desirable characteristics of both materials: silicon for its mechanical and thermal properties and germanium for its speed. Its superb ability to handle high-frequency signals and processing multiple streams of data and

Advanced Materials Innovation: Managing Global Technology in the 21st century, First Edition.
Sanford L. Moskowitz.
© 2016 John Wiley & Sons, Inc. Published 2016 by John Wiley & Sons, Inc.

information makes it the chip of choice to power a growing number of wireless products, including personal computers, smartphones, and iPads.

The common process that links these two seminal technologies—laser and silicon–germanium chip—is called epitaxy, the technique of growing one type of semiconductor material onto another so that both lattice structures fit neatly together. The MOS-SG and epitaxy processes employ very different techniques to make transistors, integrated circuits, and microchips, Whereas the MOS-SG process creates electronic composites by taking a piece of semiconductor material (silicon) and modifying it on and beneath its surface—through such operations as oxidation, doping, implantation, and photolithography—epitaxy does something very different: it creates electronically active composites by "layering," that is, constructing a series of sheets or films of varying thickness and composition upward from the surface. Whereas the MOS-SG process typically creates field-effect transistors, epitaxy is most closely linked to the making of very advanced junction (bipolar) devices.

The story of these two technologies brings into focus the role of large established companies in advanced semiconductor materials. Whereas we have seen, the start-up and small and midsized firms take the lead in the silicon semiconductor revolution (Fairchild and Intel); the older, bigger companies directed the rise of the heterostructured materials and their products. But the two companies involved did not, in fact, fare equally well in their pursuit of superlattice technology. As we will see, Bell Labs (and its corporate parent AT&T), while pioneering the semiconductor laser for fiber optics, could not in the end take full advantage of its own creation and in the end did come in second place behind a small and aggressive start up. In contrast, IBM not only innovated the silicon–germanium chip but has remained the major innovator and supplier of this breakthrough technology up to the present day. We will now consider the story of Bell Labs and its semiconductor laser; we will then, in the chapter to follow, take up how IBM triumphed in its quest for the silicon-germanium chip.

9.1   BACKGROUND: ADVANCED MATERIALS, THE EPITAXIAL PROCESS, AND NONSILICON-BASED MICROCHIPS

The technique of arranging thin layers of materials onto surfaces (or substrates) has been around for centuries. The manufacture of household mirrors, for example, relies on evenly placing a thin metal coating on the back of a sheet of glass. Electroplating of one metal on top of another is another commercial type of "layering." In recent decades, thin film design has been part of a number of well-known products such as antireflective coatings on windows and on lenses to correct eyesight. Within the electrical products industry, thin films have come to be applied to the surface of the incandescent and fluorescent light bulb. Since the 1960s, epitaxy and thin films entered the electronics industry, particularly in the making of memory drives for mainframe computers.

As time went on and it became clear that MOS-SG and not epitaxy was going to be the process of choice for the most important commercial transistors, scientists and engineers wanted to push epitaxy beyond the world of just silicon and move into completely new fields of semiconductor research. By the 1980s, physicists and materials scientists had to face the severe limitations in what silicon could do; for despite its very real advantages—and its obvious commercial prowess in information technology—it also presented problems when attempting to apply it to noncomputer-related uses. For example, one could not extract from it an intense, focused light beam, which eliminated it as a player in the growing fiber optics industry, nor could it respond reliably to high-frequency signals, which ruled it out as a material for integrated chips needed for cellular (and later smart) phone devices.

The issues before those in semiconductors who wanted to push their technology beyond the familiar world of silicon was twofold: (i) finding the right starting materials to do the job and (ii) selecting and carrying out the appropriate epitaxial process. From scientific work conducted in the United States and Europe during the 1950s and 1960s, it seemed theoretically possible to create new types of semiconductor materials that could do what silicon could not. All that one needed to do was to properly layer films of different semiconductor substances on top of one another. Those elements, compounds, and composites that should be used for particular applications depend on carefully exploiting the principles of what came to be called "band-gap engineering."

The band gap of a material is determined from the principles of quantum physics. In simplest terms, it is a measure of the minimum amount of energy that needs to be applied to a material to force the outer electrons of its atoms and molecules to break free from the confines of their prescribed orbitals and enter what physicists call "conduction bands," where they can all gather together and carry a current. When they do so, moreover, the spaces they leave behind in their parent atoms and molecules create pockets of positive charge—the same holes described by Shockley in the late 1940s and that are such an important part of silicon-based devices. These holes can move and transport current within their own ("for holes only") conduction zones. In small band-gap materials, these moving electrons and holes are extremely plentiful. Together, they create a highly dynamic electronic environment that allows the semiconductor—in the form of transistors and integrated circuits—to do things the more sluggish, higher-band-gap substances cannot, including reacting to high-frequency signals, producing high-intensity light, and processing information and data at extremely high speeds.[1] A major question then was finding the right combination of materials that produced composites with a lower band gap than silicon but acceptable mechanical and thermal properties that would make it relatively easy to manipulate and commercially produce devices made from them.

But what epitaxial process to use? In actuality, epitaxy is not one process but two distinct subprocesses. One is carried out in liquid phase whereby layers of materials are deposited (built up) on a substrate placed in a solution saturated with the substance in question. The second possible process takes place in the vapor phase. The substrate is placed in a vacuum, and the atoms and molecules to be layered are introduced so that they can travel to the surface and take their proper place in the heterostructure. The process selected—whether liquid or vapor—depends on the composite being made. The major difference between the two is their level of precision in connecting up materials with different lattice profiles. Precise matching is often very important since a misalignment means electrons would not flow easily when traveling from one material to the other, and the device would either prove too slow or possibly not work at all. This was not as serious an issue in attempting to match up the atoms, say, of gallium and arsenic (to make gallium arsenide for semiconductor lasers) since the lattices of the two elements are closely similar to begin with. A less than exacting process, such as liquid-phase epitaxy (LPE), proved to be good enough to produce working lasers. This was not the case at all with silicon and germanium; these two elements have significantly different lattice structures, and forming a useful product from them required a more advanced form of layering technology, that is to say, vapor-phase epitaxy.[2] The arrival of a technology called molecular beam epitaxy (MBE)—an extremely accurate way to make ultrathin films in the vapor state—in the 1970s, signaled loud and clear that the age of thin film technology had arrived. With this tool, engineers for the first time could make devices and components never before possible. Within a few years, advanced vacuum techniques (or ultravacuum technology) revolutionized vapor-phase epitaxy and delivered new and breakthrough types of microchips for wireless communications.

9.2 BELL LABS AND THE SEMICONDUCTOR LASER

The semiconductor laser was a very different sort of project for Bell Labs. Unlike the case with the transistor in the late 1940s, where Bell had the field to itself, there were other companies and organizations hot on the heels of the elusive semiconductor laser. Within the United States the giants of American technology—General Electric, RCA, and IBM—were already in pursuit, as was one of the country's other premier research organization, MIT Lincoln Labs. Nor was the effort confined to the United States. In England, the British Post Office (which ran the country's phone system) and in Russia the well-known Ioffe Physico-Technical Institute had already made great progress toward a working device.

By the time Bell began its research into the technology in 1966, it trailed its competitors in the field by a number of years. The person in charge of the Bell effort was the then director of the solid-state research group, John Pierce. Very much like Mervin Kelly in the late 1940s, Pierce's vision for the near future was to expand the capacity of AT&T's network, in his case development of advanced optical communications technology. Much was at stake since if AT&T's network could not keep up with rising demand, it could expose the company to loss of its status and even potential standing as a regulated monopoly. The threat was certainly real, for by the late 1960s, it was clear that the nation's demand for long-distance telecommunications was growing very rapidly and, on the supply side, the existing technology in the form of the coaxial cable and point-to-point microwave transmission "could not keep up the pace." Pierce and his bosses at AT&T were well aware of the situation and addressed the coming problem in the way they usually did: by giving Bell its marching orders to find a technological solution. But, as typically the case with this company, AT&T could not ultimately exploit to the fullest its own creation. This resistance exposed the telephone giant to the more innovative start-ups that came along in the 1980s.

One cannot understand the rise of the semiconductor laser without considering how the laser itself evolved. The first lasers built in the early 1960s did not operate on semiconductors but got their light from crystals (ruby) and gases (carbon dioxide). The high-frequency light they emitted first alerted Bell (and other research labs) that the laser could be a means of transmitting vast amounts of calls, data, and information over long distances and in ways that would likely make the old copper wiring and even microwave systems then in place obsolete.

9.2.1 The First Lasers

A seminal event in the history of the laser took place on a beautiful spring morning in 1951.[3] In Washington, DC, to attend the annual meeting of the American Physical Society, the Columbia University physicist Charles Townes was sitting in Franklin Park and mulling over a problem that had been haunting him for quite a while. In doing so, he hit upon what would become the fundamental principle of the laser. In simple terms, Townes wondered how he could create a system—say a container of gas—that could internally generate an intense beam of radiation (such as light). He knew from an early theory of Einstein that certain atoms or molecules that have been pre-"energized" (so-called "excited" atoms) and then are hit by a photon of light will give off energy greater than contained in that photon. This theory of "stimulated emission" would seem to open the way to the possibility of creating an apparatus capable of producing a very powerful ray of light, as long as the system was composed of a large number of the right type of preexcited atoms and molecules. However there is a problem known as the second law of thermodynamics that says that, under normal conditions, the vast majority of atoms or molecules in any container must be in a lower—or unexcited—state. It is certainly true that an external light shined on this collection of particles would stimulate the few atoms already and naturally in an excited, highly energized state. However, this small burst of light and heat would have nowhere to go, for it would simply be absorbed and muffled by the much larger number of lethargic particles in the group. Thus, the laws of physics dictate that only an infinitesimal amount of energy could ever leave that system, and consequently no powerful beam of light could ever come out of that or any other assemblage of atoms.

But Townes had an idea. Suppose one could find a way to isolate from, say, carbon dioxide gas, only those few special gas molecules that are already and by nature energized, and place them in their own container. Then repeat the process with a new sample of gas, placing their naturally excited atoms in that same special container to join the others previously placed there. Eventually you would have captured and placed in that container a large number of very excited molecules. He called this condition a "population inversion" where now the energized atoms dominate. This situation is one of abnormal instability (or disequilibrium), and so now the second law of thermodynamics no longer applies, thus opening up the possibility of something very interesting to happen. If you inject (or "pump") a small amount of additional energy into this inverted population of already animated particles—such as shining light into it—something akin to the chain reaction should take place as in a split second one atom, stimulated to emit more energy than it receives, sends out photons that incite two or more atoms next to it to do likewise; these in turn energize four more atoms and so on. Then one could expect a massive burst of light to come out of that highly dynamic system.

When Townes returned to Columbia, he worked out a way to extract and isolate the energetic molecules from a gas. This work resulted in an apparatus that could generate an intense beam of microwaves—which are invisible—with a single frequency. He referred to this device as a "maser," which stood for "microwave amplification by the stimulated emission of radiation." As far as Townes was concerned, the maser was essentially a scientific instrument for the study of atomic and molecular structure. It also could amplify faint signals from space and from orbiting satellites. For example, it played an important role in Bell Labs satellite communications experiments in the late 1950s and early 1960s and in radio astronomy up to the present day.

The next big leap in the technology was to apply Townes' approach to the amplification of electromagnetic radiation in the visible—or "light"—portion of the spectrum. Since light waves have naturally much shorter wavelengths, considerably greater frequencies, and far more energy than microwaves, a device that could go beyond what the maser could so and actually magnify light and concentrate the resulting bursts of energy into a focused and coherent beam would have great commercial potential. In addition to its possible use in medical procedures and as a cutting tool, it would be a radically new means of communications since the higher the frequency of a signal, the greater capacity it has to contain and transport data and information.

In truth, the commercial advantages of such a beam were not totally clear to scientists or engineers in the 1950s. Nevertheless, the possibility of amplifying light was a very tempting prospect to researchers, and a number of attempts were made by different groups in the United States, Europe, and the former Soviet Union to create such a technology. Townes and his colleague (and brother-in-law) Arthur Schawlow were the first to theoretically explain the possibility of creating what came to be known as the light amplification by the stimulated emission of radiation (laser) and its probable application to communications. While their published paper on the subject proved very influential, they themselves never successfully built a working laser. Others, however, using the Townes–Schawlow paper as their take-off point, vigorously pursued this goal. Two troubling issues plagued these early efforts: identifying the right kind of material to stimulate and finding a way to create a focus and concentrate the light given off by the lasing material. A former student of Townes at Columbia, Gordon Gould, solved that second problem by using two parallel mirrors at either end of a tube within which the stimulated energy bounced back and forth, eventually forming a thin coherent line of intense light. This "parallel mirror" approach soon became standard design in all lasers.

The question of materials was another matter. The various research groups employed different media in their machines, depending on their understanding of the internal quantum nature of the substance. The first functioning laser came in 1960 and used a pink ruby as the working material and a "flash of bright light" to stimulate its molecules. It was built by Ted Maiman, an engineer working for Hughes Aircraft in Malibu, California. While an important breakthrough, Maiman's laser had serious limitations as a practical device. The amplified light that it emitted occurred in "short and powerful pulses"; but a continuous stream of light was needed for the most important applications, such as communications. Just a few months after Maiman unveiled his ruby device, a team at Bell Labs succeeded in designing and operating its own laser using a gas medium to produce a "steady and continuous beam." As a harbinger of things to come, the researchers tested out the laser as a potential medium of communications—they used the concentrated pencil-thin ray of light given off by the laser to transmit a telephone call. It became clear that the laser might prove a powerful and economic means of simultaneously transmitting voice, data, and TV signals.

However, problems persisted with lasers that bedeviled researchers at Bell and elsewhere around the world through the 1960s. It was not enough to find a material that created an intense laser light; it had to be the right type of material that "lased" at just the right frequencies appropriate to carrying calls and data. And, even when such a material was found, there was the even more complex problem of finding a way to, as Jon Gertner tells us in his history of Bell Labs, "modulate voice and data signals and then 'impress' those signals upon the laser beam." Another crucial issue was physically transporting the light beam from one place to another. Phone calls of course then traveled through copper wires; this was easy to grasp. But what about light pregnant with messages and information of all types? Could it be sent over long distances as well as short? And how would it travel, through the air, in some sort of wireless communications network? or some other way not yet understood? Evidence of atmospheric interference soon scotched this notion. It was clear that, if the laser was to be used for communications, it would have to travel in a sheltered environment. Bell was at the time quite enamored with the idea of using hollow pipes, or "wave guides," buried underground to carry phone messages. But it was soon evident lasers wouldn't work well in such enclosures: a major problem was keeping the light focused in such spaces over hundreds of miles—different signals would just bounce off the container walls in all directions and not be able to come together easily to obtain a coherent signal at the other end.

It eventually became apparent that to transmit such light over long distances and keep it focused required it to travel within very thin strands or fibers made of special glass material. Such optical fibers were already in use in the medical field to allow doctors to peer into parts of the human body. But could such a technology work in an actual phone network and over long distances? Two big issues had to first be solved. First, a new and very pure glass fiber had to be found clear enough to allow transmission of light over long distances without significant light attenuation (contaminants in unclear glass collide with the light, thus weakening signals as they traveled through the fiber). The second problem was the laser itself. If very thin glass strands were to be used, a whole different sort of laser had to be invented than then existed that had to satisfy four requirements: the ability to emit light over long periods of time, operate at wavelengths that were suitable for communications, work at room temperatures (existing lasers generally operated only at super cold temperatures), and be very small so that it could emit its light into the tiny glass fibers (lasers then generally were the size of table tops). It wasn't long before researchers began to examine semiconductors as a medium for such a specialized laser.

9.2.2 Early Research on the Semiconductor Laser in the United States

By the time Ted Maiman demonstrated the first operating laser in 1960, the transistor was 13 years old, and the integrated circuit had just been born.[4] It made sense that thoughts of using semiconductors as the functional material within lasers would have been a great enticement for scientists and engineers. Semiconductors had an almost magical appeal as the technology of the future; they were the real celebrities of the scientific world in the 1960s. Why not explore these materials for lasers? It was well understood that, with the right materials and under appropriate conditions, electrons "fall into" the holes when the two charge carriers come into contact at p–n junctions producing concentrated bursts of light the intensity of which could be controlled by the amount of the current. It was obvious from the start of these research programs that silicon was not the right material to use. Silicon made sense when an amplified current or rapid switch was called for, but it would not produce an intense beam of energy. Quantum theory of solids predicted that the compound semiconductors, especially those composed of elements from Groups III and V of the periodic table, were far more promising as light-producing materials. It didn't take long for a "proof of concept" prototype to emerge. In 1962, in a stunning scientific effort, General Electric beat out its competitors by building and demonstrating a working semiconductor laser. The semiconductor used was a chip of germanium arsenide, which, through a doping procedure, had been divided into "p-type" and "n-type" sections joined together in the middle by a junction layer. A battery operating on this bipolar "diode" forced the electrons (from the "n" section) and holes (from the "p" section) to come together and generate light right at the junction. IBM and Lincoln Labs came in with similar versions of the semiconductor laser just a short time later.

It was at this point that work on the semiconductor laser stalled. Neither IBM nor even GE had a deep commitment to the device, and difficult problems remained before it could ever become a commercial product. In order to be useful in optical communications, such a laser had to emit light in a steady beam, for extended periods of time, and at room temperature. But even the best semiconductor lasers in the early 1960s fell far short of these goals. The beams they generated came in very short bursts, and this could only be done with any regularity at the extremely low temperatures of liquid nitrogen (a very cold −371°F). Jeff Hecht, in his history of the fiber optics industry, gives us an idea of how primitive and impractical these lasers really were at the time. In 1964, the most advanced semiconductor laser could fire a single pulse of light at room temperature lasting only 50 billionth of a second using a "trigger" of 25 A of current, which is more current than a standard refrigerator draws. As seriously, the laser could not produce a thin, focused beam, as could, for example, table-sized gas lasers. Rather, the beam spread out too much creating a "fuzzy blur" rather than a sharp pinpoint of concentrated light.

These were crucial issues, and their solution did not hold out much promise of being uncovered any time soon. The problem was in the material used. Gallium arsenide by itself—even when doped—would not be enough to coerce the semiconductor laser out of the laboratory. A more complex substance seemed to be the direction to take. But what sort of material and how a laser using that material ought to be designed and built remained a mystery.

9.2.3 Bell's Semiconductor Laser: Initiation and Research

For many years, Bell Labs did not favor fiber optics as the likely technology with which to expand AT&Ts communications network. This honor belonged to signal transmission through waveguides. But Bell certainly considered optical fibers as a possible backup route in case waveguides did not pan out. Bell's search for a workable semiconductor laser was an important part of its fiber optics research. Such a laser would be the necessary source for the coherent light beam crammed with a massive amount of voice message and digitalized data that would have to travel long distances through the tiny glass filaments. By 1966, the time Bell had begun its research into semiconductor lasers, the only ones that existed were no more than laboratory curiosities and very far from commercial reality. A great deal more work was needed before the technology could be of any practical use to the nation's phone system. The director of Bell Labs assigned Bell's solid-state electronics research at Murray Hill to invent this elusive device. The project passed into the hands of two extremely capable solid-state researchers: the chemist Mort Panish, an authority on gallium arsenide, and physicist Izuo Hayashi with expertise in atomic particles. Their assignment was clear and straightforward: to conduct basic research to determine why semiconductor lasers have to be at such low temperatures to work. The implied goal was to find out what conditions were necessary for lasers to be able to operate at higher temperatures, the *sine qua non* of a practical device, and to try to make a laboratory model of one.

Their first step was to begin growing their own lasers and to study the conditions for suitable luminescence. Around this time, Panish had a sort of epiphany at a semiconductor laser conference in Las Vegas. The speaker was from IBM and his talk concerned "lattice matching" in creating new types of heterostructured semiconductors. For the Bell scientist, the presentation appears to have opened up a whole new world of possible materials for a semiconductor laser. The trick was to make sure that the various lattice structures within and between the compounds composing the heterostructured composite were carefully connected at the junctions, so electrons and holes could flow freely and meet up at the boundary. Panish tested many different types of compound materials in various combinations. Improved performance told the researchers they were on the right track. For further insight, they turned to quantum theory that pointed to optimal lasers containing three semiconductor layers encasing two junctions. Since light formed at the boundary—the active area—this "double heterostructured'" laser produced twice the light output. Panish determined that his device had to consist of a low-band-gap material (a good conductor) sandwiched between two outer layers of high-band-gap material (poorer conductor). Through empirical methods, Panish found that using gallium arsenide as a base upon which to layer various other semiconductor compounds and adding small amounts of aluminum to the outer layers of the structure significantly improved laser performance.

It was one thing to know how a material would be designed and what materials needed to be used; it was quite another to actually make a composite structure that would work. This was far from easy to accomplish, even in the laboratory. Panish and Hayashi had to look outside Bell for a process that could be used. Here is the second important instance of outside information that played a role in Bell's work. RCA's liquid phase epitaxial (LPE) process developed by Herbert Nelson in the early 1960s for the purpose of making his own laser devices seemed to be very promising. Panish and Hayashi knew about the process but had to be careful to modify it so as not to infringe on RCA's patent position. A radical makeover of the process was necessary in any case to accommodate their particular requirements.

By early 1970, Hayashi had grown more than a thousand LPE heterojunction layers and tested a multitude of lasers cut from these wafers. He made real progress as he could now demonstrate continuous action lasers at close to room temperatures. This was a real measure of success, and Bell's senior management mounted a publicity campaign to announce the great achievement. But more work still needed to be done. Even the best of these lasers could not operate continuously for more than a couple of hours—with the vast majority of the devices being able to work for only a few seconds at a time. These problems presented a serious threat to the future of the semiconductor laser, which would have to be able to operate continuously and reliably for years on end before they could be useful as a light source in any fiber optics network. But neither Panish nor Hayashi was particularly troubled by this difficulty for, after all, they had done their job: they showed that it was indeed possible to build a continuous semiconductor laser that functioned at room temperature. This was the interesting scientific problem after all, and they had solved it. It was now up to the development people to "tinker" with the basic design they created to nudge it to work for a longer period of time. The close relationship we found that exists, say, at Intel between research and development did not exist at Bell. It is fair to say neither man could in any sense be called champions of the semiconductor laser. They were asked to do a task by their superiors, and they did it and now they were done and wanted to move on to other things. It was now up to the engineers to create a working prototype that could be scaled up to a commercial product.

9.2.4   Bell's Semiconductor Laser: Development

In truth, Bell had much work to do to achieve a practical laser. It is not clear that, despite the progress made by Panish and Hayashi, Bell management felt comfortable enough with the semiconductor laser to proceed to development. The millimeter waveguide still held the attention of Bell's senior managers, and the future of both the semiconductor laser and optical fibers remained in doubt. But work continued on under the leadership of a well-known engineer at Bell, Barry DeLoach. Somewhat ironically, DeLoach's claim to fame at Bell up to this point was inventing a high-frequency semiconductor device that proved crucial in the millimeter waveguide project. But there was no doubt that the semiconductor laser had demonstrated real progress, and Bell management believed that if anyone could see the project through a successful conclusion, it would be DeLoach.

9.2.4.1 Toward a Working Prototype DeLoach relished his new responsibilities and soon became Bell's champion of the semiconductor laser. He and his team had to immediately face the laser's "reliability problem" and, specifically, why most of them stopped working in only a few seconds. A pressing difficulty was that his development group had virtually no detailed information on how Panish and Hayashi made their lasers. This lack of communication reflects the general barrier existing between research and development at Bell. This communications gap delayed the project since DeLoach had to practically reinvent the wheel in coming up with his own workable process to make semiconductor lasers. A second issue was that Bell placed the project on a very short leash. DeLoach and his team could continue toward a working prototype on a step-by-step basis; they had to continually show progress toward a stable and functioning device, or else they would be forced to abandon their work, and Bell would then concentrate all its resources on the waveguide effort.

DeLoach replicated as best he could Panish's process and then looked for the exact causes of failure of these multilayered semiconductor structures. There were two possibilities. They either burned out suddenly or, more probably, gradually degraded until they went dark. While he couldn't be sure which of the two mechanisms it was, DeLoach knew that the first problem could be easily dealt with by applying a special coating to the laser chip. The second possibility however—and the more likely one—would be more troublesome. Degradation appeared to be related to flaws in the gallium arsenide substrate and in the depositing of the delicate layers upon it. It was clear that what was needed was a far more sophisticated and precise mechanism to grow crystals so as to form extremely thin, precisely structured layers, one upon another. Bell was not the only R&D lab struggling with this issue; RCA in the United States, Nippon Telegraph and Telephone (NTT) in Japan, the British Post Office, and other agencies and companies were trying to get reliable working lasers that did not "dim out." By the end of 1972, the sad fact was that "thousands and thousands of lasers lived brightly and briefly in labs around the world."[5]

But DeLoach and his team—now up to near 30 researchers—proceeded on in their work to see if they could slowly and incrementally improve the lifetime of the devices. Unfortunately for them, there was no magic bullet, no grand template that could help speed the process along. The process that they had to work with—that is, LPE—lacked an inherent elasticity. DeLoach found working this process very challenging. It proved difficult to adjust from one experimental setup to another and a real problem to scale. The presence of impurities was a constant concern. One of the troublesome issues that had to be fully understood and accounted for in developing the laser using LPE was the thermodynamic makeup of each experiment. This took a lot of time and slowed down how quickly any experiment could be setup, completed, analyzed, and acted on. There were no shortcuts as many experiments had to be designed and carried out, each offering up its own set of unique problems.

The researchers worked empirically, in a stepwise fashion, adjusting the composition and physical dimensions of the gallium arsenide substrate and of the various strata built upon it. They continuously modified and improved Panish's LPE process and measured these changes against device performance. They used the blunt technique of slow, painstaking iteration. When improvements were observed, they focused on further perfecting those procedures they used to obtain these positive results. Progress came slowly, but it nevertheless did come over time as lifetimes of the lasers increased from only a few minutes to hours and then days.[6]

9.2.4.2 Resource Problems and Creative Bootstrapping   But then, in late 1972, AT&T did something it almost never did in the past: it reached its long corporate arm into Bell's ivory tower and ordered DeLoach's team to stop their work. There were serious reasons for making this decision. The US economy was on a downturn, which meant long-distance traffic suffered significant losses. This contraction in demand meant less revenue and reduced pressure on the phone network, two conditions that militated against an ambitious growth strategy using new and untested technology. But most importantly, after 7 years of research effort, Bell still had not been able to produce a workable laser. Without that critical source of light to carry the optical signals, there could be no fiber optics system. Not only had the device not yet been perfected but there seemed to be a long way to go before this would happen, if at all. While progress had certainly been made, it was a slow and expensive slog. Most problematic were the continued complications involved in the layering process. While certain laser samples were made that performed quite well, some operating continuously hundreds of hours, what was required was for each device to run for millions of hours continuously without fail. Given that Bell wasn't even close to this as yet, the critical question of how to standardize the production of commercially useful lasers wasn't even on the drawing board. AT&T management had in fact no real hard evidence that its researchers—even DeLoach himself—would ever find a solution. Bell had very little history with this technology; there were really no other projects on which the laboratory had worked—not transistors, nor waveguides nor wave tubes, not the underwater cable, nor even the maser—from which DeLoach and his team could draw upon to improve the performance of these complex heterostructures. This technology was, in short, not really part of Bell's intellectual DNA. So, to AT&T management, it seemed that Bell was doing little more than groping in the dark. Finally, it decided to stop throwing good money after, what it deemed, "unpromising research."

This should have been the end of the story of Bell Labs and the semiconductor laser—which would also have been the end of AT&T's venture into fiber optics. However, like a pit bull that, despite repeated blows to the head, keeps charging ahead, DeLoach was not yet ready to capitulate. With his boss' permission—and possibly even encouragement—he went directly to Bell's upper management to ask them to keep the project alive, if only in a severely truncated form. His argument was that progress had been made, his group had clearly come closer to success than ever before, and there should be at least a few people to keep it going in case there was a breakthrough. Essentially, he was telling management that matters could be arranged that significantly reduced the risks of continuing on and that AT&T and Bell would not have to spend much additional money to maintain a skeleton crew and on-site equipment that has already been paid for. Nor would the researchers spend all their time on the laser and so would be available to work on other projects that Bell felt were more immediately promising. In short, AT&T would not be giving up much but would possibly gain a lot. It is likely that DeLoach's reputation at Bell played a role, and possibly important one, in the laboratory acceding to his request. He had done well for AT&T in the past and may very well hit upon a solution, and it wouldn't cost the company very much to find out. And Bell didn't want to lose DeLoach who might leave out of frustration and go somewhere else, a possible current or future competitor, and do important work for that company and, in turn, significant damage to his present employer.

This reasoning won the day and he and two other engineers were allowed to continue their laser experiments. This small group of intrepid researchers kept at it, slowly and incrementally finding their way to a better laser. Being under the radar, as it were, had its advantages. They were free to experiment in directions they might not otherwise have taken if they were being more closely watched by the nervous eyes of AT&T executives. As it was, the attitude of the latter was simply, "it would be great if they succeed but no great loss if they don't." That the group was now so small and the resources at their disposal severely limited meant that the engineers had to very carefully think about what experiments to perform, about the steps taken and equipment to be used to conduct them, and be more effective than ever in extracting the most information and insights from every experimental procedure.

These constraints on how they worked stimulated their creativity as they solved the many problems they had to confront in their quest for a practical semiconductor laser. The heart of their research was in the improvements they made to LPE technology and their ability to create precise and alternating layers of gallium arsenide and gallium aluminum arsenide, the major components of Bell's semiconductor laser. While they invented nothing radically new in terms of technique or equipment, they simply got progressively better at laying down thin semiconductor film. For example, the team could create sharper boundaries between layers and make each layer more uniform and chemically stable. Over time, they further improved the mechanical and thermal characteristics of the heterostructured devices that they made. DeLoach's success came at a pivotal time for Bell and AT&T.

9.2.4.3 Development of the Semiconductor Laser Gains Importance at AT&T/Bell Labs   In 1974, when DeLoach and his team of two were in the throes of their research, the US government filed an antitrust lawsuit against AT&T.[7] A number of factors can be cited for this action by an agency that had for decades protected Bell's telephone monopoly. A generalized reaction against centralization settled in the United States with the intensification of the Cold War in the 1960s and 1970s. With a totalitarian USSR at the center of American anxiety and uncertainty of the future, a monopoly as extensive and powerful as AT&T began to be looked upon as a rather sinister presence that needed to be brought under greater control. And at this time there were upstart companies that were beginning to make some noise and demanding to be allowed to exist and operate within America's telephone network. In particular, Microwave Communications Inc. (MCI) wanted to set up a system of microwave towers to offer cheap long-distance telephone service for businesses between Chicago and St. Louis. The reaction against the monopoly seemed to cross political lines, at least in the executive branch, for Presidents Nixon, Carter, and Reagan all supported greater deregulation of the nation's telephone system.

The 1970s then was a period of deep uncertainty for AT&T and the future existence of its monopoly. The company and the Department of Justice spent years battling it out in extended and costly litigation. AT&T's intensity and sheer viciousness with which it attempted to thwart competition proved to be self-defeating. AT&T took such tried and true defensive measures as political lobbying, predatory pricing, and industrial sabotage to their furthest limits. These tactics were not only ineffective this time around but, by further confirming the fears of the justice department that AT&T was now an out-of-control monopoly that would stop at nothing to push competitors out of their way, actually further hardened the hearts of government regulators against the company. By the late 1970s, it is likely that AT&T finally realized that the game was up and a breakup was inevitable. As Tim Wu tells us in his excellent book, *The Master Switch*, the highest levels of management at AT&T understood very well what was coming: "John deButts, the most Bellheaded chairman in the firm's history, resigned in 1979, by some accounts because he realized that a breakup was inevitable and he wanted no part of a Bell divided."[8]

But Ma Bell was nothing if not resourceful. If it lost the monopoly wars, it had every confidence it could beat any sort of competition that stood in its way. After all, it had controlled the communications network from the earliest days; it knew more about how it operated and how it could be used as a competitive weapon against any newcomer. And it had Bell Labs, AT&T's ace in the hole in the technological battles that were soon to come. Arguably, the most important of Bell's developments at this time was the semiconductor laser and the optical fibers through which its light would travel. With this technology, AT&T could at least hope to beat MCI and other newcomers to the punch who were planning to sidestep the existing network; it now needed the capability to build its own fiber optics system alongside, and supplementing, its long-distance lines, and it would do so in the most congested—and lucrative—market arena: the businesses communications network along the Northeast Corridor.

DeLoach and his band of diehard visionaries brought Bell Labs and its corporate parent this important leverage at just the right time. True enough, if AT&T had actually closed up shop on the laser in the early 1970s, it might have reinstated this research in the late 1970s when it saw it would have to face real competitors in a few years and better get its act together on developing advanced communications technology. But by then it would have been hopelessly behind its competitors, both large (RCA) and small (MCI), in the race for the fiber optics laser. DeLoach and his men kept the project going, slowly and painfully, making progress until they found themselves with a device that would do the job. Eventually, their progress toward a working laser and their maintaining project continuity until AT&T's external context changed convinced the company that they needed to reinvigorate the now small but still game laser program with additional personnel and cash and take the device to the finish line.

9.2.4.4   The Million-Hour Laser   Even at that, Bell's fiber optics laser was not an unalloyed success, and, in the more competitive climate that Bell found itself, the shortcomings of the invention played havoc with its ability to make AT&T the major force in the fiber optics industry. The fact that Panish and Hayashi had left the scene to pursue their scientific interests meant that DeLoach and his crew had to start on the project without the personal input from the original team that had created the device. Even worse, the conflict that often arose between the research scientist and development engineer appeared to have been played out to the hilt in the case of the semiconductor laser. In a 1984 interview with the American Institute of Physics and its Niels Bohr Library, Eugene Gordon, a major figure in Bell's laser work, talks bluntly about the difficulties that arise when there is conflict between research scientists and development engineers, manifested in the case of the semiconductor laser by Panish's desire to keep vital information from the development group:

> Art [DaSaro, from Development] tried hard to collaborate. Hayashi tried hard too. But Parish insisted [to Hayashi] that no material [used in their research] was to be given to Art and Hayashi obeyed ... Unfortunately, the denial by Parish and his desire not to share the glory with the device people cost Bell Labs a clear first.[9]

But, assuming that Gordon's account holds up—and given his detailed and objective description of the semiconductor laser in his interview, there is no clear reason to challenge it—another issue helps us to further understand the reason for Bell's difficulties, that is, the nature of the technology involved. LPE is a somewhat intractable and imprecise process. Researchers who worked with it deem it a very troublesome process to deal with; it is difficult to control and to obtain reproducible results.[10] It took DeLoach time and effort to get the handle on how to work it, its extent and limits of operation, and what types of modifications would prove most effective. Added to the barrier that came between research and development, the technical limitations of LPE became that much more severe. In this learning process, DeLoach had to redo many of the steps taken by the two scientists earlier in the research effort. Moreover, the team really had no other group within Bell to turn to. Other than Panish and Hayashi, no other scientist or engineer at Bell had much prior experience with the use and application of complex epitaxial processes. There were no shortcuts for DeLoach as he struggled inch by inch to achieve his laser. And because LPE, even in the most experienced hands, is such an inexact and unyielding process to say the least, many of the lasers that he produced fell short in performance.

In the end, it was not until 1977—that is, a full 11 years after Bell first authorized work on its semiconductor laser for communications—that the Labs had a laser that could work reliably for a million hours (circa 115 years) at room temperature, showing that it was possible to make semiconductor lasers with very long and reliable working lives. But this also was only a prototype, one that was closer to commercial respectability than that turned out by Panish and Hayashi to be sure but one that still needed time to be adapted to the real-world conditions of a practical fiber optics network. In the final analysis, it turned out to be uncompetitive to other, more advanced lasers emerging from other quarters.

9.2.5   Bell's Semiconductor Laser: Scale-Up and Commercialization

As Bell's semiconductor laser project moved from development to scaling up and commercialization, it more attentively shadowed the lab's work on optical fibers.[11] The reason for this closer relationship between light source and transmission medium is clear: laser and optical fibers would have to work very close together in any large-scale test of the system intended to carry the full load of commercial traffic.

By 1977, AT&T was preparing for the first major trial of its fiber optics technology in Atlanta, Georgia. AT&T's system of laser and optical fiber was specifically developed to operate at a light wavelength of 850 nm (1 nm equals one billionth of a meter). Bell designed its semiconductor laser to emit light at this wavelength and its fibers (called "graded index" fibers) to conduct this light—carrying hundreds of millions of bits of information per second—over long distances. While signal loss occurred, it did so at a reasonably slow rate, one that required repeaters (which bolstered or reenergized the signal) installed every 6 miles. AT&T thought this to be an acceptable rate of loss. The laser itself was finally approaching useful lifetime and consistency. The fact that Bell's Atlanta trial of 1976–1977 went off successfully showed that Bell's effort was paying off—clearly fiber had a future—and AT&T seemed poised to command that technology in the decades to come.

9.2.5.1   The Semiconductor Laser Advances to Higher Wavelengths   The fact it had taken Bell nearly a decade to get this far with the laser—with improvements still needed—testifies to the grit and determination of the people involved and not least to the aggressive and skillful championing of DeLoach. On the other hand, it also attests to the inflexible nature of the process needed to make this particular laser. Bell had spent considerable resources on getting this obstinate technology to this point; even its champions flinched at the prospect of having to modify it any further, for to do so meant more time, money, and manpower, and few (if anyone) in the organization had the will to do any such thing. Now that Bell was close to their laser and that competitors like GTE were breathing down its neck, it had little desire to make any significant changes to it at this point, for attempting to do so would surely, given past experience, add another decade to the effort. The laser they had finally developed worked well enough at its appointed 850 nm wavelength. The fact is, this laser system suited very well AT&T/Bell's "externally static" culture. It did not challenge AT&T's existing network, as it was going to be part of a totally new parallel system, one made of optical fibers rather than copper cable. Then too, considering how long it took for Bell to come up with this laser, it had every reason to believe that, once it built its new network, it would remain as is—and not require costly revamping—for many years to come. Finally, by constructing this fiber optics system—and by granting licenses to electronics companies wanting to make the 850 μm lasers—AT&T would be able to meet any future increase in demand and show the government that it is sharing its technology with the outside world, both of which would go far in maintaining favor with federal authorities.

What AT&T/Bell's culture could not accept was a rapidly changing technology involving its communications empire. But this is exactly what it had to face. For other companies could and did continue to search for even better laser optical fiber systems. Japan's NTT's Electrical Communications Laboratory worked particularly hard on finding superior fiber optics systems. The laboratory found that it could further reduce signal loss by removing residual water from the fiber (water molecules tended to absorb light energy). It discovered as well that light with longer wavelengths (especially over 1000 nm or 1 μm) experienced extremely little signal loss when traveling within such low-water fibers. Not only could the longer-wavelength light travel further through the fibers without weakening but it could also carry much more information in the form of voice and data.

As this knowledge spread throughout the global fiber optics community, the hunt began for a new and improved semiconductor laser that could deliver the longer-wavelength light. Within the United States, MIT Lincoln Labs put considerable effort and resources into finding such a laser. Its research uncovered that a high-wavelength laser requires a semiconductor material composed of gallium, arsenic, indium, and phosphorous mixed in the right proportions and with the base substrate (the ground floor in the structure) made up of indium phosphide, for which there was very little chemical or mechanical information available. Making Bell's trielement gallium–arsenic–aluminum laser was difficult enough, but trying to build one using a four-element material threatened a host of additional problems, not least of which was attempting to match the lattices of such different substances. But researchers at Lincoln Labs persisted; the fact that they had access to government money and did not face a hard deadline helped to ease the burden. They eventually managed to synthesize the complex compound indium gallium arsenic phosphide (InGaAsP) and grow "double heterojunction composites" based on this unique amalgam. Constructing these complex devices meant jettisoning the intractable LPE process in favor of a more precise vapor-phase method called metal-organic chemical vapor deposition (MOCVD). Despite its greater accuracy compared to LPE, it was still a very difficult technology to control and is what has been described as a very "contorted fabrication process."[12] Nevertheless, with time and money, researchers in the United States, Japan, and the United Kingdom managed to tame the beast to obtain "very high-yield, high-performance, high-reliability lasers."

Little by little, through trial and error experimentation, researchers managed to design and build devices that could emit light with wavelength of 1.3 μm (where 1 μm = 1000 nm) and that proved to more reliable over time than Bell's gallium aluminum arsenide lasers. NTT and British Telecom quickly turned their attention to the new and improved technology. The British further extended the technology by making additional refinements to the fiber material in the form of single-mode threads through which light coming from the new laser could carry even more information over much longer distances, thus requiring far fewer repeaters. Bell's gallium arsenide laser-graded index (multimode) fiber optics system did not measure up to the new technology with respect to duration, reliability, speed, or capacity.

AT&T and Bell, as usual, held back. In fact, at first Bell wanted nothing to do with this outside technology. It stuck with what its researchers had invented themselves in house. Nor, as mentioned, would Bell risk the time and costs of developing its own improved laser system. If DeLoach had remained an effective champion during this phase of the innovation process, things might have turned out differently. But, not familiar with the commercialization process for fiber optics networks, he no longer exerted much influence from here on. In any case, he would have had a difficult task. By the early 1980s, AT&T management had begun to take a more direct role in decision-making at Bell. Those in charge were businessmen not scientists or engineers; they could not see the potential in spending more time and money pursuing the 1.3 μm laser. Problems soon multiplied. Bell's field trials in Chicago followed those in Atlanta with decidedly mixed results. The major issue was the gallium arsenide laser. Attempts to mass produce it for this and future trials fell short. Devices failed after only a short time, in part due to overheating. Bell had to add complex—and expensive—cooling system to keep this from happening. Clearly, Bell's laser was not ready for commercial use. By 1980, AT&T moved forward despite these problems and began installing their version of an 850 nm fiber optic network.

9.2.5.2 Bell Faces Competition   As AT&T entered into the 1980s, it faced both a breakup of its long-distance monopoly by the government and increased competition. The latter came in the form of MCI and its president Bill McGowan. Founded in 1963, MCI deployed advanced microwave transmission technology to become the second-largest long-distance provider in the United States. By the 1970s, it wanted to move into something new: land-lined communications using fiber optic networks. To be clear, MCI had no more desire to take risks than AT&T. But, like many small start-ups, it possessed a very different type of organizational culture, one that was externally oriented. McGowan, who was always interested in learning about the newest technologies, believed that the best way for a small firm to compete was not to try to innovate itself but rather to shop around at outside companies and buy—or obtain the rights to—the best technology it could find. Like Ken Iverson's Nucor, Bill McGowan's MCI decided to avoid the risks of research and development altogether. Also like Nucor, MCI could build up this new technology without having to break into or even tear down an existing infrastructure. MCI engineers traveled extensively during this period. They went to England to visit British Telecom to learn about the advantages of single-mode optical fibers. They talked with Japanese companies manufacturing laser systems producing 1.3 μm wavelength light capable of carrying 400 million bits of data and information per second, an order of magnitude faster than anything AT&T had to offer. They were sold: MCI would stake its claim in telecommunications within the United States and in competition with AT&T by building a fiber optics network using Japanese semiconductor lasers in conjunction with single-mode optical fibers, the latter to be produced by Corning. As head of a small, hungry company within which information traveled easily between departments, McGowan quickly marshaled his resources and moved aggressively. He leased the right of way along the Northeast Corridor between New York and Washington, DC, from Amtrak and rapidly put in his fibers. When completed, MCI's fiber optics system outperformed that of AT&T in every category: it was 50% faster and much cheaper to build and operate (MCI needed only one repeater for every four required by AT&T). MCI's success proved to be a seminal event in US telecommunications; it "marked a turning point in America.

Bill Mcgowan's gamble was right on the money; ... almost overnight all the long distance companies ... switched to single mode fiber [and 1.3 μm lasers] for their nationwide backbone system."[13]

In fact, by the early 1980s, the writing on the wall was disturbing enough for AT&T to take notice. The future clearly belonged to the 1.3 μm laser. The once-powerful monopoly could no longer fight the inevitable and finally began to build these more advanced lasers to be used in future fiber optics systems. Bell now was a follower, a second mover, in the fiber optics game. The fundamental research and development work had already been done elsewhere. All it could do was to make some modifications that branded the technology as coming from Bell and move into scale-up and production, a difficult and time- (and resource-) consuming endeavor in itself. Eventually it began making them in volume for upgraded fiber optics systems starting in the mid-1980s, that is, after MCI had already been capturing markets with the very first 1.3 μm single-mode fiber optics system.

As with Nucor Steel and Intel, the case of the semiconductor laser shows how the small, nimble start-up outmaneuvers the large, intractable, and more phlegmatic establishment. But we tell a very different story in the next chapter when we consider another significant semiconductor technology. As with the case of Union Carbide and its UNIPOL process, we are about to embark on a tale of an old, tired warhorse of a company seemingly down for the count but demonstrating an ability to find its second wind and become a major innovator once again. IBM's pursuit of the silicon–germanium chip is one of the major advance material success stories of the late 20th and early 21st centuries.

REFERENCES

1. Orton, J. (2004), *The Story of Semiconductors*, Oxford, UK: Oxford University Press, pp. 5–10. For further discussion on the science and technology of band-gap engineering, see Capasso, F. (January 1987), "Band-Gap Engineering: From Physics and Materials to New Semiconductor Devices" *Science*, 235: 172–176.

2. Orton (2004), p. 6; 200–201. See also Capasso (January 1987), "Band-Gap Engineering," pp. 172–176.

3. Two recent histories of the laser are Hecht, J. (2005), *Beam: The Race to Make the Laser*, Oxford, UK: Oxford University Press and Taylor, N. (2007), *Laser: The Inventor, the Nobel Laureate and the Thirty-Year Patent War*, Lincoln, Nebraska: iUniverse, Inc.; Charles Townes, Nobel Laureate for his work leading to the laser, wrote a detailed history in Townes, C. (1999), *How the Laser Happened: Adventures of a Scientist*, Oxford, UK, Oxford University Press. See also Bromberg, J. (1991), *The Laser in America, 1950–1970*, Cambridge, Massachusetts: The MIT Press.

4. Hecht, J. (1999), *City of Light: The Story of Fiber Optics*, Oxford, UK: Oxford University Press, pp. 147–159.

5. Hecht (1999), *City of Light*, p. 157.

6. Hecht (1999), *City of Light*, pp. 158–159. See also Dixon, R. (May 2012), "Remembering the Million-Hour Laser" *Optics and Photonics News*, 23(5): 44–47.

7. This part of the story is taken from Gertner, J. (2012), *The Idea Factory*, New York: Penguin Press, pp. 271–279 and Hecht (1999), *City of Light*, pp. 176–200.

8. Wu, T. (2011), *The Master Switch: The Rise and Fall of Information Empires*, New York, New York: Vintage Books, p. 193.

9. Interview of Eugene Gordon by Joan Bromberg on June 5, 1984, Niels Bohr Library & Archives, American Institute of Physics, College Park, Maryland, pp. 9–10. www.aip.org/history-programs/niels-bohr-library/oral-histories/4637-2. Accessed November 25, 2015.

10. The many difficulties with liquid-phase epitaxy is a common theme in the technical literature. See Miller, L. and Mullin, J. (1991), *Electronic Materials: From Silicon to Organics*, New York, New York: Plenum Press, pp. 130–131; Ghandhi, S. (1983), *VLSI Fabrication Principles: Silicon and Gallium Arsenide*, New York, New York, John Wiley & Sons, p. 265; and Francombe, M. (1994), "Historical Perspective of Oriented and Epitaxial Thin Films" *Journal of Vacuum Science and Technology*, 12(4): 930–931. This last citation even notes that liquid-phase epitaxial techniques are " ... being supplanted by vapor phase approaches, which offer greater inherent uniformity."

11. This section on the scale-up and commercialization of the semiconductor laser comes from Hecht (1999), *City of Light*, pp. 176–200 and Dixon (2012), *Remembering the Million-Hour Laser*, pp. 47–48.

12. Dixon (2012), *Remembering the Million-Hour Laser*, p. 47.

13. Hecht (1999), *City of Light*, p. 198.

10

THE EPITAXIAL PROCESS II

IBM and the Silicon–Germanium (Sige) Chip

The need to serve the explosion in data bandwidth demand ... has driven transistor performance requirements beyond the reach of conventional silicon devices.

Bernard S. Meyerson, 2000

The silicon–germanium chip, developed by International Business Machines (IBM) in the late 1990s, introduced into the electronics industry a new advanced performance microchip with unprecedented switching speeds, up to four times that of traditional silicon chips, and with extremely low-power requirements, making it the ideal chip for the 21st-century wireless market. As early as 2002, Auburn University semiconductor researcher Guofu Niu spoke about the surging importance of this technology:

Silicon-Germanium (SiGe) is the driving force behind the explosion in low-cost, lightweight, personal communications devices ... [it] is the fastest growing semiconductor process ever, and is poised to continue to grow ... enabling higher than ever bandwidth communication.[1]

Thus one finds silicon–germanium as the primary material in chipsets used in wireless local area networks (LANs), cellular and smart phones, and optical (e.g., laser) data transmission systems. In fact, by the first decade of the 21st century, it has become central to modern mobile computing and communications.[2] By the

Advanced Materials Innovation: Managing Global Technology in the 21st century, First Edition.
Sanford L. Moskowitz.
© 2016 John Wiley & Sons, Inc. Published 2016 by John Wiley & Sons, Inc.

early 2000s, a number of major electronics firms were developing silicon–germanium chips based on the IBM technology or directly licensing IBM's chip. These included Hitachi, Philips, Lucent, Nippon Electric Company (NEC), Texas Instruments (TI), and Infineon. In 2004, silicon–germanium integrated circuits generated over $1 billion in revenue globally. By 2008 this figure had exceeded $2 billion and in 2011 over $3 billion, with sales of downstream products estimated to be 20–30 times these amounts.[3]

The story of IBM's silicon–germanium chip is one of the most interesting in recent advanced materials technology. The fact that IBM—"Big Blue"—came out with it tells us that breakthrough innovation still can come out of the large, established companies. Yet, as we will see, the birth of the technology was far from easy; it took many years, the seismic shift in the culture of a massive organization and the persistent, Herculean effort of one man to see the project through to the finish line.

10.1 IBM AND ITS RESEARCH

Early research at IBM reflected the vision and personality of the company's founder.[4] Thomas Watson Sr. was first and foremost a salesman who spent his early working life in Buffalo, New York, selling organs, pianos, and sewing machines for local businesses and then cash registers and tabulating machines for National Cash Register (NCR). He was also an excellent manager of people. These qualities helped propel him forward in the world of business, first working for NCR and then the Tabulating Machine Company. As he moved up these organizations, he became familiar and then fascinated with the internal workings of the machines he was selling and with the men who invented them. His acquaintance with the automotive inventor Charles Kettering, whom he liked and admired, while both were working at NCR pushed him further toward embracing the inventive process in the mechanical and electrical fields. When he transformed the Tabulating Machine Company into IBM in the 1920s, he hired creative engineers and established the first full-time engineering department in the United States for the development of office tabulating machines. For Watson, engineering had one goal: produce new or improved products that the customer wants and can use. He had no time or interest in science or theory. Inventive engineers are only as good as the products that they create and that are accepted by the market. Watson, ever the pragmatist, demanded his engineers undertake projects that were doable—in other words, could be accomplished in a reasonable length of time utilizing a reasonable amount of the company's resources. To Watson, this meant making sure the creative process was not far removed from the marketing department on one hand and the manufacturing function on the other. By the 1930s, Watson brought his engineering research and development (R&D) staff to work beside plant engineers and personnel at IBM's Endicott, New York, facility.[5]

World War II spurred technological change across industries at a rapid pace. This was especially true in the field of electronics, and by 1946, a great deal of innovative activity was reenvisioning electronic calculators and computational machines. But to keep up with this technology required a firm like IBM to start hiring university-trained scientists and engineers, many with doctorates in physics, electrical engineering, and mathematics to apply scientific principles to longer-range problems. In other words, it meant that IBM would have to start looking more like Bell Labs. None of this—not the science, the hard-to-grasp concepts, the long-range R&D, and the academic nature of the whole thing—sat well with Watson Sr., this 19th-century salesman with the keen ear for what the customer wanted, the practical problem solver using common sense and gut feelings, and the lover of mechanical gadgets and devices you could hold in your hand and fix with a toolbox and an oil can. At the same time, Watson was a savvy and cautious businessman and knew that the electronic computer was something to be watched very carefully; while it was not yet (in the early 1950s) in demand by IBM's rank-and-file customers, it was important that the company remain ahead of the competition in this field for two reasons: first, to show both customers and other companies that IBM remained in the vanguard in advanced business machine technology and second to be prepared if and when companies do start stamping their feet for electronic solutions to their business problems.

The year 1956 marked a turning point for IBM. Thomas Watson Jr. took over the reins of leadership from his father. In contrast to his father, Watson Jr., a pilot during World War II, was technologically sophisticated and sensitive to the potential importance of electronics to computers. He of course knew about transistors and their advantages over vacuum tubes. He envisioned the time when electronic computers would depend on semiconductors and knew that if IBM didn't quickly embrace this new technology, it would have a difficult time keeping up with its more forward-looking competitors. Watson Jr. revamped IBM's R&D using Bell Labs as his model. He pushed for the hiring of scientists and engineers with doctorates and wanted to publish in top ranked journals and pursue basic research in specialties of interest to IBM. He also split this research from product development and manufacturing and by 1961 established its own facility, the Thomas J. Watson Research Center, in Yorktown Heights in Westchester County, New York. Rather than being under any particular department, IBM Research at this facility was controlled by corporate management—by Watson himself—so that the day-to-day requirements of a product division did not place restrictions on the longer-term goals of the Research Center. Watson selected as director Emanuel ("Mannie") Piore, a prominent physicist with a Ph.D. who had been the chief scientist at the Office of Naval Research (ONR) during the war.

The Thomas J. Watson Research Center undertook a number of major projects during the 1960s and 1970s. These involved such diverse areas as artificial intelligence, linguistics, superconductors, linear programming, and solid-state lasers. Within a few years, with a growing number of published articles in top ranked scientific journals coming out of the Center, IBM gained a reputation of being an important research organization. It soon attracted top research scientists—experts in these various disciplines—from around the world. The crown jewel of IBM's research organization was its semiconductor research department, headed by G. Robert Gunther-Mohr, with a Columbia Ph.D. in physics, who had studied under two future Nobel Prize winners: Columbia's Charles Townes who pioneered the laser and Harvard's John Van Vleck, a leading solid-state physicist.

While the laboratory had its successes—most notably its own work on the semiconductor laser—it also failed to develop anything of commercial importance from some of its most important projects, including microwave and cryogenic computing. IBM research soon acquired the reputation of being unable to deliver anything practical to the company. By the 1970s then, the Watson Research Center in general and its semiconductor division in particular had become marginalized as far as IBM's product development was concerned. Academics and industrial scientists continued to be attracted to the Center and fundamental research continued to be conducted here, but the really important new components and products involving computer design and production issued from those departments outside the laboratory.

To avoid becoming irrelevant to IBM, the Research Center looked to undertake projects of more immediate interest to the company, which did not duplicate what other departments were doing. No one else at IBM had much interest in MOS technology, so this is where the Center began to stake its claim. The result was the eventual scale-up and commercialization of MOS chips for mainframe computer memories. This was certainly an important victory for IBM research. An even greater triumph for the Center involved a very different type of chip.

10.2 IBM AND THE SILICON–GERMANIUM CHIP

In the early 1980s, IBM had begun to align its research programs, including those carried out at Yorktown Heights, closer to real-world problems. In doing so, it had already started to rein in its scientists at its laboratories so that they undertook research more directly related to its businesses. For years, IBM had made in-house integrated chips composed of bipolar junction transistors. While neither these transistors nor integrated chips based on them could be considered particularly elegant—certainly not compared to Intel's superb MOS-GATE technology that made the field-effect transistors popular in personal computers—IBM did not really care. These bipolar chips, which could operate at very high speeds, although fairly power hungry, were perfectly acceptable for its large mainframe computers, the 360 and 370 series that for all intents and purposes was the great computer maker's bread and butter at this time. And business was very good indeed. In fact, 1984 turned out to be a landmark year for the company in terms of revenues taken in.[6] Then a new sort of bipolar device emerged from IBM's Research Center in Yorktown Heights. This technology would eventually lead IBM in a radically new direction, one that had little to do with the mainframe business.

10.2.1 The Silicon–Germanium Chip: Initiation and Research Phases

In the 1960s, IBM had selected bipolar technology for its logic components in its mainframe computer line.[7] The main reason was their speed; they were much faster than field-effect transistors made with MOS processes. The increasing speed and power of these devices depended on making sure electrons and holes could travel as quickly as possible across the junctions, separating the "N" (electron-rich) regions from the "P" (hole-rich, electron-deficient) regions. This meant making the middle section of the "NPN" sandwich progressively narrower. For close to 20 years, IBM could do this by using a technology called ion implantation. First developed by the TI spin-off and semiconductor memory company Mostek in the 1960s, ion implantation accelerates charged particles (ions) in an electric field and propels them into a semiconductor (such as silicon). Ion implantation proved to be especially effective in being able to accurately implant within the slab of silicon the requisite charged particle exactly where it needed to be. This gave IBM engineers the ability to construct the middle positive region thinner and thinner and thus to produce new generations of faster bipolar junction transistors and chips for their mainframes.

However, by the early 1980s it was becoming all too clear that ion implantation had its limits; one could only go so far before problems arose. To remain competitive, IBM needed to make its mainframes operate faster than ever before. This meant creating bipolar transistors with the thinnest junctions ever produced. But now ion implantation faltered badly. When IBM researchers attempted to create such a razor thin section using this technique, serious leakage of electrons across the junctions occurred, leading to disruptions in the current and catastrophic power loss. IBM had reached a roadblock that threatened the very existence of its vaunted leadership in business computers, and so "a major disruption of IBM's technology roadmap was in the offing."[8] This problem riveted the attention of IBM researchers. One approach was to go back to basics, as it were. Instead of ion implantation, they experimented with improved doping processes in furnaces under different temperatures, pressures, and concentrations. But no combination of conditions prevented the dopants from wandering uncontrollably and thus producing middle sections that were too wide and diffused to be useful.

At this point, a young IBM physicist working at the Research Center, Bernard Meyerson, saw an opportunity and began addressing this problem. Meyerson, like a number of technology champions highlighted in the story of advanced materials, can be described as an intrapreneur: he had both business experience and great technical expertise. It would take both skills to successfully navigate IBM's bureaucracy to see his project to a successful conclusion. Meyerson grew up and was educated in New York City. Prior to entering college, he helped run a small furniture factory. By all accounts, he enjoyed the work and demonstrated a real instinct for how business operates and what it takes for one to grow. But when he entered college, his interests gravitated toward the sciences. He majored in physics at the City College of New York (CCNY). Often referred to as the proletariat Harvard, CCNY became a mecca for some of the best and brightest of New York's students. For Meyerson, the school offered a physics student an inexpensive but rigorous and highly respected course of study.[9] He went on to obtain his doctorate in physics at the school and then found work at IBM specializing in semiconductor research. With traditional avenues for improved transistors foundering, Meyerson began exploring a different approach altogether and that was based on vapor-phase epitaxy. The company had already developed vapor-phase technology to create thin films for important applications, such as the development of thin-film magnetic recording heads used on disk drives having higher densities and capacities. IBM was also well acquainted with molecular beam epitaxy (MBE), which made films one atom or molecule at a time in an ultra-high vacuum chamber. Invented by Bell Labs as a scientific tool, IBM pioneered its use in the making of commercial products, such as junction transistors.[10]

The fact that Meyerson's research involved vapor-phase layering, a technology very familiar to IBM, meant he could count on support from top management at the Research Center, as long of course as his work involved practical research that would lead to useful products in the short term. Meyerson's approach in making faster transistors was simple enough. He would use vapor-phase epitaxy to build up a three-layered (e.g., "NPN") junction transistor with an extremely thin middle ("P") section, just the thing needed to keep IBM in the lead as a producer of the most advanced mainframes on the market.

10.2.1.1 A Question of Temperature While the idea sounded like a good one on paper, there was a major hurdle. In order to be able to build up such a complex structure, the surface upon which it would be grown had to be extremely—that is, atomically—"clean." In other words, it could not contain any extraneous and unwanted contaminants that could be mixed in with the carefully structured and delicate layers and cause electrical problems, such as short circuits, and in turn device malfunction. But creating and maintaining a silicon surface clean of virtually all impurities was no easy matter. It required subjecting the substrate to treatment of very high temperatures (in excess of 1000°C) in a hydrogen atmosphere in order to force the contaminants already on the top of the substrate to acquire enough kinetic energy to "jump" off and also to create a protective film over the surface to insulate it from being pummeled by particulates in the air. But putting the silicon substrate (one of the "N" layers) through such a thermal beating has its costs. Under such high temperatures, the surface structure becomes disorganized and chaotic so that it is hard to perfectly "fit" the lattice of another semiconductor layer (the "P" layer) on top of it. Such a lattice mismatch would result in poor flow of electrons and holes at the junctures, thus causing devices made of these transistors to be very slow or not work at all. Moreover, at these temperatures, the desired dopants—the ones that cause a semiconductor region to be either negative or positive—cannot be placed exactly as the layering process requires; with the additional energy from the heat, they tend to wander off in different and unexpected (and unwanted) directions, thus also compromising the integrity of the transistors.

Meyerson and his team searched for an alternate way to prepare a surface for layering that did not require such high temperatures. Their experiments showed that the negative impact of thermal stress to the surface and to the unwanted migration of dopants was significantly minimized at preparation temperatures of less than 800°C. This then was the uppermost limit that the IBM team was willing to go; the question remained how to prepare a silicon surface at these lower heats. Once again, a breakthrough at the research phase came from a source outside of the project. Meyerson and his team discovered important research done in the 1970s by a physicist at Rensselaer Polytechnic Institute in New York State. The author of these seminal articles, Sorab

K. Ghandhi, showed that the silicon surface could actually be "fixed" or stabilized by dipping silicon wafers in a solution of hydrofluoric acid at temperatures below 800°C. The acid dissolved off pollutants and also bonded with silicon atoms in such a way that kept contaminants from the environment from lighting and remaining on the surface. In short, the process kept the silicon surface clean and ready to become a foundation upon which to construct very precise and intricate semiconductor layers.

10.2.1.2 A Question of Layering: Molecular Beams Versus Chemical Vapor Deposition   Research then proceeded to the question, once the surface had been cleaned in this way, of what technique to use to erect the epitaxial structure. It was clear right away to Meyerson and his team that liquid-phase epitaxy, the technique employed by Bell Labs to make semiconductor lasers, could not be used in this case. It was simply too inexact a technique and did not give the scientist enough control over the layering process with the precision needed. But a possible solution already lay in wait in IBM's Watson Center: molecular beam epitaxy. Bell Labs invented it but never saw MBE as a commercial technology; but the people at IBM certainly did and here was the perfect test case. Using MBE technology, a semiconductor layer could be built up with the greatest precision, one atom or molecule at a time. IBM scientists already used it in their hard drive work and in their semiconductor research. Meyerson, well aware of this early work, began exploiting its unique capabilities to see if it could make extremely thin layers.

MBE technology demonstrated that it could make functioning epitaxial transistors based on silicon. Meyerson's team learned much from these earlier experiments, particularly how to go about constructing transistors and other devices using the technology. The researchers also discovered something else: MBE did not appear to be a promising commercial avenue due to the expense of securing and operating the equipment and the length of time it took to build layers in such an exacting manner. Clearly, Meyerson needed to find another approach. His experience with MBE was the crucial stepping stone to the next phase of research. MBE design depends on creating and releasing the electron beam in a very high vacuum. The next logical step was to take this expertise gained in high vacuum technique and superimpose it on another process that was well known in the chemical industry: chemical vapor deposition (CVD), whereby a gaseous element or compound causes a desired film to be formed on a surface through a chemical reaction. The great advantage of CVD under ultrahigh vacuum is that it is chemically selective and can form films in this way much faster than the molecular beam approach. Meyerson proceeded to make experimental epitaxial transistors using CVD. He placed silicon wafers in quartz tubes filled with silane (a silicon-containing gas and then positioned these tubes within a furnace under ultrahigh vacuum conditions. The silane gas reacted chemically with the silicon substrate to form accumulating layers of silicon crystals. These crystal strata were transformed into N and P regions by adding the appropriate dopants into the tube at the right times.

10.2.1.3 The Germanium Solution   Meyerson's work on the application of epitaxial processes in making silicon bipolar transistors proceeded slowly and empirically over a 4-year period, from 1984 to 1988. As the team made obvious progress,
IBM management began to take more interest and lent out more people to join the
effort. At one point, nearly 100 people were working on the project. These were
mostly technicians who performed different types of laboratory tasks, all under the
Meyerson's guidance and direction. His ability to continue to sell the project to senior
management, especially in light of the successful progress made by the team, made
all the difference to securing the support of IBM to the effort. While Meyerson's
ability as a salesman for the technology clearly cannot be dismissed as trivial, it was
very important indeed: it is also true that the nature of the project itself embodied
a technology—thin-film processes—with which IBM executives were most familiar.
As long as progress continued to be made and as long as IBM continued to back bipolar technology, Meyerson and his group could continue to count on IBM support. But
making functioning bipolar transistors was not enough. These transistors had to be
much faster than any IBM had made before. Making that middle section of the sandwich very thin helped a great deal. But it turned out that while necessary, this in itself
was not sufficient to produce the high-speed devices needed for the next-generation
mainframes.

It is at this point that Meyerson began looking at the possibility of doing something
radically different with that middle section; he thought of the possibility of putting in
there a semiconductor alloy composed of silicon and germanium. Back in the 1950s
and early 1960s, researchers preferred germanium over silicon in transistors because
it was much faster; its electrons didn't need as much energy to become mobile and
conduct current within the material. Silicon, as we know, became the chosen material
because it was fast enough for what was needed and its thermal and mechanical properties made it the superior material commercially. But now Meyerson realized that a
solution to his problem was to combine the two semiconductors, that is, make the critical base region—that middle section of the junction transistor—an alloy of mostly
silicon with a small amount of germanium mixed in. Doing so involved a trade-off;
the alloy will not be as fast as germanium but much faster than silicon alone; and its
mechanical and thermal properties would not measure up to silicon by itself but be
superior to that of germanium. Meyerson felt the advanced materials process he had
created (ultrahigh vacuum chemical vapor deposition (UHV-CVD)) would be ideal
for making that thin middle silicon–germanium section. It had proved itself with all
types of epitaxial silicon transistors already. But the other method with which IBM
was familiar was MBE; this approach could not be ruled out: its performance working
with a silicon–germanium alloy was still not known.

To hedge their bet, the researchers attempted to make and test silicon–germanium
bipolar devices using both methods. In both cases, they started with a slab of silicon as the substrate and one end (or "piece of bread") of the bipolar sandwich.
They cleaned it using the low-temperature hydrofluoric acid process. Then, upon this
now pristine surface, they built up the critical middle portion made of an alloy of
germanium and silicon—the meat of the sandwich—followed by building on this

another layer of silicon, the end piece of bread of the structure. The most difficult task was constructing that middle layer, which contained many sublayers carefully placed so that there would be just the right proportion of silicon and germanium at just the right distance from the surface. This entailed an extremely delicate and exacting operation, which, in turn, meant the need "to control the proportion of germanium atoms in hundreds of successive atomic layers of silicon." Moreover, as they constructed these sublayers composing that middle region, they worked into them carefully selected dopants—such as boron—to make sure each layer had the proper net charge to make the device, whether NPN or PNP in construction, work as it should. Once again, it became clear that MBE was simply too costly and labor intensive to be commercially viable. Even more seriously, the MBE silicon–germanium transistors performed poorly. It was at this point that MBE finally reached the end of the line in its bid to become an industrial process, at least for this technology.

10.2.2 The Silicon–Germanium Chip: Development Phase

Meyerson and his team had made great progress in the research phase but were still a long way off from creating a working prototype. While they battled the scientific and technological problems, they had to deal with some difficult obstacles to their work from both inside and outside of the company.

10.2.2.1 Internal Competition   Despite their clear progress, as they entered upon the development stage, Meyerson and his team faced a series of problems emanating from within IBM that threatened the existence of the project. The first involved resistance on the part of an important faction within the Thomas J. Watson Labs. This "old guard" within Watson Labs semiconductor group had cut their teeth on the tried-and-true method of doping silicon with semiconductors like phosphorus and arsenic in order to make the "N" and "P" regions of the junction transistor. Earlier in their careers, these scientists used furnaces to thermally dope the silicon. More recently, they, as the semiconductor industry as a whole, replaced thermal doping with the more precise technique of ion implantation. To them, this was the way forward to faster junction transistors, not the untested and very dubious route being taken by Meyerson. They openly questioned Meyerson's work on a number of points. As he himself reports, this large and influential bloc greeted his work on silicon–germanium with mixed response.[11] In their heart of hearts, the opposition firmly believed that Meyerson's silicon–germanium devices made by UHV-CVD would never leave the laboratory. One of their major concerns was that Meyerson's laboratory prototypes would not be able to be scaled up successfully on a production line and that the yields would be too low. They also pointed to the lack of stability—and therefore reliability—of silicon–germanium alloys in electronic components. IBM had spent many years trying to come up with stable MOS field-effect transistors for memory chips, and it did not want to go through that again by playing around with Meyerson's unusual and complex technology, especially since they would have to work in logic chips, a technology that is an order of magnitude more complex than memory. As far as they were concerned, Meyerson's transistors would never amount to more than specialized devices for highly exotic, niche markets and so would be of little value to IBM.

Meyerson had to act cautiously. A strong and imbedded contingent of IBM researchers were talking against him and his project. He was also still fairly new to the organization. While his reputation was growing, he still had not proven himself as yet, especially in showing that his work had relevance to the company. But at IBM, much like at Intel, the way to handle such conflicts was through a creative dialectic, or, what Meyerson calls, "friendly competition." Through his encouragement, the two groups agreed to compete against one another to see at various points in time who was ahead. This competitive research went on for a number of years and became institutionalized at IBM. Meyerson's task was to convince top management in the labs (and senior executives in the company) that they would not be taking a great risk in supporting his technology over the better known ion implantation. The great advantage of this competition was that it kept Meyerson and his work in the game. It also muffled the complaints by the old guard and put his work on an equal footing with the ion implantation group. Ultimately, of course, Meyerson had to win that competition by showing that his process could produce a wide range of commercial-grade epitaxial (UHV-CVD) silicon–germanium transistors that were faster, more stable, and more scalable than anything ion implantation could offer.

From 1988 to 1990, the team managed to build progressively faster transistors. By 1990, they could make epitaxial silicon–germanium transistors that were twice as fast as conventional silicon devices. Meyerson reports that their success helped turned the tide in acceptance of his work:

> The survival of high-quality devices largely debunked the previously widespread notion that the SiGe base was 'fragile' ... In fact, a SiGe layer designed to be uncondition- ally stable could be treated in the same manner as a conventional silicon layer ... [this progress] changed the mindset of IBM and attention turned toward using the technology in practical circuit applications.[12]

Also attractive was the fact that, while other companies turned out scientific publications on the technology, only IBM was anywhere near a commercial product. In other words, when it came to epitaxial electronics, "Initially, IBM was pretty much on its own." At this point, Meyerson's superiors began thinking about the silicon–germanium chip as it would evolve from a working prototype to full-scale production. IBM executives also wanted to know that Meyerson's device could be made on familiar, existing production technology—in other words that tons of money and resources would not have to be spent designing, building, commissioning, and testing a whole new plant. Meyerson then dictated to his team that all further silicon–germanium work be developed in accordance with the same production technology as used by the silicon transistor side and that results be measured by exactly the same standards. This requirement kept the research from going in directions that were far afield from IBM's traditional know-how. Most importantly, Meyerson's team could modify their UHV-CVD process to be used on the existing production lines then in place to produce silicon bipolar devices, and they could do so, as Meyerson tells us, "with minimum change and retooling." This requirement "set a new and important direction for all silicon-germanium epitaxial base work: full compatibility with the silicon technology base of the day."[13] This, in turn, would allow fairly rapid scale-up and automation of the process.

10.2.2.2 Grappling with a Shifting Context and Shrinking Resources

In its heyday, around 1990, Meyerson's silicon–germanium transistor program involved more than 100 scientists, engineers, and technicians. The promise of silicon–germanium transistors, supported by the progress made by Meyerson and his team, kept the project moving forward. It was, in fact, no longer the dark horse in IBM's semiconductor work but had taken the lead as the most likely route to faster semiconductor chips. Never before had a company come so close to finding an alternative to silicon in making practical transistors for computers. But then, in the early 1990s, the project confronted a major crisis, not from technical setbacks but from a company-wide financial emergency and from a game-changing managerial decision. The former stemmed from a downturn in the US economy, which impacted many semiconductor companies, including IBM, and from IBM's particular problem of losing business to the young start-ups on the West Coast who were beating the older company at making and marketing personal computers. The belt had to be tightened at IBM, especially within its R&D programs. In 1992, IBM cut funding for certain types of projects including materials research that directly impacted Meyerson's silicon–germanium work. On top of this, and even more seriously, IBM decided to stop work on all bipolar semiconductor technology for both computer memory and logic and shift their efforts and resources to MOS field-effect transistors, that is, the technology favored by Intel and Silicon Valley. IBM management already believed that they had waited too long to make this move and that it had hurt their bottom line, and to a large extent, they were right. IBM executives not only were losing the personal computer battle but also faced shrinking markets in their primary business, mainframe computers. They concluded there was no place for Meyerson's work on silicon–germanium bipolar chips in the new order of things.

The project should have died there. But Meyerson would not be thrown over and hung on by his fingernails. The energy, passion, and will power that he marshalled in keeping it alive and progressing touches on the heroic. Meyerson and his now greatly reduced team had to work more or less under the radar. If the project had now become a vanity project for Meyerson, he did not particularly care, as long as research could continue in some form. This is, of course, a similar situation we have seen in other advanced materials innovations, such as DeLoach and Bell's semiconductor laser. And, as in these cases, Meyerson and his team had to work fast and smart. They had to be creative and efficient in designing and carrying out their experiments, for they had virtually no more money coming into them from the company. This means they were forced to use the equipment they already had on hand and before senior executives co-opted their facilities for other projects.

While the situation appeared dire, Meyerson actually had quite a bit of "invisible" support behind him from a number of quarters that bolstered his resources and

allowed him and his skeleton staff to proceed in their work. The three-man team organized itself to be most effective and made sure the members covered the major areas
of activity: a technologist who designed and ran the experiments; an applications
engineer who ran, maintained, and operated the UHV-CVD systems; and an IBM
manager, Meyerson himself, who championed the technology, established and maintained a critical network of contacts throughout the company for support and political
leverage, and interfaced between research, development, and, when the time came,
production. Most importantly of all, he did all he could to keep IBM from ditching
the project. In doing this, he not only had to demonstrate technical progress, but he
also had to think of ways to integrate his bipolar chip into what had suddenly become
a MOS-based company.[14]

As he had from the beginning, Meyerson led and coordinated the effort. He had
previously established extensive contacts throughout the company and had amassed
an enormous amount of goodwill and respect from his colleagues in many departments. It was now, at this highly critical juncture, that he called in his markers from
the many people who owed him favors. And, in truth, they did so not only out of a
sense of loyalty and obligation but also because they believed Meyerson could really
accomplish something with his work, something that was likely to greatly benefit the
company even in the new MOS regime. Calling on his network that spread throughout IBM, Meyerson bootlegged equipment and materials as well as technical advice
and suggestions. To a large extent, Meyerson's influence rested on the fact that his
work extended the reach of IBM's core competency, epitaxial technology. Even if the
original point of his work—to make bipolar transistors—no longer applied, the process remained a powerful one and likely to lead to new routes in the future. A forceful
statement made in support of Meyerson at this time by James McGroddy, corporate
vice president for Science and Technology, reflects this view. In an interview on the
shrinking R&D resources at IBM in the early 1990s, McGroddy criticizes certain
projects that IBM never should have supported because they did not align with the
companies' past accomplishments. However, he singles out Meyerson and his work as
particularly important to IBM because it advances the companies' already impressive
work in epitaxy:

> As an example of a researcher he would especially like to keep [at IBM], McGroddy sin-
> gled out for praise the work of Bernard S. Meyerson, the most recent IBM researcher to
> be named a fellow ... According to McGroddy, Meyerson showed that epitaxial growth
> in chemical vapor deposition, which everybody thought had to be done at very high
> temperatures ... could be done [at much lower temperatures and, as a result] ... one could
> get [superior] germanium-silicon alloys [for advanced chips].[15]

Knowing that Meyerson had the support—if not the funding—from senior executives and that his work fit perfectly within IBM's technological competency, other
managers at the Thomas J. Watson Labs and elsewhere in the company participated
informally as his "shadow" team. Meyerson reports that while the program in the
early 1990s had been officially canceled by IBM, "the culture was such that a core of
management, from the first level fab managers to senior executives, always found a

way to aid and abet what at times was a remarkable guerilla operation."[16] A number of these "provided technical and management support," even to the point of siphoning off their own program resources into the silicon–germanium work.

10.2.2.3 Dealing with a Dynamic Market   Despite effectively maintaining and managing a stripped down research team, moving work ahead efficiently in the face of dwindling resources, and establishing an extended network of support within IBM, Meyerson and his team knew they had to find a way to harness their work on silicon–germanium to the direction the company was now heading; otherwise, all of his brilliant experiments and pathbreaking results would come to naught. Up to now, Meyerson geared his work toward the bipolar transistor, and this device no longer fit within IBM's strategic priority.

The saving grace for Meyerson was the great flexibility of his process. One aspect of this flexibility was the large range in the speed of the chips designed. By the mid-1990s, the team could make a silicon–germanium chip capable of processing much higher frequencies than possible with silicon alone This was of critical importance, for it meant that here was a chip that could handle the rapidly fluctuating signals that characterize wireless technology, such as used in cell phones and laptop computers (and later on smart phones and iPads). In other words, Meyerson's silicon–germanium chip did not have to compete with MOS technology for a place in mainframe computers. It had new lands to conquer all on its own, which could catapult IBM into an entirely new line of business.

10.2.3   The Silicon–Germanium Chip: Scale-Up and Commercialization

But IBM's management—and the new IBM president, Louis Gerstner in particular—had to be convinced that entering this new market did fit into IBM's strategic outlook and did not conflict with and even jeopardize the company's new push into MOS architectures for mainframes. Here is the second instance where the plasticity of Meyerson's process played an important role.

10.2.3.1 Integrating the Silicon–Germanium Chip into IBM's Production Process   Meyerson had previously demonstrated that silicon–germanium chips could be manufactured using IBM's existing manufacturing process. But this was when the company was making silicon-based bipolar transistors. With IBM's MOS devices for mainframes pushing the bipolar operation out of the way, this was no longer the case. But Meyerson could manipulate his process so that IBM's new MOS plant could also accommodate his silicon–germanium bipolar devices. He did this by conceiving of a new type of hybrid transistor that incorporated both his bipolar design and the MOS architecture. Called a bipolar complementary MOS (or "BiCMOS") transistor, it captured the best characteristics of both worlds: the lightning speed of a silicon–germanium chip needed in wireless application and the ease and economies of production typical of MOS technology. By carefully constructing the chip so that the two parts—the bipolar part and the MOS part—occupied separate regions of the chip and designing in the ability to overlay the former only after the MOS portion of the chip had been made, Meyerson made sure his BiCMOS chip could be made by using IBM's new CMOS production facility, thus saving time and money during scale-up. The researchers understood that success of their project depended on melding the CMOS and silicon–germanium processes: " ... it is important to use the same tool-set (as far as possible) and to keep the actual physical processes [for CMOS and Si-Ge] that same [in order to avoid] major process and device parameter changes."[17] This "hybridization" meant that Meyerson's silicon–germanium process, rather than competing with IBM's MOS technology, actually complemented and indeed enhanced it; such a chip made from both of them could capture a sizeable portion of an entirely new and rapidly growing market—wireless communications.[18]

10.2.3.2 Finding New Markets Meyerson's chip now had a potential market and from the production viewpoint was aligned closely to IBM's new CMOS strategy. But the silicon–germanium chip no longer seemed destined for use in mainframe computers, IBM's traditional customer, but rather as the electronic center for wireless devices. The question was whether IBM's new CEO Louis Gerstner and his senior managers would accept the challenge of entering such an unfamiliar market space. Success in getting the company to embark on such a journey depended in part on Gerstner's long-term strategy for the company and in equal measure on Meyerson's relationship with upper management and his ability to nudge his contacts there in the right direction.

There is no question that Gerstner's plans for IBM provided a welcoming environment for Meyerson and the pursuit of his wireless chip. Prior to his taking the helm of IBM, the company had begun the process of decentralizing into separate and semi-independent business units so that each product area essentially became its own fortified and isolated fiefdom, along the lines of DuPont in the 1950s. In this type of organization, competition rather than cooperation between departments was the norm. This atomizing of the company would have spelled disaster for the silicon–germanium initiative. None of the departments would have had any use for the technology. Its original application, bipolar transistors for mainframes, no longer applied for any of the business units, all of whom in one way or another organized and directed their efforts toward feeding the mainframe. Nor would there have been any central interdepartmental program that would have funding or facilities for supporting the new device. The likelihood then was that there would not have been any home at IBM for silicon–germanium technology and Meyerson and his team would have eventually had to disband or transfer their work to another company willing to pursue this line of research.

The coming of Gerstner to power at IBM changed everything.[19] Ironically, although he strongly supported the end of the bipolar program, which had closed one door for Meyerson's work early in its development stage, he ended up working with the scientist to create another even more important outlet for that chip. Gerstner vigorously opposed the decentralization of IBM. He believed the company's separate divisions should come together and work toward one central strategic goal. By doing so, he maintained, the business units would learn from one another and understand that they possessed complementary and mutually supporting competencies that, if harnessed and properly directed, could reinvigorate the company.[20] The answer for Gerstner was clear: leverage IBM's aggregate expertise to target the information and telecommunications services market. Meyerson, sensing new opportunities afoot, picked his time and then made his move. He used his still-potent network of contacts in the company to arrange to talk to Gerstner directly. He convinced him that his technology could serve as the basis for IBM's move into wireless markets, for example, making silicon–germanium chips for cell phones, laptops, and other devices that would come along the pike.

Gerstner grew interested. After all, Meyerson was one of the—if not the—most respected scientist at IBM. Moreover, Gerstner recognized three important qualities about Meyerson: he was creative and determined; he had close contacts throughout the organization, which gave him wide-ranging support; and he clearly knew how to manage R&D projects given the significant progress he had made with few resources over the last couple of years. His chip, moreover, seemed extremely promising. It clearly offered for the first time high-speed, low-noise, and low-power requirements, just the right combination of qualities needed for a new generation of wireless devices. In fact, a number of companies outside of IBM were already working with—and funding—some of Meyerson's research. Clearly, Meyerson's creation was a "world-class technology orphan" that was ripe for commercialization. Just as clearly, given the interest of outside companies in the technology, IBM had better jump on the technology, and soon, or be left behind in the dust.

And in fact, Gerstner could incorporate these markets into the company's overall strategic plan. After all, by the mid-1990s, the worlds of information and communications were converging into one integrated system. The Internet revolution beckoned, which needed mainframe computers to operate servers as well as communications systems—such as the telephone lines—through which customers could access the World Wide Web. The use of wireless technology that could play the same role without wires and cables was beginning to take hold and was growing very fast. Clearly this was the future of any company proposing to be a comprehensive service provider in the digital age. And this is where Meyerson and his silicon–germanium chip came in. If IBM could control the making of chips for mainframe computers, then it should do the same with integrated circuit components that were central to wireless products. By 1996, 2 years into Gerstner's reign, this "world-class orphan" had finally found a home, not an outside firm, but the very company where it first breathed life. Consequently, in the summer of 1995, Gerstner decided to fund the silicon–germanium project internally. A little over a year later, Meyerson's now expanded group designed and built the first commercial silicon–germanium process. Despite the past delays, IBM beat out all other companies in this market.

10.2.3.3 Creating New Strategies   The pressing question that remained was not so much technological but strategic: what sort of business model IBM should adopt as it entered upon this unfamiliar world of wireless communications? At this point, IBM did not have the experience or skills to leverage its new "analog and mixed signal" silicon–germanium chip in the marketplace. Up to this time, IBM's expertise in semiconductors was as a volume producer of standard and inexpensive memory chips. As such, "IBM remained low on the semiconductor value chain." The OEMs to whom the company sold their chips were the ones to capture the greater portion of the total value added, for they designed the final products actually sold to consumers. Meyerson knew he had a superior and important new technology, and he desperately wanted IBM to control it and the consumer products to flow from it. In other words, he wanted IBM to look downstream and become a wireless OEM based on its proprietary technology.[21]

To do this, IBM's two most important requirements were to obtain the capability of designing the new chip for wireless products and selling this technology in consumer markets. To fill this "competency gap," at Meyerson's urging, and in harmony with Gerstner's new order, IBM called upon its increasingly externalized culture. In previous years, IBM, the once self-contained and vertically integrated giant, decided to seek help from the outside. It turned to Microsoft and Intel to supply critical software and hardware, respectively, for its personal computers. It had also (as we shall later see) tapped a Japanese company in a joint venture to develop the first liquid crystal displays for computer screens. Now, with Meyerson's urging, Big Blue was going to further flex its expansionist muscles by reaching out to companies in a completely new field—mobile communications. It formed alliances and partnerships with—and acquired outright—outside companies possessing appropriate expertise in chip design, application selection, and product marketing in the area of wireless communications. Meyerson took on the job of interviewing hundreds of possible alliance partners based on his knowledge of wireless technology and its industry. He first selected Analog Devices (of Waltham, Massachusetts) to investigate the initial markets for silicon–germanium chips and the technical and commercial hurdles that needed to be faced to successfully enter this field. IBM subsequently purchased the company CommQuest that could design silicon–germanium chips for new-generation cell phone systems and hired technical and marketing people who could augment CommQuest's capabilities. It also partnered with Analog Devices and soon thereafter with Northern Telecom (Nortel) and Hughes Electronics. These firms took IBM's chip and designed silicon–germanium analog integrated circuits for various products.

Beyond this expansionism, IBM's culture had a dynamic component and the silicon–germanium chip fit very comfortably into this thinking. The real value of the new technology for IBM and its partners rested on two essential capabilities: to make progressively faster and more powerful chips and to create a broad range of innovative devices for various market applications. This dynamism—although not necessarily linked to the intense level of short-term thinking at Intel that came from obsessively shadowing Moore's laws—nevertheless produced impressive results. Through the 1990s, IBM's silicon–germanium process grew more sophisticated. The power and range of silicon–germanium alloy is reflected in how so few of the transistors made of this material were actually needed on an advanced BiCMOS chip. By 2002, the silicon–germanium portion of the BiCMOS chip contained 600 SiGe bipolar transistors, approximately 0.5% of the 1.2 million imbedded CMOS transistor elements (the former gave the integrated chip the ability to pick up and handle very large frequency—nearly 300 GHz signals—while the CMOS portion provided the logic and memory to make it a minicomputer).

Soon, this super chip began its rise as the workhorse of the wireless revolution.[22] Meyerson's UHV-CVD epitaxial process was nothing if not a promethean-like tool capable of fabricating all sorts of circuits for many types of wireless products. Nortel, for example, needed wireless chips for telephone communications; Hughes demanded a very different silicon–germanium integrated circuit element that worked with very diverse electronics parameters. The fact that these two companies wanted very specific and dissimilar types of silicon–germanium circuitry and that IBM's advanced epitaxial process could satisfy these disparate markets served was essential, as Meyerson understood very well: "Because the two companies covered such a wide range of frequencies and applications, they were able to extensively evaluate the applications of the SiGe technology."[23] By the late 1990s, Gerstner's IBM, with a culture that can be described as "externalized dynamism," fully embraced silicon–germanium as one of its major technologies, as evidenced by the fact that now it "was held to the same commercial standards met by any and all of IBM's technology offerings."[24] With the full support of the company behind it, Meyerson has remained at IBM over the years, and new generations of his silicon–germanium chip have continued to be sold. IBM retains its position as lead company in this field in terms of both innovative output and global sales.

REFERENCES

1. Niu, G. (2002), "SiGe Technology for Revolution of Personal Communications" http:// www.eng.auburn.edu/~guofu/sige_intro.htm. Accessed March 31, 2016.

2. Maiti, C. and Armstrong, G. (2001), *Applications of Silicon-Germanium Heterostructure Devices*, Boca Raton, Florida: CRC Press, pp. 21–25; 359–390.

3. Szweda, R. (ed) (2001), *Silicon Germanium Materials & Devices—A Market & Technology Overview to 2006*, Oxford, UK: Elsevier Science, Ltd. In addition, see "IBM Ships 100 Millionth Silicon Germanium Chip" EDP Weekly's IT Monitor, May 27, 2002, and IC Insights McClean Reports for the years 2002–2015.

4. Bassett, R. (2002), *To the Digital Age: Research Labs, Start-up Companies, and the rise of MOS Technology*, Baltimore, Maryland: Johns Hopkins University Press, pp. 57–66; see also Pugh, E.W. (1995), *IBM: Shaping and Industry and Its Technology*, Cambridge, Massachusetts: The MIT Press, pp. 19–182 and Watson, T.J., Jr. (1990), *Father Son & Co., My Life at IBM and Beyond*, New York: Bantam Books.

5. Pugh (1995), pp. 37–47.

6. Ibid., pp. 301–324.

7. The following narrative of the initiation, research, development, scale-up, and commercialization of silicon–germanium chips comes a number of sources, particularly those articles authored or coauthored by Bernard Meyerson, including the following: Harame, D. and Meyerson, B. (2001), "The Early History of IBM's SiGe Mixed Signal Technology" *IEEE Transactions on Electron Devices*, 48(11): 2555–2566; Meyerson, B. (1986),

"Low-Temperature Silicon Epitaxy by Ultrahigh Vacuum/Chemical Vapor Deposition" *Applied Physics Letters, 48*(12): 797–799; Meyerson, B. (1990), "Low-Temperature Silicon Epitaxy by Ultrahigh Vacuum/Chemical Vapor Deposition: Process Fundamentals" *IBM Journal of Research and Development, 34*(6): 806–815; and Subbanna, S., Ahlgren, D., Harame, D., and Meyerson, B. (1999), "How SiGe Evolved into a Manufacturable Semiconductor Production Process" IEEE International Solid State Circuits Conference, Presented at the IEEE Solid-State Circuits Conference, February 15–17, 1999, San Francisco, California. IEEE International Solid-State Circuits Conference Digest of Technical Papers.

8. Harame and Meyerson (2001), p. 2555.

9. City College of New York produced noteworthy inventors and important engineers who are featured in our stories including Intel's Andrew Grove who majored in chemical engineering. The school's science and engineering departments produced noteworthy inventors and important engineers who are featured in our stories including Andrew Grove who majored in chemical engineering.

10. Cho, A. (1999), "How Molecular Beam Epitaxy (MBE) Began and Its Projection into the Future" *Journal of Crystal Growth, 201/202*: 1–7; Meyerson (1990), p. 807. Additional insights into the importance of IBM's earlier work on thin-film disk heads for the company's growing competence in thin-film technology—such as so-called "sputtering" techniques—prior to when it embarked on the silicon–germanium chip are found in a 2005 oral history panel conducted by the Computer History Museum that included reminiscences of the major players in the development of this technology: "Oral History Panel on Advanced Thin Film Disk Heads," interviewed by Ian Croll on November 11, 2005, in Mountain View, California, Computer History Museum. An article that appeared a few years earlier on IBM and thin-film inductive heads provides some more technical details on the company's achievement in thin-film inductive heads in the 1960s and 1970s; see Chiu, A., Croll, I.M., Heim, D.E., Jones, R.E., Jr., Kasiraj, P., Klaassen, K.B., Denis Mee, C., and Simmons, R.G. (1996), "Thin Film Inductive Heads" *IBM Journal of Research and Development, 40*(3): 283–297.

11. Harame and Meyerson (2001), pp. 2557–2558.

12. Ibid., pp. 2558–2559.

13. Ibid., p. 2558.

14. Ibid., pp. 2562–2564.

15. Sweet, W. (June 1993), "IBM Cuts Research in Physical Sciences at Yorktown Heights and Almaden" *Physics Today*, p. 77.

16. Harame and Meyerson (2001), p. 2562.

17. Subbanna et al. (1999).

18. Ibid. Also see Harame and Meyerson (2001), pp. 2562–2564.

19. Leifer, R., McDermott, C.M., O'Connor, C.G., Peters, L.S., Rice, M.P., and Veryzer, R.W. (2000), *Radical Innovation: How Mature Companies Can Outsmart Upstarts*, Boston, Massachusetts: Harvard Business School Press, pp. 69–70. In their case study on the silicon–germanium chip, the authors explain that Gerstner's strategy was critical to Meyerson's goal since that strategy included "the sale of chips to external customers," a revolutionary idea for "Big Blue."

20. Gerstner, L. (2002), *Who Says Elephants Can't Dance? Leading a Great Enterprise Through Dramatic Change*, New York: Harper Business, pp. 60–61; 119–135.

21. Harame and Meyerson (2001), p. 2562.

22. Ibid., pp. 2562–2564.

23. Harame and Meyerson (2001), p. 2563.

24. Ibid.

PART IV

HYBRID MATERIALS AND NEW FORMS OF MATTER

Liquid Crystals and Nanomaterials

11

PRODUCT-ORIENTED MATERIALS I

Liquid Crystals and Small LC Displays—the Electronic Calculator and the Digital Watch

Research on liquid crystals has enabled the Radio Corporation of America to develop a new type of electronic display that could have widespread practical use
New York Times, May 29, 1968

The materials considered in this last portion of the book are neither completely structural nor functional; they reside somewhere between these two groups. Another way to put it is that they can be structural or functional in application or serve a completely different purpose. They may be totally new substances or, on the other hand, materials known for decades but only recently dusted off and reworked because they are useful in 21st century technology. Nanomaterials, such as nanotubes, illustrate the case of the new form of matter that can be applied to structures or as electronic materials. Liquid crystals, on the other hand, have been known to exist—and have been studied by scientists—since the 19th century. But only since the 1980s, because of their unique structure and properties, have they found an important place in the digital age as the prime material for electronic displays of all types and incorporated in today's TVs, personal computers, mobile devices, cameras, and a host of other business and consumer technologies.

This and the following chapter explore liquid crystal technology after which we delve into the world of the nanomaterials. Many advanced materials, as we have seen, owe their existence to well-defined and dominant processes, some exhibiting a great deal more elasticity—an ability to create many different products—than others.

Advanced Materials Innovation: Managing Global Technology in the 21st century, First Edition.
Sanford L. Moskowitz.
© 2016 John Wiley & Sons, Inc. Published 2016 by John Wiley & Sons, Inc.

However, in the case of liquid crystals, no one clear, well-defined, and prevailing process comes into play. The actual synthesis of the material itself has not been a serious issue. More problematic has been harnessing liquid crystals in the making of displays. Here no one template existed that could be used to create many products. The rules that worked for one type of display could not be used to any great effect on another. So, for example, each attempt to increase the dimensions of a display forced researchers to virtually start from scratch and, once again, to recommit significant time and resources to finding a solution. In other words, liquid crystal materials are an example of a "product-oriented" technology wherein each type of liquid crystal display (LCD) operates by its own unique set of rules. This reality greatly influenced how companies perceived the risks involved in pursuing this technology and in fact is an important factor in accounting for why the United States—the original inventor of the LCD and the place where the first entrepreneurial LCD start-ups arose—ultimately ceded leadership in the field to the Japanese.

In the narrative of the LCD that follows within this chapter, the basic research and early product development take place at RCA. But, with the company ultimately abandoning liquid crystal research and development (R&D), our attention must shift to other companies and finally another country in order for us to closely track product development, scale-up, and commercialization.

11.1   BACKGROUND

The beginning of liquid crystal science goes back to 1888 and the work of Austrian botanist Friedrich Reinitzer.[1] He had come upon a curious organic compound called cholesteryl benzoate, which he had extracted from a plant he was studying. He noticed that the substance appeared to have two distinct melting points. Between the temperatures 145 and 178°C, it was a cloudy, opaque liquid; above 178°, it became a clear liquid. This phenomenon soon attracted the attention of a German physicist Otto Lehmann who determined that the cloudy liquid appeared to have the structure of a crystal, that is, its atoms and molecules were organized in an ordered, regular pattern. Since up to that time scientists believed that crystals could only exist as a solid—the atoms, molecules, and ions within liquids were supposed to move too freely and chaotically to form crystals—this discovery greatly intrigued both chemists and physicists. In fact, they came to the conclusion that these "liquid crystals" (or, as Lehmann coined them, "Flüssige Kristalle") were not solid, liquid, or gas but a totally new form of matter. Over the next few decades, scientists conducted extensive investigations into these strange materials. They found that there were, in fact, many different types of liquid crystals. But most intriguing of all were what they called the "nematic" materials. These liquid crystals contain long, rod-shaped molecules, which, because of complex intermolecular forces, are all aligned in the same direction, an effect resembling "the schooling of fish." It is this ordered alignment that gives the liquid its crystalline properties.

In the first part of the 20th century, researchers began to dig deeper into the physical properties of liquid crystals. In 1911, Charles Mauguin studied their optical behavior by shining light into nematic-type liquid crystals sealed between two pieces of glass. He was the first to find a way to cause the liquid crystal molecules to "twist" into a spiral shape—by rotating one of the glass plates—and observe the optical effects when the cell was hit with polarized light. In effect, he stumbled upon what decades later came to be known as the "twisted nematic (TN)" liquid crystal effect, one of the important phenomena associated with LCDs. In the years following the First World War, scientific interest in liquid crystals began to wane. However a specialized group of scientists in France, Germany, Great Britain, and the Soviet Union continued investigating their behavior, including their properties under the influence of electric and magnetic fields. One of the great champions of the study of liquid crystals in the years following World War II was George Gray of Hull University in the United Kingdom. He particularly excelled in examining the molecular structures of the different liquid crystals and linking these to their optical and electromagnetic properties. He is best remembered for writing the first textbook in the field called *Molecular Structure and the Properties of Liquid Crystals* published by Academic Press in 1962.

Gray's work stimulated scientific interests in liquid crystals in the United States, which had done little work in the field up to that time. Indeed, most students of organic chemistry in the United States learned little about liquid crystals in their courses. Glenn H. Brown, a young chemistry professor when Gray's book came out and greatly influenced by it, pushed for and got funding for a Liquid Crystal Institute at his university, Kent State in Ohio. He succeeded in getting his school to build and staff the institute because he had become a very influential scientist internationally and was, not coincidentally, very successful in winning government grants. His rise in Kent's administration was very swift: he became chair of the Chemistry Department from 1960 to 1965 and Dean of Research from 1963 to 1968 and, in that latter year, Kent's first regent's professor. Through the university and the institute (which now bears his name), Brown had an enormous impact in the study of liquid crystals in the United States. He and his institute attracted the attention of world-class researchers such as Alfred Saupe, William Doane, and James Fergason. Brown also helped to establish the leading journal in the field (*Molecular Crystals and Liquid Crystals*) as well as the preeminent International Liquid Crystal Conference, with the first one taking place at Kent State in 1965.

While the science of liquid crystals was alive, well, and growing, especially within the United States, commercial interest in these materials and the devices that could be made from them had trouble keeping up. Some activity did exist in a number of large industrial R&D organizations. Westinghouse was one of the first companies to investigate possible uses of liquid crystals, such as in temperature sensors. But in the 1960s, no company worked harder and more creatively on liquid technology than RCA. RCA's liquid crystal inventions held out the promise that the United States was poised to, once again, transform a scientific curiosity into a new and pathbreaking technological opportunity.

11.2   RCA AND LIQUID CRYSTAL RESEARCH

The founder and visionary of RCA, David Sarnoff, often referred to as "The General" was a true pioneer in radio and television (TV) technology and broadcasting.[2] Since the 1950s, Sarnoff's dream was to push the TV frontier further by creating a flat-panel screen that could be hung on the wall. He did not know how this was to be accomplished but he knew two things: first, it was something that was going to happen at some point, and second, if it could be realized, RCA was clearly the company to do it. The emergence of the LCD in the 1960s was one intriguing avenue to TV on a wall. The researchers at RCA's central laboratory—the Sarnoff Labs—who first undertook this research were well aware of this and knew their work fit right into the corporate strategy, at least as set out by the "General."

11.2.1   The Liquid Crystal Display: Initiation and Research at RCA

Liquid crystals were not the only possible route to a TV on the wall. There were in fact a number of other promising avenues that could lead to a flat-screen TV such as electroluminescence (the emission of light by phosphor crystals within an electric field), light-emitting diodes (LEDs), and even thin cathode ray tubes. Over the years, research into all these technologies took place in RCA's central research facility located in Princeton, New Jersey. Like Bell Labs in Murray Hill (New Jersey) and IBM's Thomas J. Watson Laboratory in Yorktown Heights (New York), RCA's David Sarnoff Research Center was intentionally designed to reflect a university campus atmosphere. The fact that it was located near to a famous university was not coincidence, the better to form ties with an elite academic institution and have first pick of the best young scientists and engineers. By the late 1950s, RCA's laboratory was clearly one of the world's great R&D organizations. The main building of this R&D complex had three floors and hundreds of laboratory bays. It was a real mecca for advanced science and engineering in the areas of physics, electrical engineering, acoustics, solid-state physics, and optoelectronics. It was where the electron microscope, color TV, and early magnetic core memories were invented. Like the other big corporate R&D laboratories, the Sarnoff Labs was rather informally organized in order to facilitate collegiality and knowledge flow between disciplines. There were only a small number of laboratory directors and group heads. When a project started, a project leader was selected for that particular assignment. He or she could then tap experts and technical staff from any number of disciplines, depending on what was believed relevant to the problem at hand. This flexible approach to R&D allowed people with diverse interests and knowledge to work on those parts of the problem on which they would be most useful and was an important reason for much of the Sarnoff Labs' success over the years.

11.2.1.1   Richard Williams and His Liquid Crystal "Domains"   Liquid crystal research at RCA began in 1962 with the work of a physical chemist Richard Williams who had at that time been at RCA for 4 years. He decided to undertake this work because it would be a possible new route to a flat-screen TV, and so it would be a project he felt the company would support. He also knew no one else in the United States was working on this and thus offered an area in which he could make his mark. Williams knew a great deal of organic chemistry and believed a particular aromatic compound called *p*-azoxyanisole could be a potentially interesting liquid crystal to study. It was also the most readily available liquid crystal that he could lay his hands on. He sandwiched the material between two glass plates. He coated the plates with tin oxide to turn them into current-carrying electrodes. He kept the liquid crystal "sandwich" heated so that the substance remained in the molten state. He then applied increasing electric current to the glass electrodes and examined the liquid crystal cell under the microscope. When he reached a certain threshold voltage, he observed the formation of a regular pattern of long parallel stripes, which he later called "domains." Then when he turned the electric field off, these stripes disappeared. This suggested to Williams that an electric current could be used to control the optical properties of liquid crystals and so possibly create electronic LCDs. Williams soon abandoned his research in part because he could not see how a practical liquid crystal device could be made in the foreseeable future considering there were no materials that actually turned into liquid crystals at room temperature and, even if there were, no practical way to control the action of the electric field on the performance of the material. Williams in fact never made any sort of LCD. But he did advance the ideas of it and a potential avenue to it. But as a scientist, he wanted to go on to other more theoretical projects. So, he wrote up his results for publication in scientific journals and moved on but not before he got another RCA researcher interested in his work.

11.2.1.2 George Heilmeier and His Two Modes of Liquid Crystal Action George Heilmeier was a very different sort of person than Williams. An electrical engineer, Heilmeier was interested in using theory to make devices that had commercial use and that would advance his career at RCA and his reputation within the engineering community. Heilmeier was also much more of an extrovert than Williams. Whereas Williams was contented to work alone on scientific experiments publishing articles in prestigious journals, Heilmeier needed people around him with which to share his ideas and discuss possibilities. He was a natural-born leader and manager of men, something that would be very important as he came to champion LCDs at RCA. And he was ambitious with the instincts to try his hand and pioneer less populated areas of research. He began his career at RCA studying the virgin field of organic transistors but soon met Williams who convinced him to focus his talents on liquid crystal technology. Heilmeier began his historic experiments on LCDs in the fall of 1964.

Heilmeier did not start out having to push the project at RCA. The fact that liquid crystals might be the key to hang-on-the-wall TV continued to hold a great deal of currency with researchers and managers. Heilmeier himself tells us, "It wasn't a question of stumbling on something and saying, Gee, what's this good for ... It was a case of right from the start knowing what one was trying to do."[3] Moreover, Heilmeier's personality and ambition could play off the fluid structure within the Sarnoff Labs to his advantage. He readily formed a multidisciplinary team to work on the problem. As his work progressed, he also began to go directly to the top brass at RCA and bring them to his lab to show them firsthand the progress he and his team were making. So his enthusiasm transferred to a wide range of personnel within the labs, assuring him of continued interest—and financial support—for his project.

Heilmeier proved to be a very creative and resourceful experimenter. He expanded the research program that Williams had started, making important discoveries along the way. For example, he found out that the effectiveness of liquid crystal substance in a display depends on two specific properties of the material: its purity (lack of contaminants) and resistivity (ability to resist an electric current). A second important insight was that different liquid crystals have, what scientists call, different mechanisms (or modes) of operation, each offering potential benefits (as well as costs) when it came to designing commercial displays. Furthermore, he identified and characterized two of these fundamental modes. He discovered one of these mechanisms by placing different dyes into the liquid crystal cell. He found that when running current in the electrode at different voltages, he could make the liquid in the cell change color. He could also turn the colors on or off by running or stopping the current. To Heilmeier's colleagues and management, this appeared to be a real breakthrough to realizing color flat-panel TV. Heilmeier dubbed this phenomenon the "guest–host (GH)" effect because he theorized that, in an electric field, the liquid crystal molecules (the host) influenced the dye molecules (the guest) to align in the proper directions to create the observed colors.

Heilmeier's team also discovered a second mode of operation, one that would become quite important in the early design of small-screen LCDs. He demonstrated that, upon shining a light into a clear liquid crystal cell, application of an electric current to that cell turned the liquid crystal an opaque milky white and that increasing the current intensified the effect. He reasoned that the current altered the orientation of the liquid crystal molecules in such a way as to cause the photons pouring into the cell from the light sources to scatter at various angles thus making the material to appear cloudy. Further experiments also showed that the light not only scattered but did so in the forward direction. This phenomenon proved very useful. By depositing reflective material onto the side of the cell toward which the scattered photons were moving, Heilmeier caused the light particles to bounce right back into the liquid crystal and, in turn, further brighten the display. Heilmeier called this effect "dynamic scattering (DS)." Heilmeier's work established the foundations of liquid display technology. He demonstrated that voltages applied across a liquid crystal cell induces the material to act like a shudder that can allow light to appear bright via DS or, if the voltage is removed, to block light from getting through and appear dark. Because of his GH effect, there was also the possibility that cells could not only show bright-dark contrast but also do so in color.

When Heilmeier showed his colleagues and superiors in the lab what he had found, he and they understood very well how these laboratory discoveries could be translated into working displays. Given their competence in TV technology, RCA engineers were very familiar with how to convert electrical signals into pictures. One such technique—known as passive matrix addressing—could be used to translate Heilmeier's experiments into actual display images. In this method, many horizontal and vertical electrodes form a matrix that divide up a screen. The places at which these electrodes intersect are potential points of light called pixels. In aggregate, these pixels compose images—numbers, letters, faces, and so forth. The trick is to activate only those pixels that are appropriate at any point in time. In the case of a digital clock, for example, pixels are shown that make up the time at any particular instant; no other pixels should be activated or else contrast suffers, and the digital image of the time cannot be clearly seen. This is where liquid crystals come in. In passive matrix addressing, an external source shines light onto an area of liquid crystal material that is sectioned by vertical and horizontal electrodes. Around each point of intersection—representing a possible pixel—is a small liquid crystal cell. Both horizontal and vertical voltages have to be present before the liquid crystal allows any particular pixel (horizontal–vertical intersection point) to light up. Voltages are applied to all of the horizontal electrodes and are maintained as long as the device is "on." However, at any point in time, voltages are only applied to selected vertical electrodes depending on which pixels are needed to form the required image at that second. Semiconductor chips (memory and logic) determine which pixels to activate, and electrical circuitry responds by placing voltage on the appropriate vertical electrodes. When the correct liquid crystal cells receive the electrical signals, the molecules within these cells orient in such a way as to "open the shudders" and light up (DS effect). Theoretically, all the other cells, at that second, should shut out the light since their liquid crystal molecules did not receive the voltages that would force them to align themselves for activation. (I say "theoretically" because, as discussed in more detail in the next chapter, there are difficulties that arise with passive matrix technology that can limit its effectiveness, especially in the application of liquid crystals to larger displays.)

11.2.1.3 The Search for Room-Temperature Liquid Crystals

Of course, none of this could work in any practical way if Heilmeier and his team failed to find a liquid crystal material that operated at room temperature, for, up to that point, all of the known liquid crystal compounds had to be heated to fairly high temperatures in order to work. But accomplishing this meant delving deep into the synthesis of totally new organic compounds, an area in which the Sarnoff Lab had little experience. Up to the mid-1960s, RCA hired chemists to help develop and apply different types of insulating materials, sealants, and photoconductors for its electronics components, products, and systems. This was mostly fairly standard, low-tech work. Investigations into liquid crystals were something very different and required fundamental exploration into the nature of organic molecular structures.

Heilmeier understood that, if he was to continue on in liquid crystal technology, he would have to create an expanded organic chemistry capability. This sort of knowledge had to be imported into RCA. The loose organizational structure within the Sarnoff Labs once again proved advantageous as it gave Heilmeier the freedom to hire first-rate organic chemists as he saw fit and to work closely with them to find new materials that he required for his displays. Two of these chemists, Joel Goldmacher, a new Ph.D. from Purdue University, and Joseph Castellano, then working on chemical propulsion systems for the Thiokol Chemical Company, soon proved their mettle. In a series of experiments conducted in the summer of 1965, they demonstrated the vital importance of the purity of the liquid crystal. They also brought to the problem chemical knowledge not previously existing at RCA. In particular, they applied their understanding of liquid-phase interactions of different mixtures of organic compounds and studied them analytically through the use of three-dimensional phase diagrams. In this way, they hit upon the first liquid crystal materials to operate at room temperatures. This discovery was absolutely vital to the continuation of the project. It certainly reinvigorated RCA's LCD work and accelerated the creation of the first working LCDs.

11.2.1.4   The First Experimental Displays   By 1966, Heilmeier and his team had in place all the components necessary to make the first working displays using liquid crystals. Over the next 2 years, this is exactly what they did, constructing displays for calculators, digital clocks, and watches. By the spring of 1968, RCA was ready to show the world the fruits of its research. On May 28, RCA held a press conference at its headquarters at 30 Rockefeller Center in Manhattan (New York City). The press was very impressed. The quote at the beginning of this chapter comes from a *New York Times* article commenting on RCA's public announcement at the time of its achievement in liquid crystal technology. Heilmeier's group brought and demonstrated the devices to the delight of reporters. An eyewitness, the organic chemist Joseph Castellano, comments on the public reaction:

> Nearly all of the stories discussed the possibility of liquid crystal displays being the answer to the long-awaited thin television screen that could be hung on a living room wall like a painting. Also of interest was the application to all-electronic clocks and wristwatches with no moving parts as well as pocket televisions, auto dashboard displays, and electronic window shades. The publicity generated by these stories sparked a flood of inquiries … Meanwhile, back in Princeton, the RCA liquid crystal group intensified its efforts to further develop the new technology.[4]

11.2.2   The Liquid Crystal Display: (Attempts at) Development at RCA

Clearly, the press conference appeared to be a real turning point for RCA's venture into liquid crystals. This turned out to be true but not in a way anyone thought, for within a few years, liquid crystal activity at RCA was for all practical purposes dead. The inability of RCA to guide the transfer of its laboratory research through the development stage to a precommercial prototype doomed the technology at the company. Three reasons explain the end of RCA's role in LCDs: the weakening of the Sarnoff Lab's power to sidestep the industrial divisions to carry out its R&D agenda, the inability of the liquid crystal group in the Lab to latch onto an industrial department that could properly manage the product development phase, and, most damaging of all, the abandonment of the project by its most effective champion.

11.2.2.1   Weakening Influence of the Sarnoff Labs   RCA executive and researchers understood they were still very far away from actually building a large-screen TV from liquid crystals or any other materials for that matter. It is more than a little ironic that the press conference of 1968 had a downside in that it played up small-screen devices, an area in which RCA had little interest. In the months following the conference, it became clear just how distant the company was from reaching its main goal, flat-panel TVs. There were a number of questions still to be answered and that clearly would require many more years of intense research effort. For example, passive matrix addressing did not work well as the screen size increased. A whole new addressing technology had to be found. Scientists and engineers began to focus on a device called a thin-film transistor (TFT) as a possible approach, but little was known of this technology. For that matter, solid-state integrated circuitry, so important in translating electronic signals into liquid crystal images, was itself still in its infancy. And totally new types of liquid crystal materials had to be found that would work on the larger liquid crystal devices.

Even so, these difficulties should not have ended RCA's quest for a liquid crystal TV. After all, it was for this sort of challenge that the David Sarnoff Labs was created; it was supposed to tackle scientific and technical projects that did not necessarily promise short-term payoffs. But the Labs could only operate in this manner if it were independent of the approval and support of the operating divisions, which, needing to develop products quickly, did not have much patience for any project smacking of blue-sky research. Well aware of the cultural differences between research and operations, and appreciating the long-term benefits that accrue to a company from the former, David Sarnoff created his laboratory to be independent of the operating groups; it reported directly to, and received its funding from, RCA's central management.

But by the late 1960s, events had taken place the net effect of which seriously threatened the independence of RCA research. First and foremost, RCA had spent massive amounts of its resources on two strategic products: color TV and electronic computer. While color TV eventually came into its own, it took far longer doing so than RCA anticipated. RCA's foray into computers turned out to be far less successful. Unable to compete with the likes of IBM, it abandoned the field in the 1960s but not before it had laid out a considerable amount of money. These tremendous expenditures, the long time to get color TV off the ground and the abject failure of its computer efforts had, by the early 1970s, made RCA management much more sheepish about tackling another uncertain and long-winding research project. Of course, a wall-hanging TV had been the long-term dream of David Sarnoff, and he could have decided single handedly to support Heilmeier and his team, and indeed he most likely would have if he had been the man he had been in the early days of RCA. But by 1968, he was already into his late 70s and in failing health. He had begun slowing down, taking part less and less in the day-to-day decision-making and eventually gave up control of the company. He died just a few years later in 1971 at age 80. It was bad luck indeed for Heilmeier that LCD research came too late to get much help from Sarnoff.

11.2.2.2 Search for a Business Unit
The biggest problem now facing liquid crystal development at RCA was that its proponents could not control the fate of the project once it no longer had the support from the man at the top. For now Heilmeier

had to go into survival mode and demonstrate considerable leadership skills to keep the LCD project afloat. He had to show management that liquid crystal technology could attract interest from an influential source and that that source would be willing to actually fund development. Doing so would reduce the risks involved as perceived by senior executives in two ways: by showing them that there was a market for it and, more immediately, by relieving RCA, now strapped for funds, of having to pay for development. Heilmeier was able to secure a number of government contracts to further develop the technology. Federal agencies that supported this work included the Air Development Center (Rome, NY), Air Force Materials Laboratory (Wright–Patterson Air Force Base, Dayton, Ohio), and the Langley Research Center (National Aeronautics and Space Administration (NASA), Hampton, Virginia). These agencies paid RCA to research, design, and build displays for electronic components and systems for military aircraft and space vehicles. Of particular importance to the military was the use of liquid crystal materials in engine monitoring displays and ground positioning locators.

While these contracts gave a much needed shot in the arm to Heilmeier's project, they were relatively small and did not bring in enough money to keep the work going over the longer term. Much more important for the future existence of liquid crystal technology at RCA was getting a corporate division or business unit interested enough to take the project under its wing and support its development and commercialization. Not only did operating units control their own funds but a vote of confidence from one of these divisions would signal to corporate heads that LCDs had a future as a commercial product and would most likely shake more money loose from headquarters. The most likely division at RCA to be interested in developing display technology was the Solid State Division (SSD), whose operations were then centered in Somerville, New Jersey. Heilmeier approached SSD to take on LCD development and appears to have been very persuasive, especially in conveying the excitement of the field and its great potential. But soon thereafter, Heilmeier ran into a typical problem faced by a research department when attempting to transfer a laboratory device from research to development: culture clash. In this case, such dissonance was particularly strong. At this time, each division at RCA had developed its own unique culture. This was certainly true of SSD. While engineers at SSD did not reject outright taking on LCDs for development, many influential people in the division had very little regard for the future of the technology. They did not consider it very sophisticated or a viable way to create radical product innovation. At best they viewed it as a possible way to quickly and cheaply enhance the performance of portable and consumer electronic devices that could bring in some cash to the division in a short period of time. The problem, as Joseph Castellano, who worked on LCDs at RCA at the time, tells us, is that SSD, not unlike Bell Labs, had a strong internal orientation as manifested in a severe case of "not invented here" syndrome causing it to ignore advice, recommendations, and alternative solutions from any one working outside the division:

> ... the SSD management was interested in fabricating semiconductors and integrated circuits using processes they were familiar with and resented being directed to engage in a totally new and untested technology, they did not believe in—or have interest in.[5]

A major issue that engineers at SSD had with liquid crystals was that they believed that the material was not sufficiently stable to be used in complex electronic products. Castellano relays how one SSD engineer thought that LCDs would malfunction if subjected to elevated temperatures, such as when people brought their watch to the beach or left it in a hot car. What engineers at SSD did not understand was that different types of liquid crystal materials were being synthesized very quickly so that these problems would have been resolved within a year or two. Heilmeier and his team knew this very well—they were in fact creating these new liquid crystals themselves—but couldn't convince SSD researchers to go along. The result was that SSD ignored Heilmeier's suggestions of pushing forward into calculators and digital watches and in 1970 went ahead instead with developing LCDs for point-of-purchase displays for advertising firms. These displays did turn out to be technical successes—they were very easy to make—but they had a very limited market (an advertiser wanted new displays only about four times a year). Without the ability to generate hundreds of millions of dollars in revenues, a requirement for SSD management to seriously recognize the technology as an important product, the liquid crystal project, while continuing to limp along, was in serious trouble.

11.2.2.3 Loss of the Champion   In retrospect, it seems clear that RCA managers chafed over the difficulty of creating different LCDs. A lot of resources had been spent and still nothing of great importance had come out of it. They could not see light at the end of the liquid crystal tunnel. The company simply scaled back on what appeared to be a losing proposition. The final blow came when Heilmeier applied to become a White House Fellow, a move that introduced him to a new career as a high-level administrator for government. Whether Heilmeier left because he had given up trying to sell his technology to the company or whether he would have taken the DC position even if he had been able to make progress at RCA is unclear. Whether he could have turned the LCD project at RCA around if he had stayed is also unknown. What we do know is that his departure sealed the fate of the project once and for all. Other members of his team no longer had any reason to stay at RCA. Castellano had little doubt the project was in serious trouble:

> With the departure of Heilmeier the LCD project [at RCA] lost its great champion as he was the main interface and promoter of the technology to RCA management.[6]

The team members dispersed to find positions in other companies; a number of them were hired by—and even founded—start-up firms focused on doing what RCA failed to do: turning LCDs into the central components of successful commercial products, particularly digital watches and electronic calculators.

11.3 SMALL LCD DEVELOPMENT, SCALE-UP, AND COMMERCIALIZATION I: US START-UPS SPIN-OFF

By the time Heilmeier left RCA to go to Washington, no other large, established firms in the United States were still involved in liquid crystal research (Westinghouse is something of an exception, as I will discuss in the next chapter).[7] Certainly no US firm had come as far as RCA in pursuing liquid crystal device technology.

It is no surprise that a number of start-ups and spin-offs were formed to develop and commercialize LCDs once the big firms backed away. Those who started these companies typically worked on LCDs in larger organizations that had marginalized or closed down these projects. These entrepreneurs were true believers in liquid crystals and set out to prove this technology had legs and that money was to be made from it. In the northeast, the most important of these upstarts were formed by former scientists, engineers, and managers at RCA. Two of the more notable ones were Princeton Materials Science, cofounded by RCA researcher George W. Taylor and located in Princeton, New Jersey, and Optel Corporation, created by Zoltan Kiss, an RCA research engineer and inventor in 1969. Both companies designed, manufactured, and sold small displays for various products. Optel, in fact, made the world's first digital watches. It became a public company in the early 1970s. In the southwest, Texas Instruments, which itself had conducted significant research into liquid crystals, spawned a number of spin-offs devoted to LCDs. One of these, Micro Display Systems based in Dallas, was funded by the Japanese firm Seiko to make LCDs for its digital watches, and by 1978 the company was manufacturing displays for both watches and electronic calculators. On the West Coast, within the Silicon Valley region, Microma pioneered its own LCDs for digital watches. Founded in 1971 in Cupertino, California, Microma was one of the first companies in the region to tackle liquid crystal technology. For a while it was a leading company in digital watches and an important source of engineers who transferred their knowledge of the technology to other firms in the industry. Other start-ups established in the region in the 1970s were Solid State Time, Omron, Exetron, and Suncrux.

One of the most important and successful start-ups pursuing liquid crystal technology during this period was created to exploit the pathbreaking research of James Fergason at Kent State University and its Liquid Crystal Institute. The firm, called the International Liquid Crystal Company (ILIXCO), located in Kent, Ohio, was based on Fergason's important discovery of a third type of liquid crystal technology designated as "twisted nematic" liquid crystal display (TN-LCD). TN-LCD technology solved one of the important shortcomings of RCA's DS systems, namely, their rapid draining of power, thus making them unsuitable for portable applications such as watches and pocket calculators. The TN technique employed specific types of liquid crystals—complex organic compounds whose molecules could actually be induced to rotate, or "twist," to form a spiral shape. In this state, the optical properties of the material were improved and heightened to create greater contrast—and thus higher-quality displays—while requiring much lower operating voltages and, in turn, significantly extending the life of batteries. This meant that TN-LCD technology outperformed other liquid crystal platforms being used by competing companies. In the early 1970s, Fergason and ILIXCO produced and sold the first TN-LCD watches in volume. Other start-ups within the United States also licensed the technology from Fergason and began making their own types of TN LCDs for watches, calculators, and other devices. Within Ohio itself, ILIXCO gave birth to a number of spin-offs licensed to use the TN effect, including Crystalloid and Hamlin. We have seen how one start-up company—Intel—came to dominate the semiconductor industry exploiting a technology other larger firms could not—or would not—embrace. Yet, none of the liquid crystal start-ups and spin-offs formed in the 1970s, with the possible exception of ILIXCO, made a major impact on the international LCD market. In fact, despite the fact that these companies were small, aggressive, technically accomplished, and staffed with seasoned managers from high-tech organizations, none of them survived past the 1970s. Princeton Materials Science closed up shop in 1975 and Optel was sold to a company called Refac Electronics in 1979, the same year that Commodore, an early producer of personal computers, acquired Micro Display Systems. Microma was acquired by, and became a subsidiary of, Intel in 1977. Failing to make a success of it, Intel sold it off to Timex a few years later. The story is similar for the other start-ups within the United States (and Europe as well). As for ILIXCO, it still exists and continues to be an important innovator in the LCD field. However, the company (now called LXD) stopped manufacturing displays decades ago and concentrates now on licensing its innovations in liquid crystal and other optical technologies to other companies. It in fact never was able to make its primary process responsive to changing demands for different types of displays, especially displays for computers and TVs. The inflexibility of its technology confined it to the highly competitive and low-profit small display sector to the point that the company ceased to play an active role in the most dynamic portions of the market.

Given the perception that small firms are so successful in high-technology markets because they are more innovative, flexible, and aggressive than established companies (like RCA), what explains the utter lack of success of all these upstarts? A common complaint from those scientists, engineers, and managers working on liquid crystal devices was that there was no clear superior process that dominated the field. Instead, there were many different ways to make LCDs depending on the firm making them, and none of these were capable of turning out large quantities of high-quality products. Castellano tells us about one important example of difficulties faced by small start-ups trying to make a name for themselves in the LCD field. Problems had to be dealt with differently by different companies using the types of specialists and equipment that were on hand; it was "catch-as-catch-can" to make things work out right, and, more often than not, the hurdles were just too difficult for the start-up to overcome:

> … many problems were encountered in attempting to reach acceptable production levels … [Adopting new types of LCDs] involved a new deposition technique that required the evaporation of silicon monoxide at specific angles to the patterned glass plates. This … [technique] greatly limited the production flow … Another problem was that the patterning of the electrodes on the indium-tin oxide coated glass … could not provide the … accuracy needed for a high-yield LCD process … Needless to say, all of these problems hampered our ability to meet production schedules, resulting in negative cash flow.[8]

In other words there was no technological "magic bullet" in the form of a superior process—such as MOS-SG that powered Intel's growth—that led one or more companies to a commercially acceptable solution; none of which was clearly superior to the others. Each firm had to find its own way to a workable process, and this was no easy thing to do. These processes are described as "tedious," "time consuming," "and laborious." Start-ups could not survive under these conditions. In the case one company, for instance, the "negative cash flow" did not abate, and soon, as Castellano relates, "we were forced to release most of our staff and by the end of 1974, we had only 12 employees....By February 1975 we were out of business, and I began looking for another job."[9]

Another telling example of commercial hopes unceremoniously shattered on the rocks of technological reality comes from none other than the creator of fluidization, Exxon.[10] In the 1970s, the company established a company called Exxon Enterprises to investigate and commercialize—or invest in—new and promising ventures not necessarily related to petroleum refining. One of the most exciting prospects for Exxon at the time was LCDs. To this purpose, Exxon Enterprises established a start-up company called Kylex and staffed the firm with managers and engineers with previous experience in liquid crystal and solid-state technology. The goal was to design and build what are called "high information content" LCDs for advanced applications, such as scientific calculators and personal computers. But this proved to be quite daunting indeed, even for those who had previously worked in such famous high-tech firms as Fairchild:

> We soon found out that a 40 character display with fine lines and spaces was quite a bit more difficult to fabricate than the four digit watch displays that I had been accustomed to producing at Fairchild ... perhaps the most serious concern was the large number of open contacts from the front glass plate into the display ... the plan [for the joint venture Kylex] was to build larger displays with more lines of characters ... [this goal] was never [satisfactorily] achieved ... [as a result] Kylex never commercialized the display ...[11]

Eventually, Exxon Enterprises sold the company to 3M Company in late 1981. The latter dissolved its display operations totally a few years later to concentrate on other more promising projects, such as advanced optical disk storage media for computers.

In the end, the small high-tech start-ups that went into LCDs could not make LCD an American technology. To be sure, they exhibited great dynamism and were very willing to go outside their walls to find new ideas and talent. But all this was not enough. The big problem centered on the fact that no "killer" process along the lines of fluidization or MOS-SG existed in this field that offered a general template or model for how to create different products relatively easily and without expenditure of money and man-hours. Without this, they could not hope to a great development and scale-up expenses through either economies of scale or economies of scope.

As product-oriented firms, the risks appeared to be just too great. Venture funding dried up and there was precious little sales revenue coming in. So, by the mid-1980s, a time when the United States had created a whole new polyethylene technology and was about to embark on such advanced material successes as ultrathin steel and microprocessors, it had utterly failed to make significant progress in liquid crystals. The vacuum created by US failure in LCDs would come to be filled by European chemical (and pharmaceutical) outfits and even more importantly by Japanese electronics companies.

11.4   EUROPE AND LIQUID CRYSTALS

The United States clearly led the world in inventing the first working LCDs.[12] In the case of RCA, these devices barely made it out of the laboratory. While the company did make simple displays for advertisers, these never amounted to an important market. The small firms—the start-ups and spin-offs—did manage to enter the market in watches and, to a lesser extent, portable calculators but could not gain much traction in these markets. Europe continued to push LCD research but never achieved an important position as producers of displays. Nevertheless, European firms and research laboratories managed to advance the technology in important ways.

During the 1970s and 1980s, European research excelled in discovering and synthesizing new and advanced liquid crystal materials. One of the truly important breakthroughs in liquid crystal research occurred in 1972 when researchers at Hull University in England discovered and synthesized a whole new family of liquid crystal materials (called alkyl and alkoxy cyanobiphenyls) possessing superior optical and electronic properties that simplified production and improved response. Called a "milestone" event by many in the field, this discovery led to development of new and more efficient manufacturing techniques. Germany also contributed its chemical prowess by discovering another important new class of materials—the cyclohexanes—which gave even better performance than the cyclobiphenyls. These two types of liquid crystal materials became standard in the more complex displays and opened the flood gates to even more advanced liquid crystal synthesis. Over the next decade, a host of new and improved liquid crystal materials became available to display manufacturers, with the great pharmaceutical houses of E. Merck and Hoffmann-La Roche playing a major role in these chemical achievements.

Europe then never really entered the LCD game commercially. Instead, its companies played a role they knew best: suppliers of the most advanced liquid crystals on the market. If the United States provided the invention and Europe the liquid crystal materials, it was Asia—first Japan and then Taiwan, South Korea, and China—that brought all these technical strands together to become the regional capital of the world in the technology. Japan led the way with small LCDs and then, in the 1990s, with the large flat-panel for TVs and computers.

11.5 SMALL LCD DEVELOPMENT, SCALE-UP, AND COMMERCIALIZATION II: JAPAN

Each company—whether established or start-up—that arose in the United States and Europe had to find its way to commercialization by itself and do so without the benefit of a well-defined and robust process that could scale-up quickly and create relatively easily different products to help spread the costs of R&D over more LCD devices. As we have seen in our stories of the fluidization, silicon gate, and other major advanced material process technologies, advanced processes help champions by making them look good to upper management and to potential investors. Recall in the case of the silicones at GE, the champions of the fluidization process were able to show top executives how quickly they could scale the process up; this was essential in keeping senior executives interested and the resources flowing into the project. Similarly, in the case of the silicon gate process, Intel could predict quite accurately how rapidly it could create smaller, cheaper, and more powerful microprocessors. Indeed, Intel could create each new generation so fast that obtaining patent protection for each new microprocessor became less an issue; by the time competitors had copied that vintage device, Intel was already putting out the next-generation chip.

Liquid crystal technology had no such powerful process. Empirical hunt-and-peck method—that is, very sophisticated tinkering—was the only way to achieve success. But this was a very risky proposition for it consumed massive resources and stole away untold man-hours with no way to predict or even guess if success was just around the corner or still light-years away. And yet, despite all these hurdles the Japanese (and S. Koreans) managed to take control of the global LCD market. We previously noted the havoc Japanese electronics firms caused in the US semiconductor industry during the 1970s and 1980s when they imported into the United States (and Europe) large quantities of cheap, high-quality memory chips. They sunk companies like Mostek and nearly destroyed the likes of Intel. In this case, the Japanese were making better and cheaper devices that the United States had already innovated and produced. But even more impressive is Japan's domination of LCDs. With LCDs, they *innovated* first in a field the United States *invented* but did not succeed in commercializing. As such the Japanese were true pioneers in this field.

A number of reasons have been cited for Japan's domination in LCDs. One is that the Japanese developed over the centuries a deep-seated appreciation of, and superb manual skill in making, small intricate objects, characteristics that they leveraged in turning out transistor radios, watches, and pocket calculators. Then too, they are willing to study the work of others and then expand and extend on what they learn. Their companies sent their scientists, engineers, and managers to the United States in the early 1950s to have Bell Labs teach them how to make a transistor, and 15 years later, they went to RCA to learn all they could about LCDs. The Japanese do not excel at original research and they know it. But they are always the good students, learning from other countries and thinking how they could adapt and apply the research of others; they come to the United States and Europe, notebook in hand, and negotiate for licenses. Observers often note the attentiveness and attention to detail the Japanese show when acquiring knowledge about a new technology from their American and European teachers. For example, the Japanese were very eager to learn of a new approach in making Lcds called the supertwisted nematic (STN) effect. This novel way to orient liquid crystal molecules promised even lower power requirements and better contrast than the older TN method. While the inventor of the technology, the Swiss firm Brown Boveri & Company (BBC), did not choose to develop the technology itself, the Japanese quickly obtained a license to do so, seeing its importance to the future of computer and other complex displays. In a similar way, Japanese chemists learned from the Europeans how to tailor design liquid crystal materials for different applications. In this way, such Japanese chemical companies as Chisso Chemicals and Dainippon Ink and Chemical Company became important suppliers of liquid crystals to Japan's electronics firms.[13]

Another cultural force of importance in Japan is dedication to one's relations and colleagues. Called guanxi in China, it stems from Confucian philosophy stressing the idea of mutual obligation between leaders and followers within a tightly knit group. Within a business context, this thinking translates into mutual loyalty between management and worker by which the former promises career-long job security and the latter to work at his or her best to do what needs to be done at an assignment, not to let the group—or company—down and not strike or otherwise act in ways to weaken or embarrass the firm. Two benefits result for the high-technology company. First, the creative scientist or engineer will work however long and hard it takes to make sure an important project succeeds. Second, personnel will not leave the company—along with valuable knowledge and experience—to start another and competing firm, a major problem we have seen for companies like Bell Labs, Fairchild, and RCA.

Finally, Japanese firms enjoy resource advantages not available to most Western companies. Japanese companies tend to be much more diversified than American or European high-tech outfits. For historical reasons—and with help from government laws and regulations—Japanese companies make many different products for a wide variety of markets. Unlike a place such as Intel, which focuses on semiconductors, electronics firms in Japan make and sell not only semiconductors but also other components and final consumer products. This diversity makes it far less risky to pursue a complex and uncertain technology such as LCDs. A further resource benefit for Japanese corporations emanates from government in supplying certain companies infusions of capital. Federal law is also more lenient than the US government (and its antitrust laws) in permitting the creation of industry in Japan consortia—and other cooperative arrangements—that can supply a greater range of technical and managerial expertise and share the considerable financial risks involved in carrying out a difficult and long-term R&D project. These economic, cultural, and legal advantages came together in the 1980s for the Japanese in their successful pursuit of commercial LCD technology.

11.5.1 The Sharp Corporation and the LCD Pocket Calculator

Top managers at Japan's Sharp Corporation—formerly Hayakawa Electric Company—did not plan to make displays for portable calculators.[14] The idea for it and the championing of it came from middle management. Specifically, a chemical engineer, Tomio Wada, working as a researcher at Sharp's Central Research Laboratory and looking for a project that interested him and would help him rise in the company, became very intrigued when he saw the 1969 NHK (Japan's Broadcasting Network) documentary "Firms of the World: Modern Alchemy," which featured RCA's work in LCDs. Wada decided that he and his company should expand on what RCA had done to make small displays for pocket calculators for the Japanese market. The question that had to be resolved at the outset was which type of system to use in the calculator display. LEDs, electroluminescent devices, and fluorescent-type tubes (so-called "nixie tubes") offered three possible routes and had been used before for displays of various types. Wada concluded that liquid crystals possessed the range of qualities needed: it could be incorporated into thin lightweight devices and consumed little power. Moreover, unlike the other possible display technologies, LCDs would be highly compatible with a type of solid-state circuit technology—called "complementary metal-oxide semiconductor (CMOS)—which required minimal power and which Sharp itself had been developing to run other consumer electronic devices.

By this time, Sharp was very familiar with portable calculators. Since 1964, the company had been making them with various types of displays. Sharp understood the technology and markets for these products. All it needed in order to take the next step-up in the technology was to find a source for the LCDs themselves. Not surprisingly, Wada felt Sharp should attempt to establish a joint venture arrangement with RCA. This plan appealed to Tadashi Sasaki, head of the company's Industrial Equipment Business Division (IEBD). Whereas Heilmeier and his team could not find an industrial division within RCA to support their technology, Wada gained the backing of Sasaki and the IEBD for pursuing LCD technology in portable calculators. This was so in part because of the sense of common purpose Wada and Sasaki shared: if the project succeeded, the company as a whole would benefit. Another factor was the fact that there was no other competing technology to which the IEBD felt great loyalty combined with the obvious superiority of LCDs, assuming they could be made to work and be produced economically. There was also a sense that the risks would be low, assuming they could convince RCA to participate. Certainly, if any company could succeed in liquid crystals, it would be their hoped-for future partner, RCA. And the American company seemed to have already come so far along in the technology; surely it would not take very long for them to move into working models and then large-scale production. Sasaki then agreed to approach RCA with a proposal. He went to RCA's SSD in Somerville, New Jersey. RCA executives at SSD had little interest in Sasaki's proposal. Some of them wanted no part of a joint venture with a Japanese firm; others were none too sure about the commercial future of LCDs to begin with.

This rejection by RCA appears at first glance to have greatly increased the risks facing Sharp should it continue on in its quest to enter into LCDs for portable calculators. It wasn't at all clear at this point whether Sharp would even remain in the portable calculator market. Since 1963, more competitors such as Casio had entered the field and were cutting prices drastically to gain an extra percentage of market share. This price war—or as the Japanese called it "dentaku sensou"—wiped out profits of its participants, including Sharp. As long as the possibility existed for RCA to work jointly with Sharp on a new, revolutionary type of liquid crystal calculator, the Japanese company held out hope that it could establish a new standard for the industry and gain substantial competitive advantage over its rivals. With RCA now out of the picture, Sharp executives faced an uphill battle to create such a technology. At this point, they were uncertain how to proceed. On the one hand, being a highly diversified company, it did have other sources of income that could support an LCD project; on the other hand, there was an opportunity cost to consider since Sharp had other consumer products it could spend its money on that presented fewer perils, such as TV sets.

However, Wada and Sasaki decided to push ahead and attempt to convince the company's top management that Sharp should continue to pursue liquid crystal technology, with or without RCA's help. However, if Wada, Sasaki, and IEG were to persuade Sharp to undertake a project deemed impossible by RCA itself, they would have to demonstrate to senior executives that developing and making LCD for portable calculators was not only possible but likely to succeed. Wada in particular proved to be a very effective advocate for the project. He pointed out a technological reality that worked to Sharp's benefit. Sharp's engineers and executives had taken it for granted that the circuit and LCD would have to be built separately and then placed together in the electronic calculator. This meant bulkier devices, slower response by the system, and increased costs for assembly, in other words, a less competitive product. But Wada demonstrated this was not the case, that Sharp's expertise in CMOS circuitry and that technology's compatibility with LCDs could be used to the company's advantage. Specifically, it would be possible to place both CMOS circuits and the LCD on a common glass platform (or substrate), thus reducing costs of manufacture while making a more compact and faster calculator.

Sharp executives saw two additional barriers to success that needed to be addressed: identifying and synthesizing the right liquid crystal material and designing a liquid crystal cell that would work with a complex portable device. Wada did what he could—in a real sense, single handedly—to disabuse Sharp executives of the notion that these difficulties were anything more than minor, easily solvable problems. The first of these was the need to identify and synthesize the right liquid crystal materials that would operate at room temperature and exhibit high contrast in the display. The fact that Sharp did not have at this time an extensive expertise in organic chemistry could very well prove a powerful barrier to further work on the technology. But Wada, the chemical engineer, improvised and conducted his own research into what RCA and other US and European companies had achieved in finding new liquid crystal materials. Patent searches and analysis proved particularly helpful. Wada soon came up with promising candidates, which he shared with Sasaki and other executives.

Then there was the need to design and construct a liquid crystal cell implanted with a transparent and electrically conductive plate, a feat never before accomplished. Wada acted strategically to show his superiors at Sharp that this risk could be greatly mitigated with important assistance from the government. Japan's Ministry of International Trade and Industry (MITI) had, in fact, been investigating a similar problem in its own research laboratory. MITI found that placing a layer of indium tin oxide on a glass substrate was a promising route to such a cell. The problem was that these results and the techniques involved were highly classified. MITI allowed only small firms to see its research work (in the hope of increasing competition in Japan). Sharp, a fairly large company even at that time, did not qualify. Wada actively sought out the appropriate government personnel for permission to obtain these files arguing that, should Sharp succeed in building its LCD calculator, Japan's small businesses would profit since LCD technology would create a new industry composed of many small companies making components for a new and expansive technology. This convinced MITI and it allowed Wada and his team early access to its research results and alerted senior executives that a solution to one of the project's thorniest problems was close at hand.

Wada had a relatively easy time assuring Sasaki and his superiors that the other important process requirements—developing photolithography technology for forming the required conductive pattern on a glass substrate, holding the two glass plates a few micrometers apart and at a perfectly constant distance, and prealigning liquid crystal molecules in one direction—would shortly fall into place. Wada and Sasaki then took an important step that further eroded perceived project uncertainty and propelled the tasks forward: they created a special project team from different disciplines and management levels within the company to work on the display. They drew managers, engineers, and scientists from various departments and divisions. Each type of professional worked on their particular specialty: scientists from Central Research—led by Wada—perfected the synthesis of liquid crystal materials, engineers from the IC division worked on the electronics, and technical personnel from the Industry Equipment Group undertook systems design and manufacturing. To further cohere the group and increase team pride—as one *Fortune* article pointed out—members wore gold-colored personnel badges as a mark of distinction.

The strong sense of loyalty to one's group, as well as one's superiors and company via guanxi, enabled Wada and Sasaki to create a diverse yet harmonized and dedicated project team. This tactic—in concert with the other steps Wada took—helped convince Sharp's top managers that the project was not only feasible but that it could also be carried out on an accelerated schedule. Sharp's senior executives now believed that what would normally take 3–5 years to complete—assuming it could be accomplished at all—could now be completed in a year. And this is turned out to be a wise call. Given the go-ahead in the spring of 1972, Wada and his team completed the development of the LCD pocket calculator and constructed a working production line just before April 1973. An important factor that eased management fears as development and scale-up proceeded and that explains their ready acceptance to commercialize the technology was that there was a single market that had to be targeted. Only one type of display could be mass-produced, thus reducing the costs of production. Like Henry Ford's Model T, relatively small additional sums had to be spent modifying the product for different applications. We tend to think of a dynamic culture as one most likely to produce the great innovations. This was (and is) the case in semiconductors. In the case of Sharp, this dedication to a single device for one application that did not undergo significant change—certainly not compared to the microprocessor—characterizes the type of culture that existed at Sharp at this time. While almost counterintuitive, Sharp's static cultural tendency offered just the climate necessary for the firm to lead the world in this sort of innovation. Thus, on May 15, Sharp announced and subsequently introduced the Elsi Mate EL-805 pocket calculator, comprised of an LCD and five CMOS ICs placed on a common glass substrate. This calculator was the first commercially successful product to use liquid crystal technology.

11.5.2   The Seiko Corporation and the Digital Watch

Companies such as RCA and Sharp were intrigued with the idea of making clocks and watches with LCDs.[15] Neither of these companies succeeded in developing such devices. Watches in particular underwent rapid technological change in the late 1950s and through the 1960s. During these years, watches began integrating electronic components into their mechanism. The coming of the quartz watch was a landmark event in this revolution. But the electronics altered the internal workings of the watch; they didn't modify its outward look; it still retained its "face and hands" design. The notion of a portable timepiece that was both electronic and digital would radically change the traditional look of watches. As was the case with pocket calculators, digital watches existed before the advent of LCDs. Watches with light emitting diode (LED) displays enjoyed a brief period of popularity in the 1960s. But LEDs were difficult to view in the sunlight forcing the owner to squint to see the time. Also, as with calculators, the major problem with LEDs was that they had significant power requirements, which meant that if the device operated continuously, the battery had to be replaced every couple of months, certainly an unacceptable inconvenience—and expense—for consumers.

Watches using LCDs would clearly be superior to LED technology when considering power requirements. The TN liquid crystals in particular consumed far less power than LEDs. With LCDs, a watch could operate all the time (meaning no side button needed to turn the watch on and off). One difficulty that plagued the development of LCD watches was that liquid crystals tended to be very moisture sensitive, that is, they degraded when in contact with water. The method of sealing the liquid crystal material in display packages thus became an important—even "make or break"—consideration for companies attempting to commercialize these devices.

The Japanese watch company Seiko (more precisely Suwa Seiko) was the first company to successfully commercialize an LCD digital watch. It succeeded to a large extent through the efforts and dedication of Yoshio Yamazaki. Like Wada at Sharp, Yamazaki studied chemical engineering. He joined Seiko in the mid-1960s and, also like Wada, became fascinated with the work RCA had been doing on liquid crystal devices. Yamazaki believed that Seiko had to pioneer this area: it was a watch company with a reputation of being a technological leader in electronic timepieces. He felt that it was Seiko's destiny to embrace LCD technology and use it to produce the watch of the future and that it was his job to lead the effort. However, top managers at Seiko did not share his vision. The company was doing very well financially on their current product lines; it was not facing any intense competitive threat and so did not feel any pressure to take on a whole new type of technology, especially one that they knew was still unreliable when facing the rigorous wear and tear of daily use and, even more disturbing, one that still could not operate at all at room temperatures. They also felt that, even if they succeed in making a reliable LCD watch, it would not take long for competitors to enter the market pushing prices down to unprofitable levels.

On the other hand, Seiko's favorable financial situation—it was flush with cash from selling its popular line of electronic watches—worked in Yamazaki's favor: the company could afford to fund a wide range of projects, even those with only long-term possibilities. All that was required was to find a reason why Seiko should direct funds to such a project. Yamazaki took it upon himself to do this. Creating that glimmer of promise depended on finding an appropriate liquid crystal material that would work at room temperature and be relatively stable under various real-world conditions. Although a chemical engineer, Yamazaki did not know very much about liquid crystals. Like Wada, he did his research into their structure, characteristics, and techniques of synthesis generally through reading technical and patent literature and extending that work with his own experiments. He came upon what looked to be excellent liquid crystals for LCD watches (the cyanobiphenyl compounds). He synthesized the chemical, created various mixtures based on his knowledge of phase diagrams, and tested them in simple liquid crystal cells. As importantly, he did his research into the biphenyl industry both within Japan and globally and convinced Seiko executives that the company could obtain volume production of the material from either a Japanese company (such as Chisso in Japan) or global pharmaceutical firms (such as Merck in Europe).

With these results in hand, he won over Seiko's senior managers to the idea that these liquid crystals were the key to commercial LCD watches. Seiko authorized the project and allowed Yamazaki to form a small team. However, he was expected to produce results quickly or face the termination of the project (and possibly of his job). It was one thing to have found a reasonably good liquid crystal mixture; it was another to decide how best to design the liquid crystal cell. The most pressing question Yamazaki and his new team faced was which mode of operation would work the best in watches. That is, should the LCD be designed using the DS method, GH effect, or TN approach? Yamazaki knew immediately that the first was out of the question. It resulted in LCDs with slow response time, narrow viewing angle, and high-power requirements. While the second approach held out the hope of a colored digital watch display—a very enticing possibility for sure—as well as a wider viewing angle and lower power requirements than DS, it also produced poorer contrast, making displays harder to read. Even more damning, the GH LCDs had an unknown life expectancy, which would make the endeavor too risky for Seiko's management.

Yamazaki became aware of the TN route to liquid crystals through his technical contacts—both direct and indirect—at Kent State University. He decided that TN would be the perfect route that would significantly shorten development time and costs. It promised both low power consumption and high contrast as well as wide viewing angle and known and acceptable life expectancy. This decision to license TN technology from the United States reflects Yamazaki's knowledge of LCDs and his belief that this would reduce the risks of their venture. His understanding of this technology and ability to explain its importance went far in getting management to continue to put money into the project. Yamazaki further strategized to assure a steady and cheap supply of the critical liquid crystal by working closely with Merck and Chisso to secure volume amounts of the required biphenyl compounds.[16]

Yamazaki's efforts paid off. He and his team progressed rapidly. He kept his superiors abreast of what he was doing and the results obtained at each stage so as to maintain their confidence—and resources—in the project. As the case with Sharp's LCD experience, Yamazaki's venture was considerably helped by the fact that no creative destruction was thought to be in play—both watches and calculators not using LCDs would continue to be made alongside the newer technology—and that there was only one well-define, stable market for the technology, that is, watches. This fact significantly reduced the fears of top managers, for they knew that all the money they poured into the project was a one-time deal and—assuming there wasn't too much competition for a while—that they would be able to quickly recoup it by mass-producing watches with LCDs at low cost. As in the case of Sharp and its small LCDs for electronic calculators, Seiko's static culture saw this important innovation through scale-up and created the type of climate that could accept this type of technology as important to the future of the organization. Seiko introduced its LCD watch in September 1973. Bob Johnstone, who has written a lively account of Seiko and its digital watch venture, sums up the role Yamazaki played as champion of the LCD watch: "Yamazaki was entitled to feel proud of his efforts. He had stuck with liquid crystals through thick and thin ... beginning with dynamic scattering displays then subsequently switching to twisted nematics. He had done most of the work himself, up to an including synthesizing his own liquid crystals."[17]

By the late 1970s, under the leadership of Sharp and Seiko, the Japanese were gaining control of the world market for LCD for watches (and increasingly other consumer goods). In 1980, the number of companies in Japan greatly expanded volume production of LCD watch displays. In addition to Seiko, the important producers included Hitachi, Daini Seikosha, Citizen, Epson, Nanotronix, Optrex, Sanyo, and Sharp. By June 1981, Japanese companies held a 45% share of global market for watch-size LCDs and 80% of the market for LCDs used in other consumer electronic applications. By the mid-1980s, the five leading small digital display manufacturers were all Japanese; together they made just over 350 million units valued at $477 million, or nearly half of the total market. For these Japanese companies, watch displays were the largest product category (58%), while small TV screens (2–3 inches screens in handheld portable units) were the smallest (0.2%). The other categories were displays for calculators, instruments, computer, and auto.[18]

Japan's success in LCDs for watches and electronic calculators cannot be credited to its uncovering some "killer" process that reduced the costs of developing, scaling, and producing displays. Japanese companies—particularly Sharp and Seiko—had to work with the same reality as those in the United States that creating LCDs was no easy task nor are the Japanese greater risk takers. What they do have is a culture that, while not necessarily conducive to blindingly original innovation, is very responsive to tackling difficult development and production problems of particular products for specific markets and sticking with them—and pouring as much manpower and cash as needed to "hunt-and-peck" as long and intensely as necessary until they reach their goal. The outward-looking, community-building nature of guanxi linked units within companies and companies themselves together into tight organizational, informational, and financial networks that could be focused totally on any particular project deemed important. Knowledge and money flowed through this network and into LCD development. The Japanese also had their champions who knew how to work the network to remove or at least mitigate real or perceived risks to keep the project on track. Within this context, companies like Sharp and Seiko funneled their extensive resources into LCD technology and patiently solved one problem after another until they had working models in hand.

Within a few years, Japan—followed by Taiwan and South Korea—would begin to develop even more important liquid crystal technologies based on large displays for computers and TVs. As with digital watches and portable calculators, the United States and Europe would again be bystanders in this second liquid crystal revolution.

REFERENCES

1. The following narrative is based on Castellano, J. (2005), *Liquid Gold: The Story of Liquid Crystal Displays and the Creation of an Industry*, Singapore: World Scientific Publishing Co., pp. 1–8; Kawamoto, H. (2002), "The History of Liquid Crystal Displays" *Proceedings of the IEEE*, 90(4): 460–464; Johnstone, B. (1999), *We Were Burning: Japanese Entrepreneurs and the Forging of the Electronic Age*, New York: Perseus Books, pp. 96–105.

2. Castellano (2005), pp. 14–79; Kawamoto (2002), pp. 464–468.

3. The quote appears in Johnstone (1999), p. 97.

4. Castellano (2005), p. 51.

5. Ibid., p. 63.

6. Ibid., p. 66.

7. Ibid., pp. 67–79; 91–109; 140–148.

8. Ibid., p. 99.

9. Ibid.

10. Ibid., pp. 121–125.

11. Ibid., pp. 123–124.

12. The following narrative comes from Kawamoto (2002), pp. 476–482; Castellano (2005), pp. 71–75; 127–131; and Johnstone (1999), pp. 107–108.

13. Castellano (2005), pp. 82–83; 131.

14. The following narrative comes from Castellano (2005), pp. 82–84; Kawamoto (2002), pp. 468–470; and Johnstone (1999), pp. 137–138.

15. This section comes from Castellano (2005), pp. 84–85; Kawamoto (2002), pp. 482–486; and Johnstone (1999), pp. 105–112.

16. Johnstone (1999), pp. 105–110; Kawamoto (2002), pp. 483–484.

17. Johnstone (1999), p. 108.

18. Castellano (2005), pp. 170–172.

12

PRODUCT-ORIENTED MATERIALS II

Liquid Crystals, Thin-Film Transistors, and Large LC Displays—Flat-screen Televisions and Personal Computers

It was a very emotional thing for us, that 14-inch screen. We were really inspired, because it was the first time that any of us had seen anything like it.
Hijikigawa Masaya, (Member, LCD Research Team, Sharp Corporation) 1995

While the development and commercialization of small liquid crystal (LC) displays for pocket calculators and wristwatches were certainly important accomplishments in their own right, these successes were only a prelude to a far more significant feat by the Japanese and South Koreans: large LC displays for flat-screen televisions and personal computers. This success did not merely involve extending the dimensions of small LC displays; it required a truly radical breakthrough in advanced material technology that permitted LC displays to handle larger volumes of information than was ever necessary in running portable calculators and digital watches. If LC materials were ever to find their way into televisions and computer screens—including hand-held electronic devices that rely on wireless communications such as smartphones and iPads—a more sophisticated means of collecting and processing multiple streams of data and information and translating them into images and moving pictures on a sizeable display had to be found. Passive addressing schemes (as discussed in the previous chapter) could not operate effectively in such dynamic and complex environments and had to be jettisoned. In its place, scientists and engineers conceived of

Advanced Materials Innovation: Managing Global Technology in the 21st century, First Edition.
Sanford L. Moskowitz.
© 2016 John Wiley & Sons, Inc. Published 2016 by John Wiley & Sons, Inc.

"active" addressing, a technology depending on an advanced material device known as the thin-film transistor (TFT). These TFTs are strange electronic creatures; they are made by different processes and are composed of different materials than the traditional silicon device.

The story of TFTs, and large LC displays for televisions and computers is similar to the one we just told in the last chapter: a talented and forward-looking American scientist working for an established and well-known electronics firm pushes ahead with a brilliant new idea but, despite progress being made, ultimately fails to convince his superiors that it is a field worth developing further, and work on the technology is terminated. The Japanese, ever alert to new developments in the United States and Europe, pounce on this new technology, patiently and methodically solve the battalion of problems that arise. They eventually achieve success and corner the market for this breakthrough technology.

12.1 BACKGROUND

The thin-film transistor (TFT) first came into prominence in the early 1960s. It was invented by Paul Weimer, a physicist working for RCA. At this time, the predominance of silicon in solid-state electronics had not yet been settled. The TFT represented a possible alternative that Weimer wanted to explore. While the physics of the thin-film and silicon transistors are very similar, they are made in very different ways. As previously described, the latter are made by taking a single crystal of silicon and applying the methods of doping and etching. TFTs, on the other hand, begin with a polycrystalline form of cadmium sulfide, which is evaporated onto some suitable thin substrate such as glass or man-made film. Shortly after RCA announced the creation of this device, other companies, including IBM and Fairchild, also began researching the properties and potential of the TFT. By the mid-1960s, the battle was over and silicon emerged as the clear victor for reasons previously discussed.

However, if TFTs were not destined to power the world's computers, their possible use in advanced addressing systems for sophisticated display technologies—such as flat-panel televisions—did not go unnoticed. Silicon wafer technology could not be extended to large-area displays. Since wafers were sliced off the boule of single-crystal silicon, they could not have a diameter greater than the boule itself. Although improved processes increased these dimensions—from boules with diameters of 55 mm in 1970 to 300 mm in 2005—this growth did not come close to the dimensions needed for flat-panel displays for TVs or even personal computers. But TFTs could be made on much larger surfaces and so could, at least theoretically, be adapted to big screens. But top managers at RCA ran out of patience when Weimer and his team could not create stable TFTs of any practical size. Weimer's interest in TFT's was, in any case, more scientific than practical. He was no champion of the technology. In any case, RCA ended the TFT program in 1971.

12.2 TFTS: INITIATION, RESEARCH, AND EARLY DEVELOPMENT

With RCA out of the picture, only Westinghouse remained to carry on the research and so—spurred on by a persistent and creative champion of its own—was the last great hope for the United States to control the global large-panel LCD industry. The following narrates the important research and early development work in large-panel Lcds that took place in the United States and Europe in the 1970s and 1980s. This research and early development activity centers around the story of the thin film transistor and its role in a new type of matrix display technology.

12.2.1 The United States: Westinghouse and TFTs

Weimer's published scientific papers eventually caught the attention of another scientist, a theoretical physicist, T. Peter Brody, who had been working at Westinghouse since 1959.[1] Brody did see great commercial potential in TFTs and was one of the first researchers to predict their use in flat-screen TVs. He initiated Westinghouse's first research work into TFTs. In the 1960s, Westinghouse was a leading producer of TVs and semiconductors and committed significant resources to develop new technologies in these fields. Clearly, Westinghouse should have been receptive to Brody and his research agenda. However, the company was facing financial difficulties that dampened enthusiasm for longer-term projects. By the mid-1960s, Westinghouse's semiconductor business was losing money in the face of increasing competition from the likes of Texas Instruments, Motorola, and Fairchild. The reaction from top management was to scotch research into new types of semiconductor devices and focus on improving conventional transistors. In 1967, the company gave Brody an ultimatum: either find funding soon from other divisions or "face the axe."

The former would not be an easy thing to do at this time. The units that should have been the most interested in Brody's work—that is, the divisions dealing with semiconductors—were the ones hurting the most financially and being pressured from corporate management to turn quick profits. Even without this serious constraint, Brody had to deal with an intense conservatism that is often prevalent in units of a large decentralized organization. In the last chapter, I discussed this problem at RCA. It existed just as strongly within the Westinghouse business units. TFTs were very different creatures than the types of silicon-based semiconductors Westinghouse—and the rest of the industry—made. Scientists and engineers working on semiconductor technology did not have much faith in such a radically different design requiring totally new production methods, especially when silicon already proved itself such an effective commercial material. As Brody recalls it, Westinghouse's semiconductor people opposed TFT research as unachievable and not even necessary given the state of the art of silicon technology.[2] And then there was Brody's personality with which to contend. By all accounts, he was a first-rate scientist and an impassioned and effective advocate for the cause of TFTs. The problem was he had a reputation for being a less than first-rate manager as evidenced by his inability to deliver what he promised on schedule. Leaving aside the financial difficulties of the semiconductor operations and the "not invented here" culture of the operating divisions, managers within these divisions had little faith in Brody's ability to supply them with results when they most needed them.

Despite these hurdles, Brody, ever the aggressive and creative salesman for his cause, was in fact able to persuade two of Westinghouse's divisions to support his effort: the Consumer Electronics Division and the Electron Tube Division. Of the two, the former, with its long history in radios, TVs, and home appliances, was the more influential. This division had a strong motive for backing Brody: it had been losing market share in TV to its competitors, notably RCA. Managers in Consumer Electronics felt that being first to produce flat-panel TVs would help the company reverse its fortunes in the television market. For this reason, William Coates, a top manager in the division, became Brody's chief champion believing that his work "was going to make Westinghouse."[3] Coates was little concerned that Brody's research went against the grain of traditional silicon. Coates himself said "We could care less about what technology he used ... If he could make us a flat screen that was going to cost less than a cathode ray tube – wow!"[3]

By 1967, Brody and his team had made some significant progress. They designed and constructed a new type of vacuum deposition system for making TFTs that greatly reduced the possibility of contamination from atmospheric impurities, which traditionally had been a major cause of operational failure and lack of stability and reproducibility of TFTs. By 1968, with support from Coates and from outside military contracts, Brody became head of a new department, Thin-Film Devices, to continue research on TFTs and their possible applications to flat-panel displays. Brody began exploring the possibility of using TFT's as the basis for a totally new type of matrix addressing system for larger liquid crystal (LC) displays. As described in the previous chapter, small displays use a matrix system of vertical and horizontal electrodes that cross at pixel points. LC molecules at any pixel point are activated by the electric field produced by the intersecting electrodes. For smaller, less complicated displays, this system works well. However, as the dimensions of the display grow and the amount of data and information becomes more voluminous and complex, difficulties arise. Fugitive electric field emissions begin to "bleed" over to pixels that are supposed to remain shut at that instant. These fields partially activate the LC at these points forcing the light shutters to partially open. The unfortunate result is lack of contrast and great difficulty, even impossibility, in viewing the display.

In short, the passive matrix approach did not work with larger displays. But as he gained greater confidence in the performance of his TFTs, Brody had the idea of jettisoning the electrode matrix concept totally and placing small TFTs at each pixel point to individually control each shutter. This approach appeared promising as a way to bring far greater accuracy, speed, and image detail to larger, more complex displays. Brody coined the term "active matrix" to describe this type of addressing system. Brody's ability to convince his superiors that his ideas were way ahead of any other group working in the field did manage to keep his project alive for a while longer. But this support was not to last. Coates' impact as Brody's supporter had its limits. With Westinghouse's share of the black/white and color TV market shrinking to only 3 and 1%, respectively, the company cut its losses and decided to get out of the business of making TV sets. This decision of course made Brody's main reason for developing TFTs a dead issue at Westinghouse. While the Electron Tube Division supported Brody for a few years due to large measure to Coates' influence, this was no more than a temporary reprieve. That division also began to lose money and was itself soon closed down.

With all possible avenues of support from the divisions now gone, the only source of backing remaining for the TFT project was from corporate management. The possibility remained that Westinghouse could have reinvented itself as the first manufacturer of flat-screen TVs and gained back market share in this sector. The company had already poured millions of dollars into basic research and product development and progress had been made. But by the late 1970s, senior executives perceived that the risks in continuing the effort were simply too great. As time went on, the project demanded increased amounts of cash and personnel to keep it going. And Brody's failure as a manager—and the inherent difficulties the technology presented—took their toll as Coates recounts: "Every aspiration we had, every milestone we set, we missed. We missed time tables, and we missed cost."[4]

Here is a perfect example of a conservative management forcing the wrong criteria down the throat of a radical technology project. But knowing this does not change the situation. As it was, the risks appeared at this point very high for Westinghouse. The process seemed to be quite unworkable and at the very least not easily scalable. If Brody at least had something tangible to show for all the time and money expended, some actual contraption that even slightly resembled a commercial product and evidence that it would be relatively smooth sailing from this point on, Westinghouse may very well have let the project continue. Alternatively, it might have been possible for Brody or Coates to have found a way to tie the project to a past success, and just as the champions of Rayon and Dacron did at DuPont a generation earlier, reduce the perceived risks and keep it going. Brody also might have found a way to proceed with a vastly scaled-down effort—possibly to the level of a skunk works—one that would not have consumed any significant resources in a way that could still keep the project afloat and continue to make progress until achieving a significant breakthrough that would rouse the company to expand its support and team. In fact, if he could have kept the project alive in some way for a few years more—as Bernie Meyerson was to do in the case of the silicon–germanium chip—he would have been able to take advantage of very important scientific work that came out of Scotland and that would have provided the key to large-screen LCDs or had time to construct a new, promising process he had in mind (and that he would soon attempt to deploy in his new start-up). But at this point Brody's work appeared to be going nowhere, and the chances for success, despite all the work he and Westinghouse had put into the project and in the face of the growing demands for resources he was making on the company, seemed more remote than ever. In other words, the risks seemed too great for Westinghouse to proceed any longer. But we cannot just chalk up this failure to bad luck or the organizational rigidities of a large, hidebound corporation strapped for resources. It has as much to do with the limitations of Brody as champion for his cause, limitations that would continue to haunt Brody in his future ventures.

12.2.2   Europe: New Forms of Silicon and TFTs

As was the case with small LC displays, European efforts were scientific in nature and conducted mainly within universities.[5] They played an important role nevertheless in the eventual commercialization of large, flat-panel LCDs. An important focus of research was the investigation into new types of materials to improve the performance and stability of TFTs for use in solar cells as well as LC displays. The central question behind this work was whether the materials used by Brody and others needed rethinking.

In the scientific community, attention began to shift from compounds such as cadmium selenide and back to silicon. Perhaps, it was thought, a form of silicon exists that might be the perfect material for placing on a substrate. After all, silicon had proven its worth in semiconductor chips—it was cheap, readily processed, and reasonably fast—maybe it would also be the key to the TFT. Clearly, the type of silicon that went into microchips could not be used: it is a single crystal and, as mentioned, cannot be easily deposited onto a substrate in vapor form. But two other types of silicon exist: polycrystalline and amorphous. In contrast to single-crystal silicon, which is one single continuous and unbroken crystal, polycrystalline silicon is made up of a number of distinct crystals dispersed throughout the material. Amorphous silicon is different from these two in that it is noncrystalline, which means that its silicon atoms are unordered and distributed randomly (it is generally designated by the expression "α-Si"). Experiments demonstrated that amorphous silicon showed the most promise as the semiconductor material for TFTs. It could be processed at low-enough temperatures to deposit a uniform, thin film on a large glass surface. TFTs made of this material were more stable than any TFTs made up to that time. The sticking point was low electron mobility meaning TFTs made of α-Si were very slow to respond to signals, a problem that limited the material's usefulness in large-screen displays. It was therefore imperative that some way be found to increase the electron activity of amorphous silicon. The breakthrough came in 1980, shortly after Westinghouse shut down its LC display project. Two scientists at the University of Dundee in Scotland—Peter Le Comber and Walter Spear—discovered that electron mobility in α-Si could be significantly increased by hydrogenating (adding hydrogen to) the material. This substance—represented by scientists as "α-Si:H"—turned out to be the key to the economical mass production of active-matrix (large-area) LC displays. LeComber and Spear built prototype displays in the laboratory and were able to demonstrate that amorphous silicon TFTs could be used in 1000-line LCDs, which was sufficient detail to build a TV. As scientists, however, and not being particularly entrepreneurial by nature, they had little interest in attempting business ventures to commercially exploit their discovery.

12.3   LARGE LCDs: DEVELOPMENT, SCALE-UP, AND COMMERCIALIZATION

Once scientists pinpointed the right material for making stable TFTs and an appropriate addressing system (active matrix) that incorporated these devices, the stage was set for the development, scaling and commercialization of large LCDs for TVs and computer screens. Once again, the American corporation gained little traction in this field. A number of large, established American companies worked on flat-panel displays through the 1970s and some continuing their research and development activities into the early 1980s. Most of these companies—including Hughes, Exxon, AT&T Bell Labs, Texas Instruments, GTE, National Cash Register, IBM, Raytheon, Control Data, General Electric, and Owens-Illinois—focused on plasma, electroluminescent, or other types of display technology. Only Westinghouse and RCA actively explored the one method that would eventually dominate the field, active-matrix LCDs. None of these companies picked up on the important work done by LeComber and Spear. With little progress made, they abandoned the field, leaving it open to the smaller upstarts.

12.3.1 Large LC Display Start-Up and Spin-Off Ventures in the United States

With the ending of work on displays by RCA and then Westinghouse, the only American companies investigating active-matrix LCD techniques were start-up outfits and smaller companies. A total of 15 such firms in the United States pursued large flat-panel display technology in the 1970s and 1980s. The vast majority of these were either closed down or sold off (with a few managing only small-scale production of displays for specialized markets). The two spin-offs of Westinghouse—Panelvision and Magnascreen—were the only ones to concentrate on active-matrix LCDs. Panelvision, founded by Brody, was launched on November 1980. Brody located it in a suburb of Pittsburgh, Pennsylvania, near the Westinghouse R&D Center. He secured venture capital funding to the tune of $4 million to develop and commercialize active-matrix LCD products. Because of Brody's poor track record as a manager, his backers demanded that the company be controlled by seasoned executives.[6]

The company faced difficulties from the start. Friction between Brody and the new management team caused disruptions and delays. The location of the firm in Pittsburgh, rather than in a more technologically dynamic region as Silicon Valley, limited the company's ability to take advantage of a growing venture capital market, talented scientists and engineers, expert equipment suppliers, and the newest ideas and technologies in semiconductor technology. But most importantly, Brody's manufacturing process presented a host of problems. At the center of the trouble was Brody's specially designed machine that, by using a sequence of "shadow masks" to deposit vapors in a desired pattern directly on the substrate, avoided having to employ the difficult-to-master photolithographic technique relied on by the conventional semiconductor industry. However, it turned out that the machine, while operating splendidly at less-than-rigorous resolution of 30 lines per inch, failed badly when pushed to greater than the 50-line-per-inch limit that would be demanded in a commercial display. The result was, " ... the equipment could not produce what Panelvision needed, and ... a [radically different] process was required. The shadow-mask problem forced the company back into development mode, delaying the project [for a number of] years."[7]

By convincing potential investors that the process would be relatively simple to fix, Brody and his managers were able to obtain significant additional funding over the next few years. This influx of capital allowed Brody to improve the process enough so that, by 1984, Panelvision became the first company to market active-matrix display screens. But they had only a few customers—80 in all—operating in 12 industries. With a troubled process and nowhere close to achieving economies of scale production, the company had difficulty breaking even. Panelvision attempted to create a practical manufacturing process that could provide volume production. But it was now too late. By this time, the Japanese had entered the United States market with their low-cost, high-quality displays. Potential investors, realizing Panelvision had little chance of catching up with the Japanese, backed down. In 1985, Panelvision was sold to Litton Industries, which used Brody's technology to make display products for its own avionics systems. As for Brody, he started his second company—Magnascreen—in another attempt at making a go of his technology. A similar story can be written for Magnascreen: managerial friction, foreign competition, and, most damaging of all, an intractable and expensive technology that made it, once again, impossible to " ... move active-matrix technology across the divide from R&D to large-scale production."[8] Brody began Magnavision in 1988 and resigned as CEO less than 2 years later. Like Panelvision, the company ended up concentrating on small, very specialized markets for flat-panel displays. As for Brody, he went on to form a consulting firm and come up with more inventions, all related to TFTs and active-matrix displays. Like James Fergason, Brody was recognized with many awards for his contributions to LCD technology. This does not change the fact that, just as they had done with small display technology, it was the Japanese who ultimately won the LCD flat-panel wars.

12.3.2   Japan Enters into Large LC Displays

By the early 1980s, no US or European firm, large or small, was close to controlling flat-panel display technology. While discovery of hydrogenated amorphous silicon as an important material for making TFTs was now fairly well known, no company would (Westinghouse) or could (Panelvision) devote the resources needed to turn this scientific insight into a commercial product. As the case with small LCDs, this left a vacuum—and therefore an opportunity—that the Japanese and South Koreans were soon to exploit. As with small LC displays for watches and pocket calculators, the Japanese were the first to commercialize large LCDs for TVs and computers. Once again they had the great advantage of not having to spend time and money on basic research. This had already been done by the United States and Europe; Japanese companies borrowed, adapted, and synthesized this knowledge as they saw fit.

12.3.2.1   Flat-Panel (Hang-on-the-Wall) TVs   The first important thrust into flat-panel TVs took place in the early 1980s with Sony's introduction of the handheld TV set.[9] However, these devices employed very thin cathode ray tube (CRT) technology rather than LC design. While Sony succeeded in commercializing this product, these products did not work very well at all. While the picture was clear, it consumed far too much power to be practical—the batteries had to be replaced every few hours. Then, in late 1982, Seiko extended its LCD watch capability to introduce the first TV wristwatch. The monochrome display was one of the first commercial applications of an active-matrix design. Soon other Japanese companies, including Casio and Citizen, came out with their own version of either a TV watch or portable LCD TV.

The First Hurdle: The 14-Inch LC Display While the market for TV watches peaked early and then fell sharply, the achievement itself—that such products could be made at all—sharpened Japanese interest in attempting larger flat-panel TVs. Japanese companies, especially the large electronics outfits, took TFTs seriously and used amorphous and hydrogenated silicon in making them. In October 1983, Sanyo Electric Company demonstrated a 3-inch color LCD TV using α-Si TFTs on an active-matrix structure. This success, and the growing market for portable LCD TVs, inspired other companies in Japan (Fujitsu, Hitachi), Europe (Philips), and the United States (General Electric) to seriously investigate flat-panel TVs. However, it was Sharp Corporation that made the biggest advances in the field and was the most aggressive of the competing firms to develop an active-matrix/TFT display for TVs.

In the early 1980s, Sharp had been concentrating on developing flat-panel TVs using tellurium TFTs with mixed results. It had extensive experience with TFTs as applied to solar cells. Following a series of lectures given by Walter Spear during a visit to Japan, the company decided to abandon tellurium in favor of α-Si TFTs. Here is yet another example of an important piece of external information galvanizing a company's advanced materials program. Sharp created a LC Division and appointed Isamu Washizuka—physicist, inventor, and executive—as general manager. Washizuka pushed the technology very hard throughout the organization, pointing to the great promise of α-Si TFT and active-matrix technology. He set a very ambitious goal for his team: to accelerate development to make working 14-inch displays using active-matrix design powered by TFTs made from amorphous silicon. Washizuka, now champion for the technology, believed this goal could be reached for a number of reasons. Sharp already succeeded in commercializing LC displays years earlier for its line of portable electronic calculators, and it now controlled the most advanced LCD TV display technology. Surely, Washizuka thought, jumping to a 14-inch display would be merely a cut-and-dry extrapolation of existing technology. Also, because of his influence in Sharp, he could readily tap the resources needed. But as it turned out, making this leap was no mean accomplishment. Washizuka found out that the process Sharp used to make the small TV displays had to be completely revamped to obtain a working 14-inch model. For example, Sharp's proprietary process using multiple masking steps had to be significantly reworked to be applicable to the larger screen. The effort consumed far more resources and man-hours than originally anticipated. Sharp finally announced such a display in early 1988. This was an important event since a 14-inch display is the smallest that can be used in a TV receiver. This accomplishment immediately turned Sharp into the technological leader in the field of large-display TVs.

Toward the 40-Inch LC Display and the Flat-Screen TV Once the important 14-inch prototype had been built and operated, Sharp faced the formidable task of enlarging and expanding the capabilities of that prototype to match shifting market demands. This market instability—the fact that different displays had to be designed and built for different types of TVs—significantly increased the risks of development, especially in the face of a highly inflexible process. In short, new displays for different applications would not come easily or cheaply.

So now the question of how to proceed came to a head, not only for Sharp but also for Japan as a country. From the tortured experience in increasing the size of LCDs to 14 inches, Washizuka understood the difficult road ahead in pursuing even larger LC displays for mass produced TVs. Displays had to be in color and made in volume amounts at low cost. Unlike the case of the silicon gate process and the microprocessor, there was no one common process technology that could easily create new and more complex—that is, ever larger—displays and do so on a mass production basis. Each new step demanded different and new innovations in product design and process technology. There would be no low-hanging fruit to enjoy with this project. Each advance forward—each attempt to extend the size of the display—would necessarily involve a whole new universe of problems requiring extensive modification of the previous method for designing and manufacturing LCDs. As the size of the display grew beyond 14 inches, for example, the basic number and sequence of the masking process had to be radically modified, and the architecture of the active-matrix display had to be drastically reworked. Reaching the goal of LCDs for TVs would be a slow and costly slog up a very steep and treacherous learning curve. The risks would be great, in fact too great for one company, even one as large and creative as Sharp.

Japanese companies faced this daunting risk in ways US companies did not. They made it a question of national pride, for here was the one area in advanced technology that Japan knew it could dominate. Government played a key role. It had to because the effort required talent and resources that exceeded one company. The Japanese government now became the champion of the project. It brought together and coordinated a large-scale government-industry effort to take on this task and so spread the risks involved among the different players. In 1988, a government-sponsored consortium was organized by MITI; it involved 12 major companies all for one purpose: to develop a 40-inch diagonal display.

Since the companies worked together, there was no rush by anyone of them to beat competitors to the market nor were other companies in other countries breathing down Japan's neck in pursuit. Time was not of the utmost essence in this regard. MITI played the puppet master in this outing. The tasks, objectives, and expected difficulties were described in a September 1988 report from MITI entitled *Development of Fundamental Technologies of Giant Electronics Devices*. MITI made sure tasks were divided up between the companies according to what each did best. Sharp, the first among equals in this enterprise—along with Hitachi, NEC, Casio, Seiko, Epson, and Sanyo—concentrated on the electronics, Toppan Printing and Dai Nippon Printing on developing printing techniques for the color filters, Asahi Glass on making available the large glass plates, Chisso Chemical on supplying the LC material, Nihon Synthetic Rubber on furnishing other specialized materials, and ULVAC on designing and supplying process equipment.

MITI anticipated that the consortium would come out with its first flat-screen TV in 1995, not anticipating the enormous difficulties and costs involved in increasing the dimensions of the display and in finding an acceptable manufacturing process. Whereas MITI and the participating companies initially anticipated that the effort would cost in the tens of millions annually, they found themselves having to invest billions of dollars annually. By the mid-1990s, when project costs were clearly becoming far greater than expected and when it was clear success would not be achieved in the time expected, MITI increased funding as necessary—and pushed the participants to dig deeper into their pockets as well—to keep development and scale-up going. The project had gone past the point of no return for a number of reasons. Massive amounts of cash had already been spent and it would be very difficult for MITI to explain how and why all that money had been wasted. National pride was at stake as well. Admitting defeat in commercializing the one major technology that the Japanese could call their own would bring shame—loss of "face"—on the participants and the nation as a whole. In a world where guanxi reigned and saving face was paramount, failure was simply not an option. The onset of what has been called the Japanese recession also played a role. The country needed to control this promising technology in order to revitalize some of its largest and most important companies. Moreover, its response to chronic deflation and low growth was to stimulate the economy by throwing money into it. Within the context of this policy, the government could do nothing else; it had to spend increasing sums on the flat-panel TV project, a technology that, once perfected, was sure to bring in much needed capital into the country.

The materials, components, and processes developed by these companies moved development and production forward. By the turn of the 21st century, in one of their greatest feats of innovation, the Japanese succeeded in their quest for a flat-screen TV. The same cultural component that pushed small LCD displays in Japan and that offered a friendly, accepting environment for the technology during and after scale-up applies here as well. Predictably, it did not take long for companies in South Korea and Taiwan to gain knowledge of Japan's efforts, often by the transfer of know-how and information by people traveling between the two countries. These newcomers learned quickly. Through the 1990s, a contest existed between Japan and South Korea for the production of the largest TV display. After 2000, the center of battle moved to South Korea as Samsung and LG Philips (a joint venture between Samsung and Europe's Philips Corporation) fought it out for control of the international LCD TV market.[10]

12.3.2.2 Computer Displays: Joint US–Japanese Cooperation

12.3.2.2 Computer Displays: Joint US–Japanese Cooperation A number of electronic companies in the 1980s savored the opportunity of exploiting the rapidly growing personal computer market by designing advanced flat-panel displays.[11] This was especially true in Japan where the premium placed on office space was particularly keen and the additional desk area that would result from the substitution of the large, bulky CRT monitor by a smaller, thinner flat-panel display would be eagerly embraced. Firms such as Sharp, Kyocera, Matsushita, Epson, Hitachi, and NEC worked intensely on advanced PC displays by modifying the twisted nematic (TN) and super twisted nematic (STN) LCD design. But supremacy in this area required nothing less than a major joint effort by two large companies centered in Japan and the United States.

IBM and LCDs for Personal Computers: Research Phase   IBM clearly had a strong interest in PCs—it came out with the first successful personal computer—and thus in the LCDs that went into them. The company needed a secure source of supply of LCDs that could operate effectively on their new personal computers. IBM in fact had been conducting research into LCDs since the 1960s. Like RCA's Heilmeier, IBM had a champion of LCDs—Webster E. Howard, a Harvard Ph.D.—who was able to keep Big Blue interested in the technology. While Heilmeier left the project before we can know for sure how effective in the long run he would have been in pushing LCDs at RCA, Howard stuck it out and found creative ways to keep LCD research going even in the face of a skeptical management. Also, in sharp contrast to Brody at Westinghouse, Howard had an instinct for managing people and understood that political and even psychological maneuvers were essential in pushing a difficult and resource-intensive venture through an established and hidebound organization.

Howard joined IBM and its Thomas J. Watson Research Center in 1961. Webster's research interests included thin-film technology and epitaxy, especially as applied to semiconductor superlattices. In his first 12 years at IBM, Howard conducted research and invented semiconductor devices involving the use of epitaxial systems to layer extremely thin films of different semiconductor materials under high-vacuum conditions. He was, in fact, one of the first scientists to expand IBM's capabilities in this field, thus helping to set the stage for the creation of IBM's semiconductor laser and Meyerson's silicon–germanium bipolar chip a few years later. From his interest in epitaxy and thin films, Howard came to appreciate the importance of TFTs. But IBM's top management strongly resisted pursuing this technology since it did not appear to have anything to do with improving the company's major business in mainframe computers. What hurt Howard's cause even more was the fact that he couldn't prove to his immediate superiors at Watson Research that TFT's—even using the most advanced epitaxial techniques—could be made to work in any sort of reliable fashion. Performance stability remained an issue with these devices. In 1974, IBM decided to terminate the project. Nevertheless, Howard remained firmly committed to the technology and was able to keep the project alive, in part because he had a superb reputation within IBM as a scientist and inventor. He also, like Meyerson a decade later when championing the silicon–germanium chip, persuaded management that he could continue his research even in the most stripped-down state. IBM allowed him to continue to work on it in his spare time and using a technical assistant when the latter was not otherwise engaged with more pressing responsibilities. In allowing Howard to continue, IBM managed to avoid alienating a respected member of its staff and without committing serious resources in the process. The vague possibility that something may actually come of the research directly or otherwise must also have been an incentive.

IBM and LCDs for Personal Computers: Development, Scale-Up, and Commercialization   Howard now could move forward in the work and managed to keep the project going until new developments occurred—either within or outside of IBM—that would force the company to reassess the importance of TFTs. That break came in 1979 with the publishing of the paper from Dundee University on TFTs using hydrogenated amorphous silicon TFTs. Here at last was a way to make stable, workable TFT displays. Howard further strengthened his hand by convincing IBM to hire outside consultants on the future of LCD technology in computers; they, as it so happened, strongly supported Howard in his belief that IBM would have a competitive advantage in terms of costs and performance if they were to adopt the active-matrix LCD technology. This positive assessment further alerted IBM that it must not wait too long since, if it didn't get into the game with their own LCD products, they would lose their customer base to competitors, including the Japanese.

Howard and his team soon crafted working prototypes. The question of scaling up the device and making it on a mass production basis had to be addressed. In doing so, a problem surfaced that threatened the entire enterprise. It became very clear to IBM management that the scaling of the active-matrix LCDs was going to be quite daunting and would consume considerable resources. Epitaxial know-how did not transfer very well to this problem. In fact, as we noted in the discussion of the work done on flat-panel TVs, there really wasn't a single process or technique that provided a blueprint—as Moore's law did within the context of the silicon gate process—that could be used to predict the trajectory of development of LCD devices from laboratory to market. In short, IBM saw a long, expensive, and uncertain road ahead before active-matrix LCDs would, if ever, their way into its commercial PCs.

At this point, Howard came up with a bold plan. He convinced IBM to bring in an outside company to form a joint partnership to share the risks of development and scale-up. IBM approached Toshiba Corporation, which had had extensive experience in the development of LCDs, as a possible partner. That joint venture, called Digital Technology Incorporated (DTI), involved a 50–50 share in ownership and the construction of a $110 million manufacturing plant near Tokyo. IBM was correct to assume it would take a number of years before manufacturing could begin. By partnering with a large Japanese company that lead the field in LCDs, IBM felt confident it was mitigating three risks at the same time: sharing the costs, decreasing time to commercialization, and preventing the Japanese from beating them to the market. IBM's expansionist culture, already mentioned in our discussion of the silicon–germanium chip, clearly played a large role in the company's willingness to commercialize a foreign joint venture operation. But IBM also operated in a fiercely dynamic culture, and the computer LCD represented to the company a perfectly vibrant technology worthy of its effort and continued support. Much more than LCDs for watches and electronic calculators, and even more than large LCDs for flat-screen TVs, the LCDs for computer-powered devices are dynamic in the sense that they have had to continually improve in order to keep up with the ever-growing power and decreasing size demands of PCs, laptops, iPads, and other mobile devices.

DTI produced its first active-matrix (TFT) LC displays in 1991. This was a landmark event for it was the first large-scale, cost-effective manufacture of active-matrix color LCDs for portable computers. By the late 1990s, both companies got what they wanted: a secure supply of advanced LC displays for their computers. Their joint achievement compelled other computer companies, especially in the United States, to secure their own LC displays from Asian companies:

> The fact that IBM and Toshiba had insured themselves an exclusive supply of color TFT-LCDs prompted other user firms—Apple, Compaq, Zenith—to develop their own sources of high-quality color flat panel displays to remain competitive and they looked toward Japan and Korea to supply these vital components . . . . The impact of this [building of new plants] on the LCD industry was enormous and by the end of the [1990s] millions of portable PCs with color active matrix LCDs were being sold annually.[12]

Ironically, this heightened competition in computer LCDs—along with diverging market goals between IBM and Toshiba—led to the dissolving of DTI by the two parent companies in 2001. So, in this case, as with Union Carbide's Unipol process, great technological success did not necessarily lead to long-term competitive advantage for the innovators.[13] Still, DTI did bring into being a new and vibrant industry. By the early 2000s, active-matrix LCDs began to infiltrate a new type of PC, the notebook. This type of LCD technology made possible thinner portables requiring little power and offering a high-quality image that exceeded that of even the CRT desktop monitor. In time, the manufacturing costs—and selling price—of active-matrix LCDs declined to the point that, by the second decade of the 21st century, they have become the dominant PC display technology in the world.

REFERENCES

1. The following is based on a number of sources including published accounts by a central player in large screen LCDs: Brody, T.P. (1984), "The Thin Film Transistor—A Late Flowering Bloom" *IEEE Transactions on Electron Devices, ED-31*(11): 1614–1628; Brody, T.P. (1996), "The Birth and Early Childhood of Active Matrix—A Personal Memoir" *Journal of the Society for Information Display (SID), 4*(3): 113–127. Another important source is Florida, R. and Browdy, D. (1991), "The Invention That Got Away" *The Journal of Technology Transfer, 16*(3): 19–28. See also , Johnstone, B. (1999) *We Were Burning: Japanese Entrepreneurs and the Forging of the Electronic Age*, New York, New York: Basic Books, pp. 122–126; Kawamoto, H. (2002), "The History of Liquid Crystal Displays" *Proceedings of the IEEE, 90*(4), pp. 493–494; and Castellano, J. (2005), *Liquid Gold: The Story of Liquid Crystal Displays and the Creation of an Industry*, Singapore: World Scientific Publishing Co., pp. 175–178.
2. Brody (1984), pp. 1616–1619; Brody (1996), pp. 114–117.
3. Florida and Browdy (1991), p. 21.
4. Ibid., p. 22.
5. Johnstone, pp. 146–148; 162–163; Kawamoto, p. 494; Castellano, pp. 178–180.
6. Florida and Browdy (1991), p. 23.
7. Ibid., p. 24.
8. Ibid., pp. 25–26.
9. The following section is taken from Johnstone, pp. 126–144; Kawamoto, pp. 494–495; and Castellano, pp. 196–202.

10. Castellano, pp. 231–236.

11. Ibid., pp. 205–220.

12. Ibid., p. 220.

13. Datta, A. (2004), "Display Technologies Incorporated (DTI): What Went Wrong?" Illinois State University College of Business: Case Study (Unpublished), pp. 1–8. http://papers.ssrn.com/sol3/papers.cfm?abstract_id=733523. Accessed November 25, 2015. See also Bennett, A. (July 3, 2001), "Toshiba, IBM to Break Up LCD Joint Venture," *ITWorld*. http://www.itworld.com/article/2796079/hardware/toshiba--ibm-to-break-up-lcd-joint-venture.html. Accessed November 25, 2015.

13

NANOMATERIALS

The Promise and the Challenge

During experiments aimed at understanding the mechanisms by which long-chain carbon molecules are formed in interstellar space ... graphite has been vaporized by laser irradiation, producing a remarkably stable cluster consisting of 60 carbon atoms.
Richard Smalley, 1985

Since the latter part of the 20th century, nanomaterials have received an enormous amount of attention, not only by scientists but also by the industry and the press, and no wonder. The economic potential of these materials is clearly enormous, for they have possible wide-ranging application in manufacturing, electronics, structural engineering, electrical technology, biotechnology, and pharmaceuticals. These materials have been held up as the exemplar of what is possible in advanced materials research and development and as one of the—if not—most important new technologies of the present century. As a result, the last couple of decades have seen a significant increase in the amount of money that venture capitalists, high-technology companies, universities, and governments have committed in pursuing these elusive products.

13.1 BACKGROUND

The following sections review the background first of nanomaterials in general and then nanotubes in particular.

Advanced Materials Innovation: Managing Global Technology in the 21st century, First Edition.
Sanford L. Moskowitz.
© 2016 John Wiley & Sons, Inc. Published 2016 by John Wiley & Sons, Inc.

13.1.1   Nanomaterials

There are in fact many different types of nanomaterials including metals, ceramics, polymers, and composites.[1] Some nanomaterials exist in the natural world (e.g., natural colloids in the form of milk and blood); some are man-made (nanotubes, quantum dots). A nanomaterial refers to a substance composed of particles that are extremely small, that is, no bigger than a few atoms in length. Nanomaterials are distinguished from "bulk" matter, such as polymers, by the fact that the latter possess properties that are independent of the size of the sample used. This is not the case with nanomaterials: as the sample size of the nanomaterial decreases beyond a certain point, the laws of quantum mechanics kick in which significantly alter the chemical, physical, and electrical properties of the specimen. A nanomaterial thus lies somewhere between a bulk substance and the most fundamental of units, atoms, and molecules. Another unique characteristic of nanomaterials is their special behavior at the surface and on interfaces (such as at the boundary between two contiguous crystals). Bulk materials have only very small percentage of their atoms at the surface or on the boundary between two materials; the vast majority of the atoms lies in the material's interior. On the other hand, a very large percentage—up to 50%—of the atoms of nanomaterials lie at or near surfaces and interfaces. This congregation of atoms results in the material having very unusual mechanical, electronic, and chemical properties. Because of these effects, one of the most important types of nanocomposite materials, nanoparticles of clay interspersed in polymers, shows remarkable strength and superior thermal properties.

Despite the long history of nanomaterials, many believe that the widespread scientific and commercial interest in this form of matter really began in 1959 when the Nobel Laureate physicist Richard Feynman, in his now-famous lecture at a meeting of the American Physical Society entitled "There's Plenty of Room at the Bottom," riveted the public's attention on the concept of nanomaterials and the real possibility of scientists and engineers creating new materials by manipulating individual atoms.[2] Feynman turned out to be quite prescient, and through the 1960s, 1970s, and 1980s, the synthesis of nanomaterials gathered momentum with the development of new and more sophisticated scientific instruments, notably the ultramicroscope, electron microscope, and scanning tunnel microscope. From this research, scientists have learned that there are two main ways in which nanomaterials can be made. In the "bottom-up" approach, the nanomaterial is constructed by assembling individual atoms or molecules—from the "ground up, as it were—into nanostructured matter. This can be accomplished in a number of ways, such as subjecting a solid possessing the desired atoms and molecules to combustion, laser action, or other "dramatic" event that loosens and releases the atoms and molecules that reside at or near the surface and collecting these fugitive species onto a substrate of some kind. In this process, the nanomaterial is built up in layers, one on top of another. The second approach is known as the "top-down" method. Here, one starts with a bulk material and, through such processes as etching and lithography, "whittles" it down to nanosize. An example of this method is the miniaturization of the silicon chip to near-atomic dimensions.

Despite all this study and attention given to nanomaterials in the late 20th and 21st centuries they still have yet to find great traction as commercial products. Prices for many of these materials, though they have come down over the years, remain high compared to established technology. The humble nanoclays—used for packaging of food, bottling of beer and cold drinks, in wire and cable applications, and in composite form in components of the automobile—are today the most important nanomaterials in terms of quantities sold. The far more famous—and hyped—nanotubes have so far found only limited markets in composites for aircraft and automobiles, as active materials in semiconductor chips, in biomedical application, and to shield rooms from electromagnetic (EM) and radio-frequency radiation. Nevertheless, over the next few decades, many analysts believe that nanomaterials, in general, and nanotubes, in particular, may finally begin to impact the global health care, energy, and electronics sectors.

While it is difficult to pin down the exact size of the global nanomaterials industry, one study tells us that it reached $1.7 billion in 2010 with an average annual growth rate of 10.4% from 2005 to 2010. There are predictions that by 2017, the nanomaterial market will go to around $6 billion and well over $10 billion by 2020. Geographically, there are a few surprises. In 2010, North America produced and sold 38% of the world's nanomaterials while Europe kept up with 37%. Asia accounted for less than 25% of the world's total, although this percentage is on the rise because of the increasing demand in countries like China, Taiwan, and South Korea.[3]

13.1.2 Nanotubes

For years, nanotubes have been seen as the crown jewel of the nanomaterials.[4] Nanotubes are embedded in the most advanced science—its discoverer won the Nobel Prize—and year after year it appears to be right on the brink of a commercial explosion. There is good reason to devote so much scientific effort and public attention to them: carbon nanotubes have been shown to have highly unique properties—its stiffness and strength greater than any other material, for example—as well as extraordinary electronic properties. Carbon nanotubes are reported to be thermally stable up to 2800°C, to have a capacity to carry an electric current a thousand times better than copper wires, and to have twice the thermal conductivity of diamond. Carbon nanotubes are used as reinforcing particles in nanocomposites. They could be the basis for a new era of electronic devices smaller and more powerful than any previously envisioned. Nanocomputers based on carbon nanotubes have already been demonstrated. The markets that are supposed to open up for nanotubes at any time include their use in composite materials for automotive, construction, and other structural applications; in such optoelectronic applications as sensing, monitoring, and automated control systems; in semiconductors and fiber optics; in the biomedical arena, especially as a material within advanced diagnostic systems; and in the energy field, such as in advanced battery technology and next-generation solar energy systems. While they have yet to achieve a mass market presence, nanotubes remain an extremely exciting—if still uncertain—new material.

Nanotubes are made up entirely of carbon atoms; they are essentially small, rolled up cylinders of the well-known material graphite. Nanotubes come in two forms: multiwalled and single walled. A multiwalled nanotube (MWNT) contains a number of concentric tubes, whereas a single-walled nanotube (SWNT) is essentially a single cylindrical tube. The difference between the two in a practical sense is in their level of purity. MWNTs tend to be significantly less pure than their single-walled counterpart. In other words, they tend to have imperfections—such as missing electrons—over their many surfaces that diminish their usefulness. SWNTs have a greater range of potential applications in high-value products and systems. They are superior materials and considered the "big game" of the nanomaterial jungle. Because of their unique structure, their optical and electronic properties can be tailored by controlling the diameter of the tubes and the orientation of the carbon atoms embedded within the walls. Through such manipulation, SWNTs can be made to order for use as additives in conductive polymers, structural composites, optoelectronic sensors and displays, energy storage systems, and a variety of "printed" electronic components. They can also be attached end-to-end to make nanofibers with extremely small diameters with a host of promising uses.

However, and here is the rub, they are also much more difficult—and far more expensive—to make than MWNTs. So, as of 2017, whatever markets exist for carbon nanotubes generally involves the multiwalled variety. An important application here is as an additive to impart superior properties to certain composite materials for making specialized parts and components for sports equipment, automotive parts, and adhesives. Some close watchers of the material estimate that the global market for carbon nanotubes approached $1 billion in 2012 and is expected to reach somewhere between $2 and $3 billion by 2020, with SWNTs overtaking the multiwalled material as the dominant type of nanotube.[4] Other experts in the field are not so sure, believing that nanotubes are not realizing their once great promise and that other new materials—notably graphene—are likely to become the next big thing in advanced materials.

Research, some of it very promising, continues to be carried out by universities and industrial R&D organizations to improve the production of, and increase the range of applications for, SWNTs. Indeed, the many and various research projects focused on SWNTs are impressive, if not downright dizzying: extremely strong fiber and yarn; bulletproof clothing; water- and germproof nanoparticle coatings; advanced batteries for electrical cars and consumer electronics; highly efficient cells for solar panels; extremely small and powerful transistors and microchips for 21st century computers; superconducting cables for the virtually instantaneous and cheap transmission of electrical energy within power grids; high-density storage systems of hydrogen gas for advanced fuel cells that can power automobiles; advanced coatings with desired optical properties for a new generation of stealth aircraft; coatings for textiles for wearable electronics and biomonitoring applications; industrial sponges with superior absorption power for use in environmental remediation (e.g., removing oil from water); smart sensors, nanoscale motors, and electromechanical devices; transparent, electrically conductive films (to replace indium tin oxide) for high-reliability touchscreens and flexible displays; more powerful and selective catalysts for chemical synthesis and energy conversion systems; nanoscaled capsules for timed delivery of pharmaceuticals; and contrasting agents for MRI systems.

13.2 NANOTUBES: DISCOVERY AND EARLY RESEARCH

The discovery of nanotubes remains an extremely controversial issue within the scientific community.[5] SWNTs have been produced and examined by chemists and physicists in the United States, Europe, and Japan since the 1950s. However, many attribute the work of Japanese scientist Sumio Iijima of Nippon Electric Company (NEC) in the late 1980s and early 1990s as particularly important. His approach to the synthesis of nanotubes—adding transition metal catalysts to the carbon in an arc discharge—no doubt expanded the field. His research brought the subject of carbon nanotubes to the attention of nanomaterial researchers worldwide.

While important in many ways, the work of Iijima and those who followed and extended his research remained of purely scientific interest. The account of the commercialization of nanotubes, however, really begins with the discovery in 1985 at Rice University in Texas of a totally new type of carbon molecule with the unlikely name of "buckyball." This intriguing story is one of brilliant science followed by a difficult—and still unfolding—journey across the "valley of death" from a university laboratory to a discerning but unpredictable marketplace.

13.2.1 Early Research

Early research into the structure and nature of these materials hinged on the convergence of a number of fields, notably astronomy, physics, and chemistry. The eventual discovery of a new form of matter shows how important interdisciplinary cooperation is to scientific and technical breakthrough.

13.2.1.1 A Question of Space Dust In the late 1960s, astrophysicists wrestled with one of the most fundamental questions of existence: how did the universe come into being and where was it heading? Elucidation of the big bang theory and the dynamics of the expanding universe; the birth, evolution, and death of stars; and the creation and properties of black holes; these were the themes that captured their attention and were the main focus of their work. These scientists did not have much chance to test their hypotheses and theories in the laboratory; instead, they had to prove or reject their hypotheses by observing what went on in the universe itself, as measured by powerful telescopes operating on earth and—beginning in the late 1960s with "Stargazer"—in space.

But this work ignored—or at least didn't pay too much attention to—what went on in the interstellar space, that is, the region of the universe lying between the stars. Those physicists and chemists who did wonder about this wanted to know something very fundamental: what, if anything, existed in space within and beyond our solar system? One of the founders of this field of study was Charles Townes, who we previously met as the codiscoverer of the laser in the 1950s and 1960s and later a pioneer in the science of radio astronomy. In the late 1960s, Townes, who had moved from New York and Columbia to become a professor of Physics at the University of California at Berkeley, worked the famous Hat Creek Radio Telescope to scan for evidence of microwaves emanating from space. He found that the different types of microwave radiation coming from a cloudy region in the center of our galaxy were identical to radiation emitted by the molecules of such common substances as ammonia, water, formaldehyde, hydrogen cyanide, and acetylene. In other words, Townes and his team showed that the interstellar space, far from being the vacuum scientists had longed presumed, contained many different gases of varying complexity. Scientists were particularly excited about the discovery of complex organic molecules in space since these could be the very stuff that combined to form life here on earth.

Through the 1970s and 1980s, radio astronomers and astrochemists searched for other materials that might exist in interstellar space. They did this by hitting different substances with microwaves in the laboratory, measuring the radiation spectrum emitted by each of them and then scanning portions of space with their radio telescopes for similar types of spectra, thus indicating the existence of those same materials in those regions. In the 1970s, one of these investigators by the name of Harry Kroto was a young chemistry professor at the University of Sussex in England. His interest was in finding and identifying complex organic materials in space. This meant he was looking for interstellar compounds composed of many carbon atoms arranged in long chains. He succeeded in discovering one of these, a molecule made up of five carbon atoms arranged in a linear grouping, with a hydrogen atom attached to one end and a nitrogen atom on the other. This was the most complex molecule found in space up to that time. Within a few years, the work of Kroto and other astrochemists identified interstellar organics with as many as eleven carbon atoms and began to grapple with one of the burning questions of science at that time: how did these organic materials find their way into deep space?

Two scientists—an American and a German—were very interested in this question. Wolfgang Kratschmer, a solid-state physicist, studying at Max Planck Institute for Nuclear Physics in Heidelberg, and Donald Huffman, an experimental astrophysicist from the University of Arizona, met one another when Huffman spent his sabbatical in Germany in 1976. Together, they developed a theory that dust particles floated around within particular regions of space and that these particles were linked in some way to the actual creation of the planets. Since very little was known about interstellar dust—and of course it couldn't be purchased anywhere—Kratschmer and Huffman had to find a way to prepare samples of dust particles as these would have been created in space. They would then measure their properties—size, shape, structure, and spectroscopic profile—and compare these results with dust particles found flying around in space in regions or interstellar "bands" using data from telescope observation. To do this, they built a deceptively simple apparatus to make their space dust. Taking a bell jar evaporator, they enclosed graphite rods mounted on copper electrodes. When they turned on the electricity, fragments of graphite and carbon atoms would "boil off" and form a vapor. They introduced an inert gas—helium or argon—to cool the hot carbon vapor created by the heated electrode and promote the formation of soot particles. They kept the pressure low to make sure the particles remained small and did not congeal into "lumps." They then removed a layer of soot that formed on a horizontal quartz plate mounted near the electrodes and studied its ultraviolet (UV) spectrum to see how it compared to the UV spectrum known to exist in the interstellar dust bands in space.

While Kratschmer and Huffman did not succeed in finding soot particles that matched those existing in space, what they did find was very intriguing: their UV spectrum for the particles they created showed a strange and unexpected "bump" in the shape of a "camel's hump" for soot particles containing precisely 60 carbon atoms. To Huffman's thinking at least, this was a possible evidence of a totally new form of carbon. But Kratschmer wasn't so sure; he believed that they had just made soot and nothing more and that their results were simply the result of experimental errors. By 1983, they declared their experiments inconclusive and moved on to other research interests. They had in fact made one of the most important scientific discoveries of the 20th century but did not know it.

13.2.1.2 Richard Smalley, Clusters, and the "AP2" Machine

13.2.1.2 Richard Smalley, Clusters, and the "AP2" Machine   Rice University in Houston, Texas, would now become the center of the nanomaterials revolution. The leader of this movement was, by the late 1970s, the university's most prized "star" scientist. A chemist, Richard Smalley, received his doctorate from Princeton in 1973. Rice was very glad to get Smalley to join the faculty. He soon came to hold full professorships in both chemistry and physics. This dual role reflected the interdisciplinary nature of his field of interest, materials clusters. Whereas the pure physicists of the 20th century probed the internal structure of atoms and the physical forces that hold them together and the pure chemists examined the special chemical forces by which atoms combine to form molecules, such as those of the organic compounds, the study of clusters fell somewhere in between. While molecules are (generally) stable, coherent, and readily identified types of matter that are (more or less) fixed in shape—just the type of matter that chemists love—this is not the case with clusters. Clusters are defined as an agglomeration—or assembly—of atoms (and other types of entities, such as charged particles or ions). Clusters vary in how strongly they cohere, how long they stick together, and whether, in fact, they can form any sort of definite, well-defined shape for any length of time. The forces holding clusters together may be physical or chemical in nature. And there are different types of clusters. Some are composed of different types of atoms, molecules, and ions, while others are made up of many atoms of just one type, such as carbon. Moreover, scientists study clusters for different reasons: for purely theoretical interest (cluster behavior illuminates certain parts of quantum mechanical behavior), for potential practical implications (clusters of semiconductor particles may find applications in the digital revolution), or, of course, for helping to better understand the cosmos (carbon-based clusters occupy portions of interstellar space). But whatever the nature of the clusters studied, scientists in the field had to come to grips with a very important characteristic of all clusters: Since the properties and behavior of single atoms changed dramatically when you begin adding more atoms, it was impossible to predict by straight-line extrapolation how larger clusters of those atoms would behave from the behavior of smaller aggregations. There were, in other words, very bumpy—and unpredictable—discontinuous progressions as clusters grew in size. Mapping out how properties of clusters changed with size and then explaining these changes occupied cluster scientists through much of the last quarter of the 20th century.

Smalley was one of the foremost researchers in the chemistry and physics of clusters. He was no theoretician but rather the ultimate experimentalist with a knack of building large machines to study complex phenomena. He was in fact the inventor of what was considered the most powerful and sophisticated apparatus for the scientific exploration of clusters. No other machine experimental or otherwise could match Smalley's brilliant contraption, named "AP2" (it was the second such machine that Smalley had built). Cluster scientists from around the world came to Rice University to study it and, assuming they got permission from Smalley, to use it for their own research purposes. Smalley and his colleague at Rice, Bob Curl, had their own research agenda, one that promised to have both scientific and commercial prospects: semiconductor clusters. They built their apparatus with money provided by the Department of Defense in order to study the properties of silicon, germanium, and germanium arsenide clusters with the ultimate goal of creating advanced electronic devices for military use.

The AP2 was a far more complex and precise experimental machine than Huffman and Kratschmer's graphite evaporator; it could create different types of clusters and molecules that could not be made by conventional means and measure them more exactly than any other scientific instrument. The heart of it was a large stainless steel cylindrical chamber, which stood atop a vacuum pump system. The chamber was filled with the inert gas helium under high pressure. A laser fired vertically down through quartz windows on one side of the cylinder and onto a circular carbon disk within the chamber. The latter rotated so that the laser blast always struck a fresh portion of the disk to make sure that there were no differences in recorded measurements because of varying surface structures. The high-powered laser created a plasma cloud of carbon atoms, ion, and molecules that was carried down the cylinder by the highly pressurized, rapidly moving helium gas. This gaseous mixture then entered a small nozzle at the end of the cylinder. Once forced through the nozzle by back pressure, the gas expanded at the other end very quickly at supersonic velocities, thus causing extremely rapid cooling and the nearly instantaneous formation of clusters of different sizes. The question now was to find out what sorts of carbon clusters had been formed and the prevalence of each type. To do this, the cluster mixture had to be broken apart into its separate cluster types according to the number of carbon atoms and a determination made as to how many of each type of species had been created. For this, a second "excimer" laser hit the mixture causing it to break up into its various constituents of different sizes and therefore masses. At the same time, it ionized these now-isolated cluster groups, thus imparting electrical charges to them. Because of these charges, these ionized particles could be deflected into—and recorded via electrical signal by—a mass spectrometer. This instrument measures the masses of—and thus the number of carbon atoms associated with—the different clusters as they enter. The strength of a particular signal indicates the frequency of a certain cluster—the stronger the signal, the more there is of a cluster with a certain number of carbon

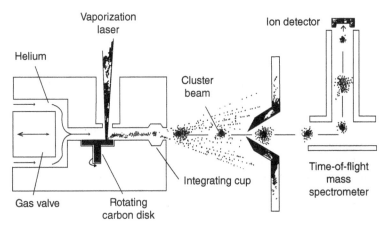

Figure 13.1   Richard Smalley's AP2 experimental machine: clusters are made by vaporizing material from a disk and analyzed in a mass spectrometer (Source: Aldersey-Williams[5]. Reproduced with permission of Wiley)

atoms. These signals were then turned into digital data and displayed on graphs on a microcomputer.

This was Smalley's machine, and he portioned out time on it to scientists around the world to conduct their own research (Figure 13.1). In the early 1980s, Harry Kroto was one of these. Kroto had earlier met Smalley's colleague, Robert Curl, at a scientific conference in Austin. Curl invited him to come to Rice to check out the wonders of Smalley's AP2. Kroto had interstellar compounds on his mind and hoped to show that specific organic compounds containing over 30 carbon atoms were floating around in outer space. The problem was chemists had never prepared these compounds in the laboratory. How then could Kroto match radiative emissions coming from space with those emanating out of his compounds in the laboratory? But Curl believed that AP2 could help Kroto by actually making these molecules possessing thirty plus carbon molecules in the form of carbon clusters. Smalley, however, was less than enthusiastic. He was not particularly interested in carbon clusters; he had his own work to do on the machine on semiconductors and didn't want a diversion from this main goal. In addition, he did not feel Kroto's work would lead to any new discoveries since carbon had been around for centuries and its forms, structures, and properties ought to have been mapped out long ago. Luckily however, a slot of time on the machine opened up in the late summer of 1985 and, at Curl's urging, Smalley decided to let Kroto conduct his research at that time.

13.2.1.3   Chance Discovery of a New Form of Matter: C_{60} and the "Buckyball"
The experiments Kroto and Curl performed on the machine in August and September of 1985 would turn out to be truly pioneering and the most important research they—and Smalley—would ever undertake. Once again, we have an example of an

outside influence coming into a project at a crucial time and energizing the research so that it enters into a very productive phase. In this case, that impetus came from researchers having nothing to do with Rice University or with Smalley's interest in semiconductors. Ironically, this landmark discovery did not directly concern any of the higher organic compounds that Kroto was attempting to study. It also had very little to do with interstellar materials of any kind. What it did involve was an unexpected measurement on AP2 that could not be explained. During Kroto's experiments, an interesting—even strange—"blip" appeared on the mass spectrometer readings. During the experiments to make complex organic compounds, AP2 created all sorts of carbon clusters from small (few carbons) to large (many carbons). But what grabbed the scientists' attention was the appearance of one very strong signal with clusters containing 60 (no more and no less) carbon atoms. At first, Smalley, Curl, and Kroto thought this C_{60} measurement was nothing more than an anomaly of their particular experiments. But even after adjusting all the experimental values they could think of—power and wavelength of the lasers used, contaminants from pump and valves, and so forth—the signal remained strong and clear. Smalley likened this impressive C_{60} peak to a flagpole on a parade ground.[6] Experiments also showed that this C_{60} signal was not only very strong but also remarkably stable. The scientists injected into the AP2 chamber other gases known to react with carbon but found that C_{60} was not a willing mate—it would not hook up with any of these active gases. If it had, that strange carbon cluster of 60 atoms would have been transformed into some other compounds, and its presence as C_{60} would have diminished—the flagpole would have disappeared. But, in fact, it refused to combine with any of the reactive gases to become something else. The "flagpole" signal remained intact through all of this.

Eventually, Smalley and his team had to conclude there really was something unique about this highly stable 60-carbon molecule (Figure 13.2). No molecule of that type was known to exist. Was this then a totally new type of substance, and, if so, what was its structure, that is, what did it look like? AP2 told Smalley and his team enough to put them on the track of its makeup and configuration. They knew that the only way a molecule of 60 carbon atoms could be so stable was if each of these atoms was comfortably attached to other carbons in the molecule. In this way, no carbons would have any desire to combine with other atoms or molecules that happened to pass its way. In chemical terms, the carbons' atoms in the molecule in question had all their valence requirements satisfied. Assuming such stability—and the experimental results allowed him to reasonably make this assumption—Smalley experimented with paper and scissors to construct the only shape that would satisfy the conditions demanded by the AP2 experiments: a closed, hollow spherical surface with carbon atoms arranged in a pattern of alternating pentagons and hexagons. Smalley immediately noted that the molecule had the shape of a soccer ball. Soon thereafter, he realized that its shape was actually that of the famous geodesic dome as conceived by the American architect, author, designer, and inventor Buckminster Fuller. In his honor, the team decided—with some resistance by Smalley who didn't particularly like the name—to call their new molecule the "Buckminsterfullerene" or "Buckyball" for short. Smalley, Curly, and Kroto soon realized that the AP2 had in fact made many sorts of closed shapes with different numbers of carbon atoms. They

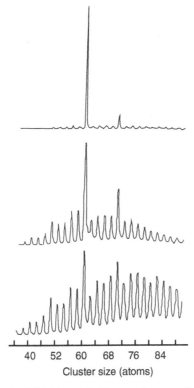

40 52 60 68 76 84
Cluster size (atoms)

Figure 13.2 Richard Smalley's highly stable 60-atom cluster peak (shown increasingly dominant by adjusting the flow of helium) (Source: Aldersey-Williams[5]. Reproduced with permission of Wiley)

termed all such molecules "fullerenes," most of which were known to science. However, they understood that the fullerene C_{60} was very special, the first among equals: a perfect geodesic sphere, which was the most stable of all of them all; its smooth symmetry meant that the carbon atoms were comfortably attached to one another and therefore had the lowest chemical activity. Its extraordinary stability pointed to it being a totally new and previously unknown form of matter.[7] By September 1985, Smalley, Curl, and Kroto felt they had enough evidence through their AP2 and mass spectrum experiments to publish their results and to declare that they had made a major discovery. The selected venue to introduce the world to their new molecule was one widely read across disciplines and even by the general public. They submitted their article to the publication *Nature* in the fall of 1985.[8]

It would be difficult to overstate the importance of this discovery to the field of advanced materials (Figure 13.3). Before C_{60}, scientists knew of only two forms of carbon: graphite (which is amorphous) and diamonds (which are crystalline in form). Smalley and his team had uncovered a third type. Even more importantly, their discovery opened up a whole new category of chemistry. For the past century

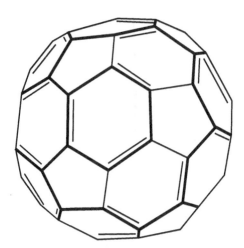

Figure 13.3   The soccer ball structure of buckminsterfullerene ("buckyball") showing alternating patterns of hexagons (6-sided figure) and pentagons (5-sided figure) (Source: Aldersey-Williams[5]. Reproduced with permission of Wiley)

and a half, there were only two types of organic molecules known: one-dimensional (such as a linear chain of carbon atoms) and two-dimensional or "aromatic" (such as closed rings of carbon as in benzene). Now Smalley had discovered a third type of molecule, spherical in shape and occupying all three dimensions. This new field of chemistry meant the birth of a completely novel type of chemical reaction and thus the possibility of creating in the laboratory a group of advanced materials never before seen by science. Soon, Smalley, excited at the prospects of his discovery, was already exploring the likelihood of using his AP2 to create new types of molecular structures by inserting metal atoms inside the C_{60} geodesic architecture.

13.3   NANOTUBES: LATER RESEARCH AND EARLY DEVELOPMENT

While the AP2 data made it likely that Smalley had discovered a new form of carbon and that his view of its structure was the correct one, he still had no direct proof of any of this. No microscope existed powerful enough to observe these assumed geodesic molecules. While other tests existed apart from AP2—such as nuclear magnetic resonance NMR—that could provide very strong supporting evidence, they were very demanding and required significant amounts of a material to work with. As far as Smalley and his colleagues knew, their AP2 machine was the only way to make C_{60} molecules, and it could only make infinitesimal amounts of it, that is, not enough for conclusive testing.

Over the next few years, Smalley and his team continued to explore their hypothetical molecule using their clustering experiments. This research increased their understanding of those factors that went into creating stability in C_{60} and tended to give support to their conjectured molecule. But without additional and airtight proof, they could not claim without some disclaimer discovery within the scientific community. Consequently, doubts continued to stalk their work and block any chance they had of capturing a Nobel Prize. The only solution was to find a way to make larger amounts of the new material. But, as the 1980s wore on, Smalley slowly drifted away from research on fullerenes, believing there was not much else to be discovered nor any new way to make greater amounts of the material. As far as he was concerned, he had done all he could do experimentally on the subject with the evidence he and his colleague amassed over the years clearly pointing toward the existence of his buckminsterfullerene. Smalley had had enough of his buckyballs and he moved on to other research interests.

13.3.1 A Small Buckyball "Factory" in Germany

But other research groups continued looking into the material and asking if buckminsterfullerenes were so stable, why are they so hard to make? One of these worked out of IBM's Almaden Research Center in San Jose, California. Here, a surface scientist, Donald Bethune, who read about fullerenes and heard Smalley speak on them, wanted to pursue the subject. Another group continuing to conduct fullerene experiments operated out of Sussex England where Harry Kroto, now professionally divorced from Smalley and Rice, pursued his own research agenda on fullerenes.

But the most important research would come from Huffman and Kratschmer's renewed work on carbon soot at the Max Plank Institute in Germany. Huffman in particular was obsessed with following through on their earlier experiments with the inverted bell jar evaporator. After word had come out of the finding of C_{60} by Smalley, Huffman recalled his own earlier experiments with Kratschmer and those unexplained "camel humps." He began thinking these could in fact be due to the same C_{60}. But was it possible that this new material could be formed from such a simple apparatus as the carbon evaporator and in such a simple process as the making of soot? His machine had no complex lasers and no intricate mechanism. Yet, if he and Kratschmer had indeed made C_{60}, this would go far to explain those strange jumps in the graph. If true, there were two very important implications: they had in fact been very close themselves earlier on finding the new molecules, and this could be a way to make measureable—even commercial—amounts of C_{60}. The prospects were too enticing not to continue on this road to see what was at the end of it.

More experiments were clearly needed on the carbon evaporator to find the exact conditions under which it would make C_{60}. This time around, however, Kratschmer and Huffman had a very useful instrument at their disposal—a scary-sounding thing called a Fourier Transform Infrared (FTIR) absorption spectroscope—that could conclusively show whether or not the carbon evaporator was indeed making C_{60} molecules. The final proof that they did have buckminsterfullerene's came when they compared the spectra obtained from two different isotopes of carbon: carbon 12 and carbon 13. If C_{60} were there, certain differences should show themselves in the infrared part of the spectra—a definite shift in the spectral lines—and indeed this is what they found, just as predicted earlier by quantum theory.

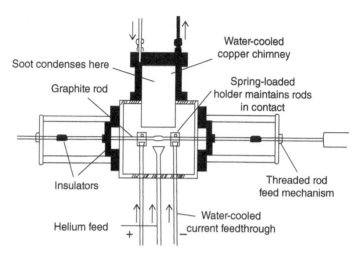

Figure 13.4   The Kratschmer–Huffman apparatus created carbon vapor when graphite rods were heated by an electric current (Source: Aldersey-Williams[5]. Reproduced with permission of Wiley)

Not only did Kratschmer and Huffman convincingly show that the buckminster-fullerene existed, they also discovered to their delight that their ridiculously simple machine was a small buckyball factory (Figure 13.4). When they raised the pressure within the evaporator above a certain point and took their FTIR measurements, they found to their surprise that they were indeed making buckminsterfullerenes—the very same C_{60} molecules claimed by Smalley—and what's more, they were making a lot of them, or at least far more than Smalley's AP2 had made. Here was a process that was simple—no lasers required—and capable of creating measureable amounts of the elusive molecule. In time, with improvements in equipment and adjustments in the operating conditions, it seemed possible that this technology could even turn out industrial quantities of the new material. No one was more excited about this possibility than Smalley himself, for when he had heard the news of the breakthrough, his " ... first response was one of delight. Here, at last, was the vindication he had sought for five years."[9] Thanks to Huffman, Smalley's buckminsterfullerene was shown to the entire world to be a real thing and could be measured and tested,

13.3.2   Smalley Reenters the Fray: An Entrepreneurial Vision

Unlike Kratschmer and Hoffman, Smalley's interest was not just scientific; he saw commercial possibilities now for his new molecule, and he very much wanted to engineer a future in which his work would have a major technological and economic impact on society. Smalley naturally enough used Rice University as the stepping stone toward this goal. As what Cyrus Mody has termed, a "star" scientist at Rice, Smalley could help shape the school's institutions to suit his goals.[10] After his landmark discovery and the increased prestige and authority that went along with it, Smalley could coax the administration to create the Rice Quantum Institute (RQI) and appoint him as director. The RQI promoted the study of nanomaterials and stressed the interdisciplinary nature of the field, both as a science and a new technology. By the early 1990s, Smalley had begun to move beyond his scientific world and to explore how nanomaterials could become a real business venture. This desire is shown in mid-1993 when Smalley testified before a Congressional committee about "how [he planned] to build on [his] success with 'buckyballs' to work toward practical, useful nanotechnology." As Mody notes, "Smalley routinely spoke of the [C_{60}] molecule's potential as, among other things, a drug delivery system."[11] To Smalley, Huffman's carbon evaporator was an extremely promising route in making his buckyballs commercial. In 1991, Smalley attempted to scale-up the apparatus to create what he called his "fullerene factory." He designed it to produce C_{60} on a semicommercial scale by using graphite rods.[12]

As Smalley worked on this technology, he realized that the QRI was not the appropriate organization around which to develop this process. Smalley decided to push Rice's administration to support another institutional venture, also multidisciplinary, but this time more geared toward engineering and technology rather than pure science and pedagogy.[13] Smalley had enormous influence over the university. Rice simply could not afford to lose their most famous faculty member to the likes of Princeton, Berkeley, and Stanford—all of whom came-a-calling on Smalley with offers to jump ship. By early 1990s, "Smalley was receiving as much in salary and research funding as Rice—or almost any other university—was paying any professor."[14] But further incentive was needed to assure Smalley's loyalty: directorship over a large, multidisciplinary, and more influential research program than QRI could provide was just the thing. The fact that Smalley helped to select the new president of Rice and that his name could attract large private donations and government grants that could be used toward building a new facility sealed the deal. This new institution, the Center for Nanoscale Science and Technology (CNST), encompassing nearly one-quarter of Rice's faculty in science and engineering, provided Smalley with the venue—and power—he wanted at Rice to pursue his ambitious program in nanomaterials.[15]

At this point, Smalley had to make a decision on the direction his new center would go in exploiting the commercial possibilities of nanomaterials. While his discovery of the buckminsterfullerene was clearly a landmark scientific accomplishment, the practical applications of these molecules, even if a way existed to make more of them than ever before, were clearly very far off. At this point, he began thinking seriously of the opportunities that might exist in the field of a closely related material, nanotube. He was of course well aware of the existence of nanotubes, and he began thinking of how close they were in structure to C_{60}. In the early 1990s, he was greatly stimulated in this thinking when reading in the scientific literature of the work on single-wall nanotubes by Iijima (at NEC in Japan) and Bethune (at IBM in Almaden). For Smalley, nanotubes were simply a type of buckminsterfullerene, which could be synthesized by adding and connecting a string of many buckyballs resulting in a highly pure carbon fiber possessing exceptional mechanical and electrical properties.[16] He believed these single-wall nanotubes—because they didn't have the electronic defects that limited the usefulness of the multiwalled variety—would replace many traditional materials in commercial applications.[17] He dreamed of very pure SWNTs with high tensile strength, one hundred times greater than steel but with only one-fifth the weight. These tubes could be combined with composite materials and made into extremely strong but light structural cables. They also have exceptional electrical conductivity and could be used in electronic devices, such as semiconductors, to increase the speed of calculations. Moreover, it soon became clear to Smalley, particularly through Iijima's work, that nanotubes—and particularly SWNTs—could be made using the same carbon evaporator that produced C_{60} buckyballs. By the mid-1990s, Smalley and his colleagues at CNST were pouring all their resources in perfecting and scaling that process. Ultimately, Smalley's goal was to form his own company around that process, selling SWNTs to other start-ups hoping to develop applications for those nanotubes and offering to help those firms to commercialize their technologies either for a fee or shares in the companies.

13.3.3   The Laser Oven Stopgap

Smalley however had a major problem: the carbon evaporator contaminated the nanotubes that were made with large amounts of useless soot.[18] To quantify the problem: each gram of SWNTs produced was covered by 100 g of soot. Separating the precious nanotubes from the carbon dust would simply cost too much as the scale of operations grew. In short, the evaporator turned out not to be a practical process. Smalley and his new Center continued to work on the problem. They revisited their old AP2 machine and thought about ways in which it might be modified to generate industrial amounts of nanotubes. In 1995, these efforts paid off with the design and development of the "laser oven" process. In this apparatus, a laser vaporizes a rod or disk made of a composite of carbon (in the form of graphite) and a transition metal (such as cobalt or nickel) placed in a heated quartz tube. The vapor is then swept down the tube, condensed, and collected downstream. Under the right operating conditions, up to 90% yield of nanotubes were obtained, most of which were the desired SWNTs. But despite attempts to improve the economics of the process, it turned out to be impractical as a commercial technology because of its enormous energy requirements, the high costs of the feedstock, and major difficulties of scaling-up the equipment. It costs several thousand dollars just to make a gram of the material. Nevertheless, the laser oven technology did at least provide enough SWNTs of sufficient purity so that they could be sold to researchers at universities and within industrial labs. These customers were willing to pay a premium to have high-quality SWNTs with which to work. To this extent, Smalley and his Center were finally in the business of nanotubes. While the operation—called "Tubes@Rice"—turned out to be a more or less break-even proposition financially for the Center, it was important in that it actually got nanotubes into the hands of researchers who could then explore possible real-world applications.

13.3.4 The "Hipco" Solution: Fluidization and Nanomaterials

In the late 1990s, Smalley and the Center continued looking for a route to a process that could compete in the advanced materials marketplace.[19] With both the carbon evaporator and laser oven processes proving to be commercial dead ends, a new approach was needed. The multidisciplinary nature of the CNST was particularly important at this point. Chemists and chemical engineers associated with the Center suggested combining two processes that they knew very well—and that we have discussed in other advanced materials contexts—high-pressure catalysis and the fluidization process. The result was a new and promising approach to commercial nanotube production called the high-pressure carbon monoxide (Hipco) process, which, under the right conditions, could turn out high-quality SWNTs at a higher yield and faster rate than any other existing technology. Hipco is a gas-phase process that makes SWNTs by intermixing carbon monoxide gas through a Ziegler–Natta catalyst under high pressure and high temperature using fluidization techniques. By the late 1990s, Smalley and CNST had taken the new process to the point of producing "high-quality [SWNTs at] high yield and high rate."[20] Never before anywhere has any process turned out more SWNTs of such purity on such a regular basis. To Smalley and his backers, the age of the nanotube appeared to be at hand.

13.4 NANOTUBES: LATER DEVELOPMENT AND SCALE-UP

The time had come for Smalley to begin thinking of moving the new technology out of the university environment and into a for-profit start-up. The first steps in this process were to transfer patent rights to the new firm.

13.4.1 Technology Transfer: From Rice University to Carbon Nanotechnologies Inc.

While Smalley and Rice held the critical patents for Hipco, the university did not yet have an official technology transfer mechanism setup.[21] The transferring of intellectual property from a university to a company was still a relatively new idea, and Rice itself had not yet been called upon to undertake many such transfers as levels of patenting within the institution were still quite low. Once again, it was Smalley who, employing his clout in the university and reminding administration that patent transfer capabilities would attract other star faculty, proactively forced the issue. At Smalley's urging, Rice brought in consultants to advise on establishing and operating a patent transfer office. Smalley himself accounted for a significant portion of the university's total patenting activity. As a result, Rice agreed to foot the bill for the services of a law firm with special expertise in the patenting of advanced materials research.

Smalley then engineered the formation of the start-up called Carbon Nanotechnologies Inc. (CNI) and helped formulate the licensing agreement. Rice agreed to license CNI with regard to "Single-Wall Carbon Nanotube Production and Application Technologies." In effect, this meant the Hipco process. The contract also outlined important milestones that had to be met. Within the first year, for example, CNI had to demonstrate definite and measureable success, including raising $1 million and continuing to develop the HiPco process as well as be engaged in at least one product development partnership. To satisfy this last requirement, CNI soon began to look at such major chemical companies as potential product partners as Dow and DuPont.[22]

Smalley further used his institutional network built up over the years to fund the company and to find executives to staff it. One of the most important contacts was Paul Howell, a donor to Rice who endowed Smalley's academic chair. According to the patent agreement documents, it appears that Smalley himself was being considered for the role as president of the new venture.[23] He however had other ideas. He understood that he was essentially a researcher with no expertise in managing a company. He was neither equipped to turn his ideas into a full-scale commercial product nor, despite his ability to attract government grants, to access the level of venture funding the new business would require. He needed to locate an experienced business guru who could do these things. It didn't take him long to find his man. Howell introduced Smalley to Bob Gower, a seasoned executive and the former CEO of Lyondell Chemical. Gower agreed to be the CEO of CNI; he also was the largest initial investor and a very useful contact to others in the chemical industry and magnet for venture capitalists who were willing to invest in a company to be run by such a knowledgeable and respected businessman.[24]

Smalley played an important, but less direct, role in the company. His great reputation and ability to attract investors in his own right clearly benefitted CNI. But he did not intrude very much on the day-to-day running of the company or even on the formulation of its long-term strategic direction. Smalley wanted to give Gower a free hand in leading the start up. Possibly more importantly, Smalley's health—he had been diagnosed with leukemia in 2002—prevented him from taking a more active part in the company's operations. By this time, Gower had become the main champion for CNI and its technology.[25]

13.4.1.1 CNI and Its Pilot Plant

The first order of business for the new company was to begin scaling-up the HiPco process.[26] Rice's CNST had built laboratory models only. Now the time had come to take the next step toward a commercial unit. The pilot plant would have to produce nanotubes at ten times the rate of the largest laboratory reactor. This meant it would have to generate 250–300 g/day of SWNTs. Gower realized that CNI could not do this alone. He also wanted to reduce the financial and technical risks to the company (which would further appeal to future funding). In 2001, the fledgling company contracted with Kellogg Brown & Root (KBR), at the time a subsidiary of the oil equipment company Halliburton. This partnership made good sense. KBR had worked with many chemical companies designing and building their plants; it had particular expertise in high-pressure and fluid catalytic processes, the perfect combination for expanding the HiPco technology. Moreover, both firms were located in Houston, thus easing day-to-day communication between executives and engineers. CNI soon leased office and laboratory space at KBR's technology center in Houston. Around this time (April 2001), CNI received an additional $15 million funding from chemical industry entrepreneurs Gordon Cain and William McMinn to create a pilot plant and scale-up to precommercial stage. The plan was for CNI–KBR to expand the pilot plant to 1 kg (1000 g) per day capacity by mid-2002, scale-up for full commercial production by mid-2003, and begin forming partnerships with other firms to commercialize and sell nanotubes for flat-panel displays and EM shielding as the nearest term applications.

The pilot plant was a larger version of Rice's HiPco fluidization setup. It consisted of a cylindrical, high-pressure reactor attached to a high-temperature furnace and gas flow controls. Carbon monoxide gas was passed through the catalyst, which was kept in a fluid, turbulent state. The SWNTs were produced in powdered form. High pressures had to be maintained in the cylindrical reactor in order for the reaction to take place. Fluid conditions assured that adequate mixing took place, heat buildup through the catalyst system was readily controlled, and fluid flowed at appropriate and regular speeds throughout the reactor.

Given the history of fluid catalytic systems (high-performance fuels, polysilicones, etc.), one would predict a rather smooth and rapid scale-up to economic size of the HiPco process followed by growing markets as prices fell in the wake of a drastic decline in manufacturing costs. One would also anticipate a flexible process producing many different types of nanotubes for a variety of eager markets. This was certainly the hope and indeed expectation of the way the enterprise would evolve. At around this time (2002), the online trade publication Electronic Engineering Times (EE Times) trumpeted the coming of the nanotube revolution:

> Carbon nanotubes are the most promising of all the building blocks for nanotechnology … Besides being stronger than steel, carbon nanotubes exhibit many fascinating electronic properties … Companies like Carbon Nanotechnologies Inc.…..are growing nanotubes to the specifications of a diverse set of future applications … Over the next five years, carbon nanotubes will likely be integrated into hundreds of applications … [27]

Predictions by a number of experts bet on an annual growth rate of between 15 and 25% for carbon nanotubes with a global market of US\$ 2.5 billion by 2020.[28] This optimistic global forecast came not only with confidence in CNI's technology but also in the progress in manufacturing methods being pursued by a number of other firms within the United States, Europe, and Japan. An advanced form of HiPco was already in the works, which employed a new and even more powerful catalyst—composed of cobalt and molybdenum—in a fluidized reactor. The advanced materials community firmly believed the so-called CoMoCAT system would improve on HiPco by delivering higher quality SWNTs at lower cost.[29]

13.4.1.2 SWNTs and Their Problems But during that first decade of the new century, progress in the commercialization of advanced nanotubes stalled. Nearly 100 firms worldwide were making and selling nanotubes and related products. But these were the less exciting, less valuable, and more easily manufactured multiwalled variety. Companies like the American outfit Hyperion Catalysis International concentrated on making batches of ready-to-mold plastic resins containing its MWNTs for very specialized use in automotive (external body parts and fuel systems) and electronics (integrated circuits, disk drives, batteries, fuel cells) applications. Other nanotube firms also emerged including Nanocyl in Belgium and Cnano in the United States, offering MWNT products or promising to produce hundreds of tons annually of SWNTs. A few established companies, such as Showa Denko in Japan and Bayer company in Germany, also tried their hand at producing highly pure nanotubes for the commercial market, the latter developing their own fluid catalytic process to make SWNTs. By 2017, the large market for nanotubes had yet to happen. While companies continued to make MWNTs for various products, these remained quite limited due to molecular imperfections that are an integral part of the material. While SWNTs should have tapped a large demand because of its special properties, they have not done so.

CNI was clearly the flagship start-up firm for single-wall nanotubes. It was the firm started by the most famous scientist in the nanomaterials field. By all accounts, CNI had one of—if not the—strongest patent portfolios among nanotube manufacturers. In fact, by 2007, CNI appeared to be quite successful. It was shipping its nanotubes to over 700 customers globally and over the years had established hundreds of strategic relationships with numerous companies, large and small. These partnerships helped CNI share the risks of innovating new nanotube products in terms of cash, knowledge, and human resources. For example, it formed strategic arrangements with such established corporations as DuPont and Motorola as well as the small start-up Nanoink (to develop special types of ink for the photolithographic process used in semiconductor manufacturing).[30]

Up to this point, we note that many of the risks that often cause management and venture capitalists to terminate a radical technology project had been dealt with rather masterfully by Smalley. During the research and early development phases, he was alert to and ably incorporated important information generated outside of his immediate scientific circle that breathed new life into his efforts. The independent discovery by Kratschmer and Huffman of a simple way to make SWNTs reenergized Smalley's efforts to take his nanotube project to the next level and gave him leverage to obtain more funding from Rice. This effort in turn led to the creation of the CNST, which allowed Smalley to tap the knowledge and expertise of many different disciplines, including chemical engineering, which steered him in the direction of using fluid technology for his SWNT reactor. He then helped ease the managerial, financial, and technical risks associated with the transition from the world of the laboratory to development by (i) establishing a partnership with the chemical engineering firm KBR, thus giving the new firm access to its production facilities and expertise in high pressure and fluid processing, and (ii) selecting Bill Gower as head of CNI, a move that eased venture capitalist concerns over the running of the company and, through Gower's connections, linked CNI with a network of additional funding sources.

But now scaling-up the HiPco process itself proved to be problematic. One of the major difficulties was the fact that the process had to take place under high pressures, a condition, we may recall, that did not exist in the fluid catalytic technologies used in the making of fuels and the polysilicone materials. CNI found it difficult to control the process under these rigorous operating conditions. One of the issues that plagued scale-up under high pressures was separating the nanotubes from the catalyst. Doing so was necessary since the catalyst atoms contaminated and thus affected the properties of the nanotubes. Another difficulty was the unfortunate tendency of the process to produce a motley collection of hard-to-disentangle SWNTs of different types and properties. But the value of the material depended on selling one variety of nanotube in a highly pure, even pristine condition. In addition to the issue of purity was also one of selectivity, that is, obtaining only that one type of SWNT desired by a customer. Solving both these problems meant designing new and more sophisticated types of separation methods, which significantly increased production costs, thus forcing prices to remain high.[31]

But this was not all of it. The production process didn't just involve the reactor. Also important was the technology to combine and disperse the nanotubes with other materials to make advanced composites with precisely specified properties. A major market—indeed what has been called the "Holy Grail" for SWNTs—was as an additive in plastic composites. But impregnating the nanoparticles within the plastic so that they imparted the desired properties to the resin was far from an easy matter. It is important to disperse the particles uniformly throughout the polymer and at the same time obtain sufficient intimate contact between plastic and nanotubes without causing the polymer to become brittle. As a recent article on carbon nanotubes observes:

Dispersion issues are a significant factor that needs to be worked out. There is a strong need for straight, defect free, high purity tubes that disperse easily during processing ... The more expensive SWNTs offer that potential, as they are straight in structure and generally have fewer defects than MWNT ... they have excellent properties that would allow you to produce a composite with a very high modulus – as much as 10 times higher than a conventional carbon fiber composite ....At best, the existing technology could provide a 25% increase in strength of a plastic using SWNTs. But this was not sufficient to make nanotube competitive with other more familiar additives. . Unless it can boost strength somewhere between the 100% and 200% level, nanotubes cannot displace the current industry favorites, carbon and glass fibers.[32]

The difficulties encountered with synthesizing, purifying, and dispersal of SWNTs kept the cost of making and applying the material at too high a level. Whereas traditional fibers used to make polymer composites sell for as little as $0.10 per gram, SWNTs sell anywhere from $50–$2500 per gram, with the typical price tag coming in at hundreds of dollars.[33] Nanotubes simply could not displace its existing competitors in the composites business, a failing that greatly limited its market potential.

A closely related problem for nanotubes is the fact that the market for SWNTs by its very nature is highly fragmented. Even within plastics, there is a wide variety of polymers, and within each polymer family, many different uses demanding different physical characteristics (stiffness vs. electrical, conductivity vs. strength, etc.). If the processes for making nanotubes had been sufficiently flexible—along the lines of the original fluid catalytic process of earlier decades—CNI could have easily adjusted the process to create the exact type of nanotubes needed for particular polymers and their applications. Costs would have been lower to start with and would have come down further over time as CNI descended the learning curve. Although market fragmentation would have prevented the possibility of controlling one single large segment, CNI could have dominated in a number of small but lucrative market arenas. In fact their aggregate demand would have kept its process technology busy and close to full capacity, thus further pushing down costs and inviting additional price reductions. As it was, the complex technology—composed of both of the catalytic synthesis of nanotubes and purification and dispersion operations—was severely unyielding in terms of range of products it could make and the efficiencies with which it could make them. While time and money might sufficiently tame the process for a small market niche by finding just the right conditions, catalyst, reactor setup, and dispersion technique appropriate for that particular application, the cost in doing so would be high thus further slicing the already small piece of the pie. Moreover, any attempts by CNI to go after other market possibilities in order to increase its aggregate demand would be another time-consuming and costly affair since so little of what applied technically to one application could be readily carried over to another. It would be back to the drawing board, inventing another wheel: different operating conditions, catalysts, separations technology, and dispersion methods. If CNI could have held the course for a bit longer, it may have come upon a new improved—and more flexible—process that would have made it easier, and cheaper, to go after and dominate different market arenas. (Over the last few years, the company SouthWest NanoTechnologies in Oklahoma has been adapting the CoMoCAT process, which appears to make specific types of nanotubes custom designed for particular applications without expensive separations and purification steps.)[34]

But, this was not to be. We have seen in previous cases that effective champions have been able to keep troubled projects alive in a number of ways. Most commonly they retrench, convince top management that they should continue with the research with sharply reduced resources. Then, as (hopefully) they notch in results to show their bosses, their team expands and more money comes their way to the finish line. But such an option does not exist for a small start-up with no other bankable products to sell. CNI had to come up with a highly pliable process for pure SWNTs and had to do so quickly in order to begin bringing in serious revenue. Gower was like an excellent pilot at the helm of a passenger jet that has lost all power only a few feet from the ground; he did not have much strategic wiggle room at his disposal.

Faced with such difficulties, it is no surprise then that by 2007, CNI ceased to exist as a separate company. In that year, Unidym (a subsidiary of Arrowhead Research and located in Menlo Park, California) acquired Carbon Nanotechnologies to help develop applications for nanotubes technology for flat-panel displays. Then in 2011, Arrowhead sold Unidym to the South Korean company Wisepower Co., which makes lithium batteries and battery packs and LED products. With this purchase, CNI was pushed to the sidelines in nanomaterials research.[35]

13.5 NANOTUBES—COMMERCIALIZATION: THE CASE OF BAYER MATERIALS SCIENCE

CNI was not the only important company to attempt to enter into advanced nanotube development and commercialization. Since the turn of the new century, the most important of the established firms to do so was Germany's Bayer Chemicals. In 2009, Bayer decided to pursue the nanotube market through its Division of Materials Science (MS). Its goal was to produce MWNTs with higher quality and lower prices than anyone else and use these highly pure materials to develop new applications for Bayer's existing products. Bayer explored possible future applications of its nanotubes, including conductive coatings, smart films, and electronic components. One important market goal was to make large amounts of MWNTs for improved plastic composite materials, in particular, modifying its polyurethane polymers for high-end cosmetics and medical products. Unlike CNI, which looked to control small market niches, Bayer saw nanotubes as an advanced additive for its mass market composites. All that was needed, Bayer reasoned, was a superior process technology that would make MWNTs by the tons for volume markets as efficiently as possible. Economies of scale would then kick in and costs and thus prices would then come down allowing Bayer to undersell competitors. This strategy assumed of course prices would be low enough to induce customers to substitute these materials for the traditional and more familiar fibers that they had been using to modify their plastics.[36]

We have seen how established companies are more willing to embark on radically new advanced materials technology if top managers have reason to believe the project will follow the trajectory of a successful innovation of the firm's recent past. Thus Bayer believed that the processes and markets for its nanotube—called "Baytube"—technology ought to mimic that of its polycarbonate polymer, a major success story for the company in the last half of the 20th century. Executives supposed a direct comparison, as a Bayer document titled "Baytubes – the Next Polycarbonate" strongly intimates. In it, a graph shows the rise in Bayer's production in polycarbonates from 1980 to 2005 and the polymer's expanding markets as prices decline suggesting that Bayer could do the same thing and have the same degree of success with its MWNTs, namely, "ramp up production with relatively little capital cost and manage to price everyone else out of the market."[37] Bayer also knew what the one killer market was for its MWNTs, that is, as the major component in nanopolymer composites to make high-strength material for structural applications.

One of the advantages of focusing on the MWNTs was that very little basic research is needed to be done. Their structure and properties were well known throughout the scientific community. In other words, Bayer was not attempting to discover a new material but rather to find a better way to make a known one. Consequently, Bayer did not have to deal with the often difficult issues that arise during the advanced materials research phase. The first important question that Bayer did need to address was what process to champion. It decided to develop its own proprietary version of a fluid reactor that could make MWNTs on a mass production basis. The company's plan was to first build laboratory-scale equipment (with a reactor made of quartz glass), move onto a fluidized bed pilot plant capable of turning out 200 tons of nanotubes annually, and then scale this up to a full commercial facility with capacity of 3000 tons per year, approximately the same capacity as Bayer's first commercial polycarbonate plant decades earlier.

Over the next two and half years, Bayer made excellent progress. In January 2010, Bayer MS, the organization created by Bayer to commercialize the technology, opened the world's largest carbon nanotube pilot facility in the world at CHEMPARK in Leverkusen, Germany, an industrial area with easy access to raw materials and power, sophisticated waste management, and a workforce highly skilled in chemical manufacture. Bayer MS spent €22 million—or only 0.2% of its annual sales—in the planning, development, and construction of the pilot facility and did so "in anticipation of significant rise in demand."[38]

But before long, Bayer faced serious problems, for it had badly misjudged the nature of the market the company had hoped to conquer. The mass markets for nanotubes never materialized, even as prices declined. Bayer found that it could only tap specialized, low-volume applications where customers were willing to pay high prices for a competitive edge, such as conductive plastics or field emitters for flat-screen TVs (which provided competitive advantage over LCDs) and for conductive plastic wafer trays for the semiconductor industry. But these markets did not consume large volumes of nanomaterials. When Bayer tried to force its process to full capacity, turn out high-quality nanotubes, and push the competition out, it ended up pumping out more MWNTs than it could sell. In fact, "nobody was buying it [MWNTs] because there were no [single large-scale] applications for it."[39] Prices for MWNTs did decline—from nearly $1000 per gram to less than $100 per gram—because of competition from hundreds of small outfits using a variety of manufacturing technologies. So Bayer could only sell a portion of what they made and at very low prices. Production costs remained high because Bayer's technology could not be employed to full capacity, thus sacrificing the benefits of economies of scale. Less than full-throttle production also compromised product quality, in part, because fluid flow conditions in the reactor were less than optimal resulting in too much of the catalyst staying on the nanotubes, thus diminishing their performance. In this condition, the materials could not begin to entice materials substitution.

High costs, declining prices due to competition, no killer market, and poor performance product that promised to sully the company's reputation for quality were not what Bayer had signed up for. It even got to the point that Bayer actually called on laboratory researchers at companies around the world pitching possible application ideas. Unlike CNI, Bayer did have other products bringing in revenue. Maybe most seriously of all, no champion of the technology appeared to guide the project through this difficult period by finding ways to reduce the risk pressures that had been building up over the months to the point that it could remain airborne, at least until the next hurdle came along. Collapsing the scope of the project into a bootlegged operation, bringing in a major player closely associated with Bayer's polycarbonate success, or finding an alternative production process compatible with one Bayer already had in operation were possible ways to handle the situation; as we know, these tactics succeeded in the past with other advanced materials efforts in other companies. None of these may have actually resulted in final success for Bayer, but we will never know: without the guiding hand of a forceful individual at the helm, none of these options were even attempted. In 2013, Bayer gave up on its nanotube project, shut down its Baytubes pilot plant, and sold off all of its intellectual property in the field.

As of 2017, the commercial future of nanotubes—both the single- and multiwalled variety—remains in question. As mentioned, improvements to the HiPco process may supply the key to commercial success, especially of SWNTs. But nanotubes now face a threat from a related, but very different, material known as graphene, which some believe will replace nanotubes as the advanced material to watch and the next disruptive technology, one that will "change the way many consumer and industrial products are manufactured."[40] Proponents of graphene—a material composed of single, tightly packed layers of carbon atoms—contend that it has a greater chance of building up a rapidly growing market demand due to its superior optical, electrical, and mechanical properties. Graphene exhibits the types of properties that make it an ideal material for use in conductive films, composites, high-strength plastics, super-protective coatings for wind turbines and ships, and batteries with dramatically higher capacity than anything available today. IBM has recently pioneered the use of graphene in building computer processors, although the company has yet to be able to integrate the material into commercial microchips and digital processors.[41] As with nanotubes, graphene has proven difficult to scale-up. In the spring of 2014, researchers at the University of Dublin reported on a major breakthrough in making graphene on a mass production basis.[42] Whether graphene pushes carbon nanotubes aside as the next big material may very well depend on what nanotubes will do in the future and whether graphene technology will also find it too difficult to negotiate that treacherous gap between laboratory invention and commercial reality.

REFERENCES

1. Baker, S. and Aston, A. (February 14, 2005), "The Business of Nanotech," *Business-Week*, pp. 65–71; Johnson, R. (September 11, 2002), "Nanotechnology Is Rebuilding Electronics One Atom at a Time," *EETimes*. http://www.eetimes.com/document.asp?doc_id=1145143. Accessed November 28, 2015; and AzoMaterials. (2015), "Nanomaterials and Their Applications," AzoMaterials Online. http://www.azom.com/article.aspx?ArticleID=1066. Accessed November 28, 2015.

2. Feynman, R. (December, 1959), "There's Plenty of Room at the Bottom," *Transcript of a Talk Presented by Richard P. Feynman to the American Physical Society*, http://www.pa.msu.edu/~yang/RFeynman_plentySpace.pdf. Accessed April 5, 2016.

3. Lucintel. (2011), "Global Nanomaterials Opportunity and Emerging Trends," *Lucintel Brief*, Las Colinas, TX: Lucintel LLC. http://www.lucintel.com/lucintelbrief/globalnanomaterialsopportunity-final.pdf. Accessed November 25, 2015. See also, Lucintel. (2011), "Growth Opportunities in Global Nanomaterials Market 2011–2016: Trend, Forecast, and Opportunity Analysis," *Market Report*, Las Colinas, TX: Lucintel LLC.

4. Loiseau, A. (2006), *Understanding Carbon Nanotubes from Basics to Applications*, Berlin: Springer; Meyyappan, M. (ed) (2005), *Carbon Nanotubes: Science and Applications*, Boca Raton: CRC Press; and Pfautsch, E. (2007), *Challenges in Commercializing Carbon Nanotube Composites*, University of Missouri, Columbia.

5. The following narrative is taken from Aldersey-Williams, H. (1995), *The Most Beautiful Molecule: The Discovery of the Buckyball*, New York, New York: John Wiley & Sons; Baggott, J. (1994), *Perfect Symmetry: The Accidental Discovery of Buckminsterfullerene*, Oxford, UK: Oxford University Press. See also descriptions by the major players: Kroto, H. (1997), "Symmetry, Space, Stars and C_{60}" *Reviews of Modern Physics*, 69(3): 703–721; and Smalley, R.E. (1997), "Discovering the Fullerenes" *Reviews of Modern Physics*, 69(3): 723–729.

6. Aldersey-Williams (1995), p. 67. See also Baggott (1994), p. 64.

7. Smalley, R. (1992), "The Third Form of Carbon" *Naval Research Reviews*, *43*: 4–6; Smalley, R. (March/April 1991), "Great Balls of Carbon: The Story of Buckminster-fullerene" *The Sciences*, pp. 24–26; and Kroto, H. (1997), "Symmetry, Space, Stars and C_{60}" *Reviews of Modern Physics*, *69*(3): 703–721. See also Aldersey-Williams (1995), pp. 91–123.

8. Kroto, H.W., Heath, J.R., O'Brien, S.C., Curl, R.F., and Smalley, R.E. (1985), "C_{60}: Buck-minsterfullerene" *Nature*, *318*: 162–163.

9. Aldersey-Williams (1995), p. 219.

10. Mody, C. (2010), *Institutions as Stepping Stones: Rick Smalley and the Commercialization of Nanotubes*, Studies in Material Innovation, Philadelphia, Pennsylvania: Chemical Heritage Foundation, pp. 5–8; 25–26.

11. Ibid., p. 14.

12. Aldersey-Williams (1995), pp. 224–225 and Baggott (1994), pp. 208–210. See also Haufler, R., et al. (1990), "Efficient Production of C_{60} (Buckminsterfullerene) ... " *Journal of Physical Chemistry*, *94*: 8634–8636.

13. Mody (2010), pp. 13–17.

14. Ibid., p. 14.

15. Ibid., pp. 19–20.

16. Nikolaev, P., Thess, A., Guo, T., Colbert, D.T., and Smalley, R.E. (1997), "Fullerene Nanowires" *Pure and Applied Chemistry*, *69*(1): 31–34. Note that Smalley was coauthor of this article and that the all of the authors worked under him at Rice's Center for Nanoscale Science and Technology (CNST). See also, Smalley, R. (1996) "From Balls to Tubes to Ropes: New Materials from Carbon," Presentation to the *American Institute of Chemical Engineers, South Texas Section*, January Meeting in Houston, January 4, 1996. (Retrieved from "Richard Smalley Papers" at the *Chemical Heritage Foundation*, Philadelphia, Pennsylvania.)

17. Ibid.

18. Mody (2010), pp. 19–20.

19. Ibid., p. 20; Bronikowski, M., Willis, P.A., Colbert, D.T., Smith, K.A., and Smalley, R.E. (2001), "Gas-Phase Production of Carbon Single-Walled Nanotubes from Carbon Monoxide via the HiPco Process: A Parametric Study" *Journal of Vacuum Science and Technology*, *19*(4): 1800–1804; Zhang, Q. (2011), "Carbon Nanotube Mass Production: Principles and Processes" *ChemSusChem*, *4*: 871–880; and See, C. and Harris, A. (2007), "A Review of Carbon Nanotube Synthesis via Fluidized-Bed Chemical Vapor Deposition" *Industrial & Engineering Chemistry Research*, *46*: 1005–1006.

20. Mody (2010), p. 20.

21. Ibid., pp. 21–23.

22. Photonics. (January 14, 2002), "Carbon Nanotechnologies Licenses Production Process to DuPont" *Photonics.com*. http://www.photonics.com/Article.aspx?AID=11544. Accessed November 29, 2015.

23. "Licensing Terms to Carbon Nanotechnology, Inc. for Single-Wall Carbon Nanotube Production and Application Technologies," Philadelphia, Pennsylvania: *Richard E. Smalley Papers, Chemical Heritage Foundation*. The document shows its uncertainty as to who the president of the start-up would be by indicating, under the line to be signed by the president: "??Gower or Smalley??"

24. Mody (2010), pp. 21–22.

25. Ibid., pp. 21–23.

26. Ibid. See also Thayer, A. (2001), "Nanotube Firm Building Pilot Plant" *Chemical & Engineering News, 79*(41): 11 and Darwin, J. (March 3, 2002), "Nanotechnology Beginning to Take Center Stage in Houston" *Houston Business Journal*. http://www.bizjournals.com/houston/stories/2002/03/04/newscolumn3.html?page=all. Accessed July 7, 2014.

27. Johnson, R. (September 11, 2002), "Nanotechnology is Rebuilding Electronics One Atom at a Time," EETimes. http://www.eetimes.com/document.asp?doc_id=1145143. Accessed November 28, 2015.

28. Marketsandmarkets. (2014), "Carbon Nanotubes (CNTs) Market by Type, Application & Geography—Global Trends & Forecasts to 2018," Marketsandmarkets.com. http://www.marketsandmarkets.com/PressReleases/carbon-nanotubes.asp. Accessed July 8, 2014.

29. Jansen, R. and Wallis, P. (2009), "Manufacturing, Characterization and Use of Single-Walled Carbon Nanotubes" *Material Matters, 4*(1): 23.

30. "Carbon Nanotechnologies Licenses Production Process to DuPont" (January 14, 2002), *Photonics.com*. See also PR Newswire. (2003), "NanoInk and Carbon Nanotechnologies, Inc. to Develop Advanced Manufacturing Solutions for Carbon Nanotube Devices," prnewswire.com. http://www.prnewswire.com/news-releases/nanoink-and-carbon-nanotechnologies-inc-to-develop-advanced-manufacturing-solutions-for-carbon-nanotube-devices-70812767.html. Accessed July 7, 2014.

31. Jansen, R. and Wallis, P. (2009), "Manufacturing, Characterization and Use of Single-Walled Carbon Nanotubes." See also Haddon, R. et al. (2004), "Purification and Separation of Carbon Nanotubes" *Materials Research Society (MRS) Bulletin, 29*: 252–258.

32. Sherman, L. (July 2007), "Carbon Nanotubes Lots of Potential—If the Price Is Right" *Plastics Technology*. http://www.ptonline.com/articles/carbon-nanotubes-lots-of-potentialif-the-price-is-right. Accessed November 29, 2015.

33. Ibid. See also Caprio, V. (July 2010), "NanoBusiness Alliance Interview with David Arthur, President and CEO of SouthWest NanoTechnologies, Inc.," InterNano.org. http://www.internano.org/node/390. Accessed January 29, 2015.

34. Jansen, R. and Wallis, P. (2009), "Manufacturing, Characterization and Use of Single-Walled Carbon Nanotubes." See also Arthur, D., Silvy, R.P., Wallis, P., Tan, Y., Rocha, J.-D.R., Resasco, D., Praino, R., and Hurley, W. (2012), "Carbon Nanomaterial Commercialization: Lessons for Graphene from Carbon Nanotubes," *Materials Research Society (MRS) Bulletin, 37*: 1299.

35. Business Wire. (March 22, 2007), "Carbon Nanotechnologies, Inc. (CNI) to Merge with Unidym, Inc. to Create a Formidable Force in Carbon Nanotube-Enabled Products," BusinessWire.com. http://www.businesswire.com/news/home/20070322005419/en/Carbon-Nanotechnologies-CNI-Merge-Unidym-Create-Formidable. Accessed November 29, 2015.

36. Johnson, D. (May 18, 2013), "Nanotube Supply Glut Claims First Victim" *IEEE Spectrum*. http://spectrum.ieee.org/nanoclast/semiconductors/nanotechnology/carbon-nanotube-supply-glut-claims-its-first-victim. Accessed July 8, 2014.

37. Ibid. and Hocke, H. (June 1, 2010), "Baytubes®-Carbon Nanotubes @ Bayer," *Bayer Material Science PowerPoint Presentation*. http://192.168.1.1:8181/http://www.lidorr.com/_uploads/dbsattachedfiles/baytubesseminartechnicalpresentation2010.pdf. Accessed November 29, 2015.

38. Barkoviak, M. (February 2, 2010), "Bayer Opens Largest Carbon Nanotube Pilot Facility in the World," *DailyTech.com*. http://www.dailytech.com/Bayer+Opens+Largest+ Carbon+Nanotube+Pilot+Facility+in+the+World/article17576.htm. Accessed November 29, 2015.

39. Johnson (2013), "Nanotube Supply Glut Claims First Victim."

40. Nanowerk. (April 20, 2014), "Researchers Discover New Method to Produce Large Volumes of High Quality Graphene," *Nanowerk News*. http://www.nanowerk.com/ nanotechnology-news/newsid=35264.php. Accessed November 11, 2015.

41. Shah, A. (2014), "IBM Shows Smallest, Fastest Graphene Processor," *PCWorld*. http:// www.pcworld.com/article/224576/ibm_shows_smallest_fastest_graphene_processor .html. Accessed November 29, 2015.

42. Press Association (April 20, 2014), "Global Breakthrough: Irish Scientists Discover How to Mass Produce 'Wonder Material' Graphene," *The Journal.ie*. http://www.thejournal .ie/graphene-irish-researchers-major-breakdown-mass-production-1424843-Apr2014/. Accessed November 11, 2015.

PART V

CONCLUSION

14

RISKS, CHAMPIONS, AND ADVANCED MATERIALS INNOVATION

> *… it was Mountbatten's championship of the [nuclear submarine] project which helped it gather momentum. By energy, enthusiasm and skillful lobbying he won the support or, at least, the tolerance of all those who might have been expected to oppose this speculative project—the only certain element of which … was that the cost would be extravagantly high.*
> Philip Ziegler in his biography of Lord Louis Mountbatten, 1985

The stories of our major innovations all began with initial ideas and insights that led to breakthrough materials technologies; they ended with the introduction of the first commercial products. It is important to note that first commercialization may or may not be the same as the firm integrating the new technology into its strategic planning. Certainly, DuPont immediately saw nylon as it golden goose and central to its long-range goals. But this was not the case with Intel and the microprocessor. While created and first sold commercially in the early 1970s, Intel did not shift from being a memory to a microprocessor company until the mid-1980s. What we can say is that our narratives end when, at the least, the new technology has at the least gained a foothold in the organization. For a number of these major innovations, if, when, and how from this point it actually becomes a star player in the firm is another story entirely. But just gaining that foothold is a massive feat for a breakthrough technology, especially one that is forced to suffer through the persistent close scrutiny of suspicious, parsimonious, and often short-sighted organizational leaders. The above quote

Advanced Materials Innovation: Managing Global Technology in the 21st century, First Edition.
Sanford L. Moskowitz.
© 2016 John Wiley & Sons, Inc. Published 2016 by John Wiley & Sons, Inc.

reminds us that such tasks can fall to politicians and soldiers at the highest levels who promote programs and agendas they value as well as to bench scientists and middle managers who want to see a promising idea be transformed into market reality. The process by which success comes is remarkably similar in both situations. This chapter examines that process more closely by finding and exploring the general themes that link our various case histories. In doing this, we distinguish what we term "above-ground" versus "underground" innovation, emphasize the critical importance of the sequence (or "gauntlet") of risks—as perceived by top management—to the former and show how successful project leaders are able to neutralize these risks and, in so doing, clear the way for their projects to proceed to eventual completion. The final part of the chapter delves into the personal and professional attributes of successful champions, those men and women who, like Mountbatten, ply their technical, strategic, and psychological insights to negotiate not only the survival but steady growth of new ideas within the conservative and often resistant corporate environment.

14.1   THE MAJOR TASK MILESTONES IN ADVANCED MATERIALS CREATION

So, what are the common themes that link our widely varying innovations as they evolve from first beginnings to winning that initial critical piece of real estate within a company? What we can say is that in each of our stories of advanced materials, the innovation process faces a series of task milestones that need to be addressed and dealt with by the project manager. Each of these landmarks is associated with a particular stage (or stages) of the research, development, scale-up, and commercialization (RDS&C) process. Table 14.1 shows these milestones in chronological order, describes them, links them to the appropriate stage of RDS&C, and provides examples. It is fascinating how innocently our projects got their start ("initiating"). In virtually every case, the researcher works on a project that aims only at incremental improvements to existing products and processes. This work is completely in agreement with existing corporate strategy. But such harmony does not last. Soon, the more creative and ambitious researchers see opportunities in pushing the envelope beyond accepted bounds ("redefining"). They are anxious to test their intellectual powers and in the process make names for themselves within the company and with their professional colleagues. Thus they redefine the nature of the project and set off in a direction at odds with what they had originally been asked to do. As they go off conducting research into uncharted territory, they must reduce to a bare minimum the amount of resources that they consume on the new venture ("scavenging"). Whether or not top management knows what they are up to, corporate leaders are not about to sink considerable time and money into such an uncertain venture. Creativity and persuasiveness on the part of the project leader in keeping the fledgling project going at this stage are absolutely vital. As research proceeds, the point is reached when laboratory thinking must begin to be translated into commercial terms ("spanning"). The gap between laboratory and commercial plant is often referred to as the "valley of death," for good reason, as many projects perish the difficult journey between

TABLE 14.1   The Major Task Milestones of Advanced Materials Creation

Major Task Milestones	Description	Phases of RDS&C	Examples
Initiating	Project started innocently, that is, involves researching an incremental change to an existing advanced materials technology. Pursuing this path accords with company's existing strategy	Preresearch	GE's silicone project began by looking for ways to produce a product already invented by one of the company's clients (Corning)
Redefining	Point reached when researcher sees opportunity to invent something truly new to company. Pursuing this novel path conflicts with company's existing strategy	Early research	IBM's silicon-germanium chip project took an important turn when Meyerson saw a way to greatly increase the speed and capability of existing chips even though this approach conflicted with company's switch to MOS design
Scavenging	Research proceeds on a new idea. This innovative direction is taken by conserving resources, a necessity because top management is wary of taking a new path	Late research	GE's Daniel Fox toiled alone in his laboratory attempting to find new carbonate-based polymer using bootlegged equipment
Spanning	Dangerous gap between the laboratory and commercial environments must be traversed	Early development–late development	GE's silicones evolve from inefficient laboratory apparatus to easily scaled-up fluid reactor
Extending	Project leader now must extend project in two interdependent directions: capture more market and build large-scale plants	Late development–scale-up	DuPont's nylon project scaled up Seaford, Delaware, plant to serve growing synthetic fiber markets
Establishing	Project leader must be sure the new technology is, at the least, acceptable to company. In this way, he/she plants flag for his/her innovation in company	Commercialization	Union Carbide's Joyce convinces top management to license out its new Unipol process, a decision that allows corporate leaders to accept the new technology as a viable addition to the company's competitive arsenal

science and technology. If the project survives this crossing, it must then grow as a commercial technology. It does so by scaling up its process and expanding the number and range of its customers ("extending"). If it has great difficulty doing this, it can—and will—be terminated, a victim of not being able to make it under the tough conditions established by the real-world marketplace. Finally, the new technology must be accepted by the company ("establishing"). Even a process or product that is a workable technology commercially may not be a good strategic fit; it may require the innovating firm to behave in a way totally anathema to its trusted and accepted business model. In this case, the organization may spin it off as a separate company or may simply dismantle it and use the pieces—and knowledge gained from developing it—to advance its existing technology portfolio.

14.2 "UNDERGROUND" VERSUS "ABOVEGROUND" ADVANCED MATERIALS INNOVATION

While all our innovations went through the same sequence of task milestones as they proceeded from research to commercialization, they differed in the degree to which corporate management was aware of, and actively participated in, these projects. In certain of our cases, project leaders worked by stealth for all or a large part of the task milestone sequence.[1] In other cases, corporate leaders were aware of, and took an active interest in, a project almost at the very start of the cycle. The decision (and it was not always a conscious decision nor even within the power of the project leader to decide) to go "underground" and conduct the project in secret (often continuing to work a company-approved effort in the open as his "day job") took place after (or even during) the "redefining" milestone. In some cases, despite a desire to carry out a project underground, the project leader could not do so; top managers knew about the work and were too interested in it to let it get out of their sight. In this situation, a project manager's superiors played an active decision-making role nearly from the beginning. Table 14.2 presents the major differences between underground and aboveground advanced materials innovation from the perspective of top management participation. When conducted underground, advanced materials innovation does not give corporate management much chance to participate. Project leaders are slowly working their way up the organizational structure, building up momentum as they achieve technical perfection and generating market steam and the size of revenues necessary for the powers that be in an organization to take notice and accept the new venture as an important part of the company's strategic context. But corporate leaders play a small role in all this; they are simply presented with the technology in full—at the end of the innovation cycle and thus a fait accompli—that they can either accept or reject totally in an "all-or-nothing" decision.

In comparison, aboveground innovation involves continual decision-making by top managers in response to the actions of project champions at each major stage or milestone. Corporate leaders have a chance to decide "Go/No Go" (NG) on a regular basis along the length of the process. Rather than a "bottom-up" process, aboveground innovation takes place "horizontally," as project managers and

TABLE 14.2   Underground Versus Aboveground Advanced Materials Creation:
Major Characteristics

	Level of Participation by Corporate Management	Decision-Making Opportunity by Top Management	Nature of Trigger for "Go/No Go" Decision	Frequency of "Go/No Go" Decisions Made by Top Management	Timing of "Go/No Go" Decisions by Top Management
Underground	Low	Little	Fait accompli based	Once	Only at end
Aboveground	High	Much	Risk based, per milestone	Many	During or after each stage/task milestone

corporate leaders negotiate regularly during the innovation cycle, with the former trying to convince their bosses that they should continue to resource the venture, at least as far as the next stage of the process. Whereas in underground innovation top managers say "yea" or "nay" to a project one time based on a completed work presented to them, they respond to aboveground innovative projects on a stage-by-stage basis depending on whether they perceive minimal risks are involved in proceeding to the next phase of the effort. They then can terminate the project at any stage along the way.

Underground and aboveground innovation offer their own costs and benefits, as displayed in Table 14.3. In the former case, while top managers get to decide on approving or rejecting a project only at the end of the process, a project leader can build his or her case for a radical new technology before having to show conservative corporate executives the work in progress, which is often not very pretty technically or economically. Premature disclosures have sunk many a project in companies. However, aboveground innovation has its own advantages; notably it allows corporate managers to participate regularly in decision-making—and so gradually buy—into the project. These participants can also offer suggestions along the way on how to improve the product or process and make it more acceptable within the corporate context.

14.2.1   Underground Versus Aboveground Innovation, Strategic Context, and the Major Task Milestones

The two major modes of internal innovation differ sharply in how they deal with a company's existing structural context and with the cycle of task milestones (see Table 14.4). The structural context is a company's current organizational and administrative system that is set up by top management to assure that only those projects that accord with existing strategic beliefs of the firm are selected for development

TABLE 14.3   Underground Versus Aboveground Advanced Materials Creation: A Comparison of the "Pros" and "Cons"

	Pros	Cons
Underground	• Can pursue major projects that corporate management would cut off if it was involved at earlier stages • Allows researchers and champions time to "get their ducks in a row" before presenting them to corporate management	• Corporate management does not get a chance to get involved with and "buy in" to project as it evolves • The entire "Go/No Go" decision is a one-shot chance decided by top management only at the end of the process
Aboveground	• Corporate management gets a chance to get involved with and "buy in" to project as it evolves • Not everything hinges on one "Go/No Go" decision by top management	• Corporate management can cut off a project at any stage even before it gets a chance to see its full potential • Researchers always under the gun to keep their bosses satisfied that the project is on track • Pushes project leaders into using deception, which can come to haunt project and company

and commercialization. The structural context then is conservative by its nature and geared toward favoring the lesser, incremental projects over the more radical breakthroughs. Underground projects, by their very nature, completely avoid living by the strict and inhibiting rules set by the standing structural context. They stay off the radar for as long as they can, slowly and methodically building their own case for viability and, when they are ready, presenting to top management a technology in full. Aboveground ventures do not have this advantage, for they are being closely watched every step of the way by the corporate powers who are ready to cut off support when risks begin to be perceived as too daunting. Under such tight control, project leaders cannot make up their own rules but rather must accept the existing criteria for success set by the company and do what they can to manipulate these to keep the project on track.

These contrasting ways of dealing with a company's prevailing strategic context—avoidance (underground) versus accepting manipulation (aboveground)—color the differences we observe across the entire task milestone sequence. While "initiating" is basically the same for both types of internal innovation (in both under- and aboveground innovations, a project gets started innocently enough by being involved in company-approved incremental research), crucial differences surface as the venture evolves over the cycle. The concept of "scavenging" or bootlegging has a different purpose depending on whether underground or aboveground innovation is involved.

TABLE 14.4 Underground Versus Aboveground Advanced Materials Creation: The Structural Context and Comparison of the Major Task Milestones

	Structural Context	Initiating	Redefining	Scavenging	Spanning	Extending	Establishing
Underground	Totally avoid	Work on incremental project with approval of top management	Done without knowledge of top management. More freedom to redefine project	Keep project moving forward by secretly turning "a new idea into a concrete new project." Delays in obtaining objective and accurate data delay being able to start building a case for eventual acceptance	Through process called "strategic forcing," continue to build case for project viability by creating novel markets (and customers) for venture. In order to remain under the radar, adopt technologies that do not consume corporate revenues	Use internally generated funds from initial sales to support scale-up and expanded market development activities. Adopt technology that is easy and cheap to scale up and that is very flexible to customer requirement and can pull in income into project in order stay under the radar	Present "fait accompli" case for project acceptance; require imposition of new strategic order
Aboveground	Do not avoid/ creatively manipulate	Same	May be done with or without approval—tacit or otherwise—of top management. Less freedom to redefine project	Show top management project can proceed without consuming corporate resources. Within these constraints, great deal of freedom to experiment	Show top management project can leverage laboratory work to increase number of customers in current markets. Prototyping important as proof project is on the way to successful scale-up	Convince top management few risks in this phase of the project by creatively exploiting company's existing core competency via linking this stage to past company successes	Convince top management of minimal risks involved in undertaking this last "acceptance" step by showing that by doing so will closely hue to the existing strategic order. No "fait accompli" strategy involved

In the former case, it is used by a project champion to turn a still unformed idea or concept into a tangible "thing," such as a laboratory setup or small pilot plant. It is needed to bring the project to the point where it has some form and to establish an early takeoff point for building project momentum and viability. For aboveground innovation, it has another purpose entirely: to show a corporate leadership breathing down the project's neck that product champions can build a small pilot plant without tapping significant corporate resources—an important risk minimization strategy since top managers do not yet feel comfortable that the project is worth pursuing and certainly not something at which to throw significant corporate resources. For both underground and aboveground innovation, "spanning" involves moving from the laboratory and pilot plant to begin interacting with potential customers. This task often leads to initial sales that can be plowed back into the project. However there are differences. With underground venturing, spanning allows the project to gain further traction; it helps to fortify the argument being constructed by its supporters for eventual corporate acceptance of the new product or process. Whereas for this underground case spanning is a question of beginning to build up market connections and sales volume, for the aboveground endeavor, the issue is one of risk reduction. The project leader must show top managers that moving from the laboratory to the marketplace can be accomplished relatively easily and with minimal risk. If the champion cannot do this, the project will come to an end at this point irrespective of the successes achieved in the previous steps. The task of "extending" takes up and continues what spanning has begun. For underground work, this means expanding markets and supplying these customers with the output from commercial production facilities. The goal is to create a juggernaut of a technology, characterized by a large and rapidly growing customer base, that cannot be denied by the corporate leadership once presented to them. In the aboveground world, the task of extending is nothing more than another hurdle to be overcome by easing corporate fears of facing uncontrollable risks. While underground managers simply avoid dealing with a company's existing, built-up, and conservative core competency—after all they are creating something totally new and need to be free of the shackles from a firm's past—the aboveground project leader cannot do this. Top management is watching closely and will cut off support even at this late date if the risks appear to be getting out of hand. Project champions must make their superiors feel comfortable in sinking considerable resources to enlarge the scale of operations. In both the "spanning" and "extending" tasks, they do this by finding some way to link their radically new project with the company's past successes based on the firm's trusted and proven core competence, to, in other words, convince their bosses that what they want to do is not very different from what the firm has already done successfully in the past. Deception may play a role at this point.

Finally, "establishing" the innovation within a company differs significantly when comparing the two varieties of internal technology creation. While both under- and aboveground innovations generate breakthrough technology, their final impact on the firm will likely differ. Underground initiatives have evolved and gained momentum outside of a company's existing structural and strategic contexts. If the new product or process has demonstrated great market success, it will be absorbed into the company, but it is likely that the only way this can happen is for top managers, who have no other real choice at this point given the clearly proven importance of the upstart technology for the future of the firm, to modify their company's structural context to accommodate the new corporate star. In the case of the underground Unipol project at Union Carbide, for example, the company had to suddenly and radically revamp how it had done things in the past (by cannibalizing its own existing plants and licensing out innovation to other firms). The same cannot be said of innovation conducted aboveground. In this circumstance, project champions have worked (more or less) within the confines of the company's organizational and administrative structure, albeit in new and creative ways that typically include deception and manipulation of the system. If, up to this point, the breakthrough material managed to pass through all task milestones unscathed, it did so by close and active negotiation between project managers and corporate leaders throughout the cycle. The final product or process will most probably not challenge the existing strategic life of a firm but fit very comfortably within it. If it doesn't, corporate powers will view the new technology with suspicion as a risky proposition, one that the company would have neither the will nor the way to make successful. The project will then fail its final test and it will be abandoned by the company or possibly spun off as a separate entity. Recall that final corporate acceptance of such aboveground breakthroughs as the microprocessor and silicon-germanium chip depended on these products conforming very closely to the corporate contexts within which they evolved: Robert Noyce's/Gordon Moore's Intel and Louis Gerstner's IBM, respectively. In summary avoidance and active and radical transformation versus manipulative deception and passive acceptance of the corporate way of doing things characterize and distinguish under- and aboveground internal innovation, respectively.

14.2.2 Underground Versus Aboveground Innovation: Firm and Project Characteristics

Of course, the innovation process in the creation of advanced materials can be quite thorny and somewhat messy. That is to say, an innovation cycle producing breakthrough materials is not usually purely aboveground or underground but often a combination of the two. It may start out as a secret project but after a few stages, word gets around the company and top managers learn of it and begin to participate in it; it has in other words now become an aboveground effort. In fact, in many cases, a sort of tension exists between the two modes of innovation, with project champions wanting to keep their venture under the company's radar until ready to unveil their work but with knowledge of the effort reaching the highest levels, especially as successes mount. (Less often, a project, once it is known by the company's leaders, can so readily go underground.)

Despite these complications, we can identify certain firm characteristics that make the company more or less susceptible to one or another of the two models of innovation. These characteristics are expressed in Table 14.5. Large, decentralized, and diversified firms tend to offer greater chances for innovators to work in isolation without interference from senior executives. In these environments, there are more places

TABLE 14.5   Underground Versus Aboveground Advanced Materials Creation: Firm and Project Characteristics

	Firm Size	Power Distribution	Degree of Diversification	Degree of Competition with Existing Major Product	Perceived Importance of Product to Corporation	Corporate Profile of Project Initiator
Under-ground	Large	Decentralized	High	High	Low	Low
Above-ground	Small or medium sized	Centralized	Low	Low	High	High

for them to "hide" and more chances of manipulating funding sources without corporate leaders being any wiser. Because top managers will be organizationally—and possibly geographically—distant from a project, and because they may not be very knowledgeable with or even that interested in that sort of technology, they are likely to keep their hands off the venture—and not even know of its existence—until the project leader can demonstrate to them that it is ready for acceptance by the company. These are the organizations most conducive to bottom-up innovation.[2] On the other hand, smaller, centralized, and more specialized firms are often run by those who pay close attention to what is going on in their shop. These executives also tend to be technically knowledgeable and interested in new ventures, especially when they involve their own areas of expertise. In addition to the size and structure of a company, other factors come into play. Certainly, whether the firm is large or small, the degree of perceived importance of the project to the present (or future) bottom line of the company helps determine how much attention it will get from corporate executives. Clearly, projects perceived by the powers that be as less crucial to the company's bottom line can likely go underground until they can prove themselves important additions to the company's product portfolio. Similarly, project champions who are proven talents in the organization and associated with high-profile projects in the past—or considered rising stars—will, whether they want to or not, command attention from those at the highest levels. It will be difficult for them to go underground on exciting new project opportunities. Another important consideration is whether an innovative effort competes directly with a technology that is proven and accepted by the corporate hierarchy. In this case, if the champion is not well known in corporate headquarters, he or she would likely need to operate under the radar until they could surface with a clearly superior product. Table 14.5 shows the nature of the firms and projects most closely associated with above- versus underground innovation.

Based on the previous criteria, underground internal innovation is most likely to take place in a large, highly diversified, decentralized firm and with a project that is led by a champion having a low profile in the company and that involves a technology directly competing with an established and trusted product or process.[3] On the other hand, aboveground internal innovation is most likely to happen in a smaller, specialized centrally controlled organization and involving a high-profile project and champion and a new type of product or process that does not directly compete with an existing technology. We find in our cases a few examples of underground innovation, particularly the two GE polymer projects (polycarbonates and silicones) and Union Carbide's Unipol venture. Both GE and Union Carbide represent large diversified firms and in the case of the silicones and Unipol, projects involved direct competition with corporate favorites (the enamels at GE and the high-pressure polyethylene process at Carbide). However, the majority of our cases satisfy one or more of the criteria characteristic of aboveground innovation.

14.3 UNDERGROUND ADVANCED MATERIALS CREATION: GENERAL ELECTRIC AND UNION CARBIDE

Certain of our case studies emerged from underground innovation activity. As anticipated, they are the product of large, diversified firms. General Electric's polycarbonate polymers and Union Carbide's Unipol process are good examples of "bottom-up" innovation. They both began innocently enough with the search for company-approved enhancements to existing technology (for GE, a way to advance existing insulating material and for Union Carbide, improvements to the polyethylene process being licensed from Philips). In each case, a creative individual from the middle rungs of the organization—specifically the R&D department—sees a possible way to do better than this, to come up with a totally new and breakthrough technology no one else had believed possible or had thought about at all. Success in this venture would greatly benefit the company and of course boost their careers. Both individuals (Fox at GE and Joyce at Union Carbide) pursued their work in secret. They had to for the same reason: they were challenging important, proven, and trusted technologies—GE's enamels and Carbide's high-pressure polyethylene process. They both understood that corporate managers existing would never support—and most likely would actively oppose—these efforts until they could demonstrate the clear superiority and market acceptance of their technologies. Both Fox and Joyce went underground to slowly but steadily build their case for corporate acceptance of their innovations. They continued to work their "day project"—the one officially sanctioned by their superiors—but, using various subterfuges, found time and limited resources to continue their secret investigations.

In reviewing the narratives of these two technologies, we notice a similar pattern in how they took shape and eventually achieved final approval for commercial production by top managers in their companies. The following table presents this sequence of activities from initiating a project to establishing a solid foothold as one of the company's portfolio technologies. This sequence of "case-building steps," adapted from Robert Burgelman's research on bottom-up internal corporate venturing, is shown on the right-hand side of Table 14.6. The table also identifies each of these steps with the appropriate stage of RDS&C and associated task milestone. Thus, for example, the case-building step of "proving feasibility of new idea" takes place during the "later

TABLE 14.6   Underground Innovation Process: "Bottom-Up" Case-Building Sequence for GE's Polycarbonates and Union Carbide's Unipol Process

Stage #	Stage Name	Task Milestones	"Bottom-Up" Case-Building Steps
1	Initiation phase	Initiating	Initiating project via incremental research
2	Early research phase	Redefining	Defining new business opportunities
3	Later research phase	Scavenging	Proving feasibility of new idea
4	Early development phase	Spanning I	Converting new idea into workable product or process
5	Later development phase	Spanning II and extending I	Initiating market momentum—early strategic forcing
6	Scale-up phase	Extending II	Accelerating market momentum—expanded strategic forcing
7	Commercialization phase	Establishing	Building a strategic context ("strategic building") within the firm

research stage" when the task milestone is "scavenging." At this point then, scavenging or bootlegging is essential to prove to the researchers themselves that they are on the right track in making their case. If scavenging does not accomplish this, they must think of a different path to take. Because top managers are not aware of what is going on—either in terms of success or failure—they do not make any decisions at this point as to whether the project should proceed or not. As long as their work stays off the corporate radar, researchers have a great deal of freedom in trying out different approaches until they are comfortably certain they have found the right one. As the table lays out, early research is the time for defining or "redefining" what the initial project ought to be. It was at this point that Fox redefined his official task to be finding a totally new polymer based on carbonate chemistry, while Joyce concluded that his most important work was to discover a much better (low-pressure) process to make the main type of polyethylene. Now with the project having a real sense of both purpose and direction, the next job is to prove, through creative bootlegging, that the great idea can be shown as achievable in a laboratory setting and with a working process demonstrating positive results. Commercial as well as technical feasibility needs to be considered. In Unipol's case, Joyce and Karol had to discover a new type of catalyst that was effective chemically at low pressures and that did not conflict with the patent position of competitors. For their part, Fox and Pechukus over at GE were able to produce enough polycarbonates using an inefficient but workable laboratory process to give to potential customers for testing. As the polycarbonate and Unipol researchers moved into and navigated across the development phase of innovation, they had to move as quickly as possible to span the difficult gap between the laboratory and commercial environments. In early development, they had to decide on and build up a functional pilot plant capable of handling semicommercial production. The goal was to get such a facility on line as quickly as possible. It had to be flexible enough to turn out different versions of their material for the various customer requirements. Fox and Pechukus selected interfacial polymerization and Joyce decided on fluid reaction technology as their major processes. In later development, they used these pliable technologies to begin to build up market momentum allowing them to bring in revenue for further development and scale-up plans. The fact that they still did not have to tap corporate funds to build and operate these facilities—and that they could convince senior managers that these units were being built in furtherance of their original, official, and less radical projects—continued to keep their real work more or less out of the sight of higher-level executives thus allowing them to retain their freedom of action. By the time Fox, Pechukus, and Joyce were ready to construct their first full-scale commercial plant, they had already secured a fairly large and varied market for their new materials through a process Burgelman calls strategic forcing, which creates a "beachhead for further development" of technology and its markets. With scaling, strategic forcing entered into a new, expansive stage. Sales of the material accelerated to the point that a compelling case could be advanced to top management that they should outwardly and officially support the now clearly successful effort. The final commercialization stage links this success in strategic *forcing* to a process called strategic *building* by which the strategic makeup of a firm is sharply—even radically—modified in order for the company to accommodate the new technology. Top management is willing to make this change because it really now has no choice: the new product or process—now in great demand by a large and growing market—promises too great an opportunity for the company to ignore. Even if the senior executives, for whatever reason, would prefer not to deal with them, they would have to explain their case to the board of directors and to investors in the company, and this might not be a particularly easy thing to do, especially with the threat that competitors might upstage a sheepish firm with new technology of their own. Corporate management's hands are now tied and the new technologies must be accommodated. In the case of the polycarbonates and Unipol, they indeed were with major shifts in corporate policies regarding patent licensing agreements and customer relationships paving the way for the establishment of these two innovations within their parent companies, specifically as central technologies in corporate divisions.

In these two cases, the "bottom-up" model seems to be particularly apt since the project doesn't really surface until the late development or early scale-up phases, that is, not until fairly well along in the innovation process. But there are other instances that are not so clear-cut. In one such case, the innovation process began in secret but soon surfaced as top management got wind of this potentially important project. GE's silicone polymer technology certainly began underground since Eugene Rochow was plainly defying his direct orders from the director of Central Research to find a process to make Corning's newly discovered material into glass fibers. But finding a totally new and important material rather than merely adapting someone else's was what he soon was determined to do, and this meant proceeding as secretly as possible. But unlike many research chemists, Rochow had the touch of the entrepreneur in him. He found a way to make his new polymer commercially acceptable by inventing a process that did not require magnesium, the production of which was controlled by a competitor (Dow Chemical), and he found a way of getting GE to start taking his new discovery seriously: he set up a dramatic demonstration that showed the superiority of his silicones over GE's existing and preferred insulating materials. These strategies not only piqued the interest of the engineering community in GE but convinced two particularly influential chemists—Patnode and Marshall—that Rochow was really on to something. In large part because of Rochow's success and ability to spread the word about this success, by the end of the research phase of his work, the silicone project caught the attention of the leaders of Central Research and, through them, corporate management. By the beginning of the development stage, the project had become an aboveground effort. By this time, Rochow was no longer in the picture, so his successors, without the freedom he enjoyed, would now have to deal closely with GE's management structure one phase at a time.

14.4 ABOVEGROUND ADVANCED MATERIALS CREATION AND THE "GAUNTLET OF RISKS"

While underground innovation accounted for a few major advanced materials technologies, it has been aboveground model that has dominated our case studies. There are good reasons for this. A number of them came from smaller firms (Nucor, Intel, Carbon Nanotechnologies), and whether large, small, or midsized, our companies tended to be deeply focused on one predominant process (DuPont and high-pressure synthesis, Intel and MOS-silicon gate, IBM and vapor-phase epitaxy). In a number of the outfits, corporate leaders were heavily involved and remained intensely interested in technical matters. Often, a new project was conducted by high-profile scientists, engineers, and managers (IBM's Bernie Meyerson, DuPont's Hale Charch, Intel's Ted Hoff) and involved products and processes perceived as significant by top executives almost from the first (DuPont and nylon, Sharp and large-screen LCDs, Nucor and thin slab casting). In these circumstances, there was little chance of a project champion proceeding without significant corporate oversight. And here we have an issue that did not come up with underground innovation. It is perfectly understandable that the latter can lead to breakthrough technology within a company that is led by cautious, risk-averse management. Toiling underground, the innovator can avoid following the oppressive and overly cautious rules and criteria that tend to select less pioneering efforts at the expense of the bolder, but less certain, projects. Only when a breakthrough innovation is in good working order commercially and has proven itself as a market powerhouse can technology champions argue their case to their corporate superiors, who up to now have been out of the loop in these efforts and at this late date are practically forced to bring the new material into the company. But what about the many more advanced materials innovations that have had to proceed with top management—mostly conservative to the core—breathing down their necks from the earliest days? How and why have so many seen a successful conclusion?

To better understand this question, let's consider a bit further the nature of the relationship between risk, core competency, and innovation. Our case studies tell us, quite simply, that companies do not create important technologies by taking great risks. Indeed, it is quite the opposite, that is, "a fundamental aspect of breakthrough innovation is managing risk." This is so because a new idea will not gain traction in an organization until what top managers perceive to be the most important uncertainties are, in their minds, greatly minimized, if not fully eliminated.[4] The main job of a project manager is to monitor, identify, and reduce the major uncertainties present at each stage of the innovation process. Management guru Peter Drucker tells us in no uncertain terms:

> The successful ones [entrepreneurs] I know all have ... one thing—and only one thing—in common: they are *not* "risk-takers" ... they try to define the risks they have to take and to minimize them as much as possible. ... the innovators I know are successful to the extent to which they define risks and confine them.[5]

As our narratives have shown, it is the perceived risk that has to be addressed and neutralized, and the most effective—and possibly the only—way to do this is by convincing top management that the project is nothing more than a logical extension of the company's trusted core competency, for, as Drucker again reminds us, "It is not size that is an impediment to entrepreneurship and innovation ... " but rather how alien a firm's leaders believe a project to be to the company's capabilities and way of doing things:

> When innovation is perceived by the organization as something that goes against the grain, as swimming against the current ... there will be no innovation. Innovation must be part and parcel of the ordinary, the norm, if not routine.[6]

Drucker's point is particularly relevant in the life of aboveground projects. In these cases management and champion must interact at regular points over the course of the project, with the latter having the responsibility of convincing his or her bosses that risks are minimal in proceeding further. Our case histories allow us to identify specific types of risks associated with the different phases of a project's evolution. Each risk has to be dealt with—that is, eliminated or at least minimized—at each phase or task milestone before the project can proceed to the next landmark. The following table links each project stage and associated task milestone with its unique risk factor. For example, intellectual risk (to be more precisely described later) is the greatest risk in being able to redefine a project during the early research phase of the project. I refer to this sequence of risks as "the gauntlet of perceived risks"—what a project that is forced to play itself out aboveground must go through before it can achieve a successful conclusion. Table 14.7 reviews the major stages, task milestones pertinent to each stage, and the type of risks involved as an aboveground project travels through the gauntlet of risks.

Regarding this table, three points need to be stressed. First, the "gauntlet of risks" applies only to aboveground innovation; for underground projects, the "case-building steps" are appropriate. Secondly, whether these risks are real or not, it is whether they are perceived to be risks by top management—which has the power to resource or

TABLE 14.7   Aboveground Innovation Process: The Gauntlet of Perceived Risks

Stage #	Stage Name	Task Milestones	Risk
1	Initiation phase	Initiating	Relevancy risk
2	Early research phase	Redefining	Intellectual risk
3	Later research phase	Scavenging	Resource minimization risk
4	Early development phase	Spanning	Prototyping risk
5	Later development phase	Spanning and extending I	Technology–market interaction risk
6	Scale-up phase	Extending II	Scaling risk
7	Commercialization phase	Establishing	Cultural risk

stop a project—that counts. In some cases (e.g., Orlon), senior executives thought (or more accurately were made to think) incorrectly that there were fewer risks involved in moving forward on a project and so gave the green light to project managers to proceed onto the next phase. This inaccurate perception turned out to be a happy one, for it kept the project on track and ultimately led to its successful conclusion. Thirdly, while each of the risks earlier relates to a particular period of innovation, they are by no means absolutely confined to that phase. So, for instance, intellectual risks could spread to and involve other stages. Similarly cultural risks may help shape early decisions in the process. The point is, however, based on what our case studies tell us, the probability is greatest that these risks will most directly and intensely pervade the phase of innovation with which they have been identified before.

So what does this "gauntlet of risks" approach look like when applied to our case studies? Table 14.8 tells the story. It summarizes the history of the innovation process for each of our advanced materials technologies within their firms of origin. The table lists the entries from least successful (terminated at very first stage) to most fully realized (successfully passed through all the risk stages). In the table, completion of a stage of innovation is indicated by an "X"; project managers were able to minimize perceived risk at this stage, thus assuring the granting of permission to move on to the next phase. The first failed stage for any entry is indicated by "NG" ("No Go"); it is at this point that project managers failed to convince their superiors that the risks necessary to complete this step in the innovation process could be overcome or were worth taking. Whatever success the project managed to attain up to this point is now of little relevance; Company leaders do not believe the endeavor will survive this next—and necessary—leg of the journey. The project is then either terminated outright or seriously delayed (for lack of interest and resources) to the point that a competitor gets to the finish line first. Those projects that failed to move ahead very far and stopped at one stage or another highlight the dangers of aboveground versus underground innovation; in the latter, the project champion has the advantage of being able to build up a compelling case for the breakthrough product without having to prove to corporate leaders that minimal risk conditions exist in every step of the way. It is interesting to consider, for example, what would have happened if a champion for a Unipol-like polyethylene process existed at DuPont and who was

TABLE 14.8 Aboveground Advanced Materials Innovation: The Gauntlet of Risk Case Study Results—Project Terminations and Completions

Advanced Materials Innovation	Company	Initiation ("Relevancy" Risk)	Early Research ("Intellectual" Risk)	Later Research ("Resource Minimization" Risk)	Early Development ("Prototyping" Risk)	Later Development ("Technology–Market Interaction" Risk)	Scale-Up ("Scaling" Risk)	Commercialization ("Cultural" Risk)
Thin slab, thin strip, and ultrathin steel	"Big steel" companies	NG						
High octane (catalytic) fuels: fixed bed	Jersey Standard (Exxon)	NG						
Memory chip: EPROM	Mostek	NG						
Microprocessor	Mostek	X	NG					
Polycarbonates	DuPont	NG	NG					
Advanced polyolefins and metallocenes	DuPont	X	NG					
High octane (catalytic) fuels: fluid beds	Sun Oil	X	NG					
MOS transistors	Bell Labs	X	X	NG				
Transistor: three-layered junction (bipolar)	Shockley Semiconductor	X	X	NG				

(continued overleaf)

TABLE 14.8 *(Continued)*

Advanced Materials Innovation	Company	Initiation ("Relevancy" Risk)	Early Research ("Intellectual" Risk)	Later Research ("Resource Minimization" Risk)	Early Development ("Prototyping" Risk)	Later Development ("Technology–Market Interaction" Risk)	Scale-Up ("Scaling" Risk)	Commercialization ("Cultural" Risk)
Integrated circuit (MOS)	Texas Instruments	X	X	NG				
Thin-film transistors–LCD	Westinghouse	X	X	NG				
Semiconductor laser (1.3 μm)	Bell Labs	X	X	X	NG			
LCDs	RCA	X	X	X	NG			
Diode: four layered	Shockley Semiconductor	X	X	X	NG			
LCDs	ILIXCO	X	X	X	X	NG		
Thin-film transistors–active-matrix LCDs	Panelvision	X	X	X	X	NG		
Nanotubes: single walled	Carbon Nanotechologies, Inc. (CNI)	X	X	X	X	NG		
Nanotubes: multiwalled (Baytubes)	Bayer Advanced Materials	X	X	X	X	NG		
Integrated circuits (MOS)	Fairchild Semiconductor	X	X	X	X	X	X	NG

Thin slab steel	Nucor	X	X	X	X	X	X
Ultrathin steel + microalloys	Nucor	X	X	X	X	X	X
High-pressure chemicals: first generation	DuPont	X	X	X	X	X	X
Nylon	DuPont	X	X	X	X	X	X
Orlon	DuPont	X	X	X	X	X	X
Dacron	DuPont	X	X	X	X	X	X
Lycra	DuPont	X	X	X	X	X	X
Kevlar	DuPont	X	X	X	X	X	X
High octane (catalytic) fuels: fixed bed	Sun Oil	X	X	X	X	X	X
High octane (catalytic) fuels: fluid beds	Jersey Standard (Exxon)	X	X	X	X	X	X
Transistor: point contact	Bell Labs	X	X	X	X	X	X
Transistor: silicon junction (bipolar)	Fairchild Semiconductor	X	X	X	X	X	X
Integrated circuit: bipolar	Fairchild Semiconductor	X	X	X	X	X	X
Integrated circuit: bipolar	Texas Instruments	X	X	X	X	X	X
Memory chip: MOS	Mostek	X	X	X	X	X	X

(continued overleaf)

TABLE 14.8 (*Continued*)

Advanced Materials Innovation	Company	Initiation ("Relevancy" Risk)	Early Research ("Intellectual" Risk)	Later Research ("Resource Minimization" Risk)	Early Development ("Prototyping" Risk)	Later Development ("Technology–Market Interaction" Risk)	Scale-Up ("Scaling" Risk)	Commercialization ("Cultural" Risk)
Memory chip: DRAM	Intel	X	X	X	X	X	X	X
Memory chip: EPROM	Intel	X	X	X	X	X	X	X
Microprocessor	Intel	X	X	X	X	X	X	X
Semiconductor laser (850 nm)	Bell Labs	X	X	X	X	X	X	X
Semiconductor laser (1.3 μm)	MCI	X	X	X	X	X	X	X
Silicon-germanium chip	IBM	X	X	X	X	X	X	X
LCD: electronic calculators	Sharp	X	X	X	X	X	X	X
LCD: watches	Seiko	X	X	X	X	X	X	X
Thin-film transistors–LCD: flat-screen TVs	Sharp	X	X	X	X	X	X	X
Thin-film transistors–LCD: computer screens	IBM	X	X	X	X	X	X	X

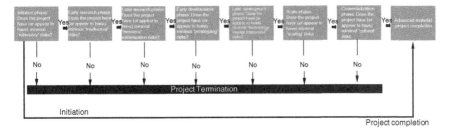

Figure 14.1   Gauntlet of perceived risk flowchart

willing and able to dive underground to create a new and powerful fluidization technology out of the probing eyes of the former Ammonia Department engineers who
now led Polychemicals and spread the gospel of high-pressure synthesis. Like Joyce
at Union Carbide, our putative hero over at DuPont may have been able to come up
with a workable process, prove its market potential, and convince a group of influential customers and colleagues to support the cause before conservative executives
could quash the venture.

We can depict what we observe in the table with a "gauntlet of risks" flowchart
(see Figure 14.1). As the managers of aboveground innovation confront each project
step, they must convince their bosses and ultimately top management that the risks
of proceeding through that stage are minimal or at least acceptable. If they cannot do
this, the project is terminated at this point irrespective of the successful traversal of
past stages.

The following sections focus on each of the seven phases of the innovation process
and the critical risks each of the project leaders must address and negotiate with corporate leaders before the venture can move onto the next step. The innovations created
by GE and Union Carbide need to be carefully considered. They are all included in our
discussion on "initiation." But beginning with the early research phase (the "redefining" task), they all go underground and we will not see them again until they resurface
at the commercialization phase, the point at which project leaders must begin again
negotiating with their superiors. In contrast, GE's silicones reemerge much sooner,
specifically at the early development phase, at which point they once again can be
observed in our table as an aboveground effort.

14.4.1   Phase I: Initiation—"Relevancy" Risks

It is surprising how quietly breakthrough innovation begins its life in a company. It
doesn't come with any crack, jolt, or clap of thunder. A progressive CEO doesn't
(generally) come up with a great idea that will shake up the company and its business
model that he or she then pushes down into the company. Nor is there some "eureka"
moment by a mid-level scientist that leads to a totally new technology and that then
bursts its way up a skeptical organization until it finds general acceptance. It also
does not usually come from an outside inventor who fights to introduce a great new
product or process into an initially unwilling company. Such drama is, in fact, quite

rare. Instead, the greatest advanced materials technologies have gotten their start in the simplest of ways, by mid-level employees trying to do what their bosses are asking them to do. These scientists, engineers, and managers are not out to set the world on fire but rather to satisfy their superiors on a particular problem that will to a greater or less degree improve the bottom line for their company. They are, after all, human and want to get ahead in the company and know they will do that by following orders to the best of their ability and not rocking the boat. Top management knows about it, approves it, and even demands it be done. It passes the first test: immediate relevancy to the firm's present business. If it doesn't meet this initial condition—if it is too alien to the organization and its customers—it dies then and there. Such was the case with the "big steel" companies and thin slab casting technology. There were those in these companies who wanted management to consider moving into this technology. But this request never stood a chance: it simply didn't make sense to company leaders given their immediate needs. Similarly for advocates within Jersey Standard of finding a superior fixed-bed process to compete with Houdry, Jersey had a much larger volume of oil in its pipeline than did Sun; it needed a process that had the capacity to refine such massive amounts of petroleum. No fixed-bed cracking process could do this job. This is why Howard and his bosses were so keen on finding a totally new and more efficient process.[7]

Table 14.9 tells this story of initiation and its need for relevancy. It recounts why advanced materials projects were initiated within their companies, whether top management approved these projects at the start, and if they didn't, whether the company allowed it to proceed or stopped it in its tracks ("NG"). The table shows that the vast majority of the projects that would ultimately result as major breakthroughs began very much within the company's existing strategic context. Out of the 46 entries, 36 or 78% involve projects that began with the explicit approval of top managers. These include both large corporations and medium-sized firms and smaller start-ups. While the percentage of those management-approved projects that involve large corporations (DuPont, IBM, etc.) is lower than that from the smaller outfits, the difference is not that great: 53% for the smaller to midsized firms versus 47% for the big firms, an indication that projects that address existing, pressing needs appeal to companies of all sizes.

As for the relatively few projects that did not get initial approval from top management, these usually got started with the discovery of some sort and were allowed to proceed simply because company leaders were not aware of the work being done. In a number of these renegade cases, sufficient discretionary resources existed within the department to keep the work going for a while (which meant the inventor's immediate boss may have approved of the initiative), or just as likely, additional resources were not really needed to any extent to kick-start the research. As mentioned, in two cases where top managers did not give their approval (thin slab casting in "big steel" companies and fixed-bed cracking at Jersey Standard), the effort ended then and there.

But even in those entries without the explicit approval by company leaders—in fact in virtually every case—the immediate interests of the company was the major impetus for the start of the project. The reasons are varied of course, but all are

TABLE 14.9 Aboveground Advanced Materials Innovation Phase I: Initiation—"Relevancy" Risk

Advanced Materials Innovation	Company	Why Project Initiated?	Initial Support from Senior Management ("Yes/No")	Result: Go/No Go	Selected Comments
Thin slab steel	Nucor	Secure internal supply of steel; expand presence in flat steel markets	Yes	Go	Nucor's president pushed this project from the beginning
Thin slab steel	"Big steel" companies	To find more efficient ways to make slab steel	No	NG	Big steel companies had grown complacent and viewed radical new approaches to steelmaking like thin slab casting as unnecessary. Each company already had less radical and less risky ways to compete with one another. Also, these companies did not believe they would be seriously challenged by small companies
Ultrathin steel + microalloys	Nucor	Help company compete with "big steel"	Yes	Go	Nucor's president pushed this project
High-pressure chemicals: first generation	DuPont	Secure internal supply of nitrogen; aid diversification strategy	Yes	Go	DuPont management supported project because they wanted a stream of new products to help pay for costs sunk in high-pressure technology
Nylon	DuPont	Looking for new fiber for DuPont's diversification strategy	Yes	Go	Nylon as the first major application of company-wide strategy of diversification through new product creation and development
Orlon	DuPont	Inventor's scientific curiosity—looking for new fiber	Yes	Go	Once DuPont management learned of the project, and thinking about the success of nylon, they were willing to see where it headed as long as no big risks incurred. Effort fit within company's diversification strategy

(continued overleaf)

TABLE 14.9 (*Continued*)

Advanced Materials Innovation	Company	Why Project Initiated?	Initial Support from Senior Management ("Yes/No")	Result: Go/No Go	Selected Comments
Dacron	DuPont	Inventor's scientific curiosity—looking for new fiber	Yes	Go	Once DuPont management learned of the project, and thinking about the success of nylon, they were willing to see where it headed as long as no big risks incurred. Effort fit within company's diversification strategy
Lycra	DuPont	Find specialty fiber that could be made by interfacial process	Yes	Go	Project fits nicely into DuPont's diversification strategy
Kevlar	DuPont	Find specialty fiber that could be made by interfacial process	Yes	Go	Project fits nicely into DuPont's diversification strategy
Polycarbonates	DuPont	DuPont looking for next polymer revolution	No	NG	No upper or middle executive thought about DuPont Plastics delving into polycarbonates even though the company's Fibers group had invented the one process (interfacial polymerization) that is at the very heart of the polycarbonate revolution
Polycarbonates	GE	Looking for superior insulating material for GE equipment	No	Go	While top management not aware of this work on polycarbonates, GE's chemical venture group was interested because of the potential relevancy of the invention and took the project over
High octane (catalytic) fuels: fixed bed	Sun Oil	Looking for cracking process superior to thermal cracking technology that would give company a competitive edge	Yes	Go	Sun's president J. Howard Pew gave his total support to Eugene Houdry

High octane (catalytic) fuels: fluid bed	Sun Oil	Looking for a catalytic cracking process superior to the fixed bed design	Yes	Go	Sun's president supported attempts by Houdry and his team to develop "moving bed" catalytic cracking unit
High octane (catalytic) fuels: fixed bed	Jersey Standard (Exxon)	Possibility of finding an improved fixed-bed process to compete with that of Sun and not infringe on Sun's patents	No	NG	No longer relevant when Jersey management realizes it will be very difficult not to infringe Sun's patents in fixed-bed technology + unlikely to find a fixed-bed technology that could avoid the same pitfalls as the Houdry process
High octane (catalytic) fuels: fluid bed	Jersey Standard (Exxon)	Wanting to avoid licensing Houdry process from Sun Oil and find a better cracking process	Yes	Go	Jersey Standard's executives supported Frank Howard in his quest to find a superior cracking process and avoid Sun's high licensing fees
Polysilicones	GE	Request from another firm (Corning) to develop silicone material for them	Yes	Go	Top management approves project for this limited result, that is, develop silicone materials invented by Corning
Advanced polyolefins and metallocenes	DuPont	DuPont wanted improved low-pressure process to make high-density polyethylene	Yes	Go	DuPont's top management supported efforts to find such a process
Advanced polyolefins and metallocenes	Union Carbide	Like DuPont, Union Carbide wanted improved low-pressure process to make high-density polyethylene	Yes	Go	Project begins on approval from top management to find a better way to make polyethylene for limited, specialized markets
Transistor: point contact	Bell Labs	At the urging of Mervin Kelly, then head of Bell Labs, AT&T wanted to find way to increase capacity of network for projected increase in future demand	Yes	Go	While not of immediate necessity to AT&T, Mervin Kelly and his Bell Labs given go-ahead to pursue new switching technologies based on advanced semiconductor technology

(continued overleaf)

TABLE 14.9 (*Continued*)

Advanced Materials Innovation	Company	Why Project Initiated?	Initial Support from Senior Management ("Yes/No")	Result: Go/No Go	Selected Comments
Diode: four layered	Shockley Semiconductor	Top management at Shockley Semiconductor (viz., Shockley himself) felt this very advanced device would revolutionize AT&T's phone switching system + Shockley felt it would give the company—and himself—the reputation of doing cutting-edge science. This was his goal for the company	Yes	Go	Project strongly opposed by the group of men to be known as the "traitorous eight" as they wanted to pursue a more practical device (the three-layered junction transistor) that could enter the market quickly
Transistor: three-layered junction (bipolar)	Shockley Semiconductor	Create most practical junction transistor	No	Go	Shockley viewed this device as scientifically uninteresting at this time and so did not support it. However, the "traitorous eight" continued to push it and keep it alive, at least for a while. Shockley allowed them to continue because he didn't want to lose the talent and their effort was not yet sucking too much out of the company
Transistor: planar silicon junction (bipolar)	Fairchild Semiconductor	This "silicon-"based device was extremely important for the new start-up as it worked better than any other transistors on the market and so gave Fairchild the competitive advantage it needed	Yes	Go	Fairchild management approved this project even though it cannibalized its mesa bipolar transistor

MOS transistors	Bell Labs	Bell looking for a way to improve upon point contact transistor. The MOS transistor would give Bell the elusive but promising field-effect transistor	Yes	Go	Bell was willing to give this product a chance, even though it competed against another promising type of transistor—the epitaxial transistor
Integrated circuit: bipolar	Fairchild Semiconductor	Solve "tyranny of numbers" problem that was of immediate concern to Fairchild management	Yes	Go	Noyce of course was both the inventor of the integrated circuit and founder—and top manager—of Fairchild
Integrated circuit: MOS	Fairchild Semiconductor	Possible future because more easily manufacturable than bipolar circuits	Yes	Go	The major champion at Fairchild for MOS circuits, Frank Wanlass, very early convinced Gordon Moore of potential importance of MOS. Almost from the beginning, a parallel research effort was utilized along with bipolar technology
Integrated circuit: bipolar	Texas Instruments	Improve TI's "micromodule" technology	Yes	Go	Top management wanted very much to have someone solve this problem of "tyranny of numbers." They in fact hired Kilby to work on this problem
Integrated circuit: MOS	Texas Instruments	Develop an improved IC—one that would be better designed for mass production—for Texas Instruments	No	Go	An "underground" group of scientists and engineers at TI took it upon themselves to explore MOS IC circuits

(continued overleaf)

TABLE 14.9 *(Continued)*

Advanced Materials Innovation	Company	Why Project Initiated?	Initial Support from Senior Management ("Yes/No")	Result: Go/No Go	Selected Comments
Memory chip: MOS	Mostek	Mostek was formed to create and sell cutting-edge semiconductor memory chips	Yes	Go	The company needed this technology as their "bread-and-butter" product
Memory chip: DRAM	Intel	Intel wanted to develop the most advanced commercial memory chip	Yes	Go	The company needed this technology as their "bread-and-butter" product
Memory chip: EPROM	Intel	Attempt to find a solution to the "floating gate" problem	Yes	Go	Intel's top managers wanted very much to solve this problem
Microprocessor	Intel	Improved way to make logic chip for Japanese client its electronic calculator	Yes	Go	Intel's top management wanted very much to satisfy the requirements of a much needed client
Microprocessor	Mostek	Improved way to make logic chip for Japanese client its electronic calculator	Yes	NG	The same. Japanese company that went to Intel also went to Mostek with similar assignment. Mostek's management
Semiconductor laser (850 nm)	Bell Labs	Bell saw this technology as a parallel initiative in case its millimeter waveguide effort failed	Yes	Go	Bell management viewed this as a backup plan

Technology	Company	Description			Notes
Semiconductor laser (1.3 µm)	Bell Labs	While Bell Labs relied on its 850-nm laser for its fiber network, it supported research on developing a more advanced 1.3-µm laser for increased range	Yes	Go	This was Bell's second-generation semiconductor laser. It was based on R&D work already done by other organizations in the United States, United Kingdom, and Japan
Semiconductor laser (1.3 µm)	MCI	To use in a new and advanced fiber optics network	Yes	Go	MCI top management supported purchase of the laser technology from the Japanese
Silicon-germanium chip	IBM	Attempt to increase speed of IBM mainframes	Yes	Go	Top management needed a solution to this problem and very much supported the effort
LCDs	RCA	Attempt to realize the dream of RCA founder David Sarnoff: flat-screen TV	Yes	Go	A flat-screen TV had been a dream of Sarnoff (and RCA) since the 1930s. Efforts to find a solution to this generally supported
Advanced LCDs based on "twisted nematic" design	ILIXCO	Create advanced LCDs for wide range of applications	Yes	Go	Top management certainly supported this. This was the very basis for survival of this start-up company. Design based on work of company's founder
LCD: electronic calculators	Sharp	Make advanced LCDs for electronic calculators	No	Go	Idea originated and pursued (semisecretly) by champion of LCDs. He started project without explicit approval of top management
LCD: watches	Seiko	Make advanced LCDs for electronic watches	No	Go	Idea pursued (semisecretly) by champion of LCDs. He started project without explicit approval of top management

(continued overleaf)

TABLE 14.9 (*Continued*)

Advanced Materials Innovation	Company	Why Project Initiated?	Initial Support from Senior Management ("Yes/No")	Result: Go/No Go	Selected Comments
Thin-film transistors–LCD	Westinghouse	Make Westinghouse the most competitive in advanced TV technology	Yes	Go	Top management initially supported the project as it wanted to be the first LCD television producer
Thin-film transistors–active-matrix LCDs	Panelvision	Make state-of-the-art LCDs using thin-film–active-matrix technology	Yes	Go	Top management certainly supported this. This was the very basis for survival of this start-up company. Design based on work of company's founder
Thin-film transistors–LCD: flat-screen TVs	Sharp	Become the first company to make large, flat-screen TVs	Yes	Go	Top management wholly supported this project from the start
Thin-film transistors–LCD: computer screens	IBM	At the time research began, no immediate relevance to IBM's immediate bottom line: improved transistors for mainframe computers	No	Go	The only reason this project got its start was the tenaciousness of its champion, who believed in the importance of the work for IBM's future and who had an excellent reputation and extensive contacts throughout the company
Nanotubes: single walled	Carbon Nanotechnologies, Inc. (CNI)	Make single-walled nanotubes to sell commercially	Yes	Go	The basis of the start-up was the making of SWNTs from technology of company's founder
Nanotubes: multiwalled (Baytubes)	Bayer Advanced Materials	Make highest-quality MWNTs more efficiently than any competitor	Yes	Go	Top management at Bayer supported the project from the start. They viewed it as an important new area to get into

directly related to current corporate goals: helping a valued client develop a material it discovered (GE and the polysilicones), searching for a way to meet the needs of projected increased demand to keep the government off the company's back (Bell Labs and the transistor), finding an improved insulating material to increase the performance of its electrical products (GE and the polycarbonates), designing an electronic calculator for a Japanese client (Intel and the microprocessor), attempting to solve the problem of why a given product is malfunctioning (Intel and the EPROM), looking to increase the speed of a family of mainframe computers it leases to clients (IBM and the silicon-germanium chip), and so forth. That in many cases the work leads in unexpected directions that often veer away from immediate company interests does not take away from the fact that initiation of the enterprise began very much with current corporate strategy and customers in mind. Thus we can say that the story of breakthrough innovation often begins innocently enough and without much noise, with a whimper rather than a bang.

14.4.2   Phase II: Early Research—Intellectual Risks

Project initiation (phase I) is absolutely critical to the total innovation project: it gets the ball rolling. This is vital if, for no other reason than, it starts the work's forward momentum. A small team often begins to get committed to the effort. Even if it is only one person, at least "something" is happening. These early champions will no doubt begin to talk to others in their circle, start getting feedback, and begin to generate interest. Even at this very early stage, it is often not that simple a thing to put a stop to the project. But how can such a venture, which began so loyal to the company's strategic vision, end up with anything close to a revolutionary technology? It is at this point that things start to get interesting. The bench scientist and engineer, sometimes with the assistance of an immediate superior, begin to chafe at the restrictions imposed by those at the top. They start to see opportunities that would take their work far beyond the confines of their more limited, official assignment. The motivation for taking on such tasks is not hard to understand: desiring a higher professional profile within their field of endeavor, obtaining increased prestige and power in the company, exerting innate creative energy, and pursuing a scientific curiosity. It may be one or a combination of these incentives. Sometime during the first phase, then, the projects begin to morph from one that hues close to the current goals and strategic mind-set of a company to one that challenges the tried-and-true, that is, from evolutionary to revolutionary.

If initiation starts project momentum, obtaining vital information from outside sources is absolutely essential to give this now newly directed project the traction needed to keep going. This makes perfect sense. If an ambitious person wants to push the project along a new course, this is the time they need to come up with a novel way of thinking or else they will continue to wallow in a traditional mind-set, and it is unlikely (although certainly not impossible) that they will find this intellectual key to the kingdom by tapping only the knowledge they have gained by working within their narrow bailiwick of their immediate group. If they cannot find that strategically essential insight or piece of information they need, their efforts will falter and appear to have no direction, and a successful conclusion will seem too far away—that is, too uncertain—to attract any further support within the company. It is then at this early research stage that searching for knowledge beyond one's proximate neighborhood is imperative as "strategic discoveries require tapping into external networks." To be clear, when we talk of finding that one important piece of information, scientific knowledge, and technical insight, we are not saying that the research phase does not involve the accumulation of other bits of information that are needed inputs into the work. But our cases do show a general pattern of early research efforts needing a specific, definable fragment of outside knowledge at a particularly perilous moment in the venture the absence of which would sink the enterprise before it has hardly had a chance to begin. In the case of nylon, for example, Carothers found his research had reached a dead end because he did not know a way in the laboratory of extending the length (or molecular weight) of his polymers beyond a certain point. If he could not find a solution, there was no way he could justify continuing. Learning about an outside device called the molecular still saved the research. Similarly, Bernie Meyerson faced a dead end in his research early on because he could not see a way of preparing surfaces for his proposed epitaxial chip without using high-temperature conditions, which also sufficiently degraded the device to render it useless. Discovering in the research literature the existence of a low-temperature way to handle surface preparation broke through this bottleneck and opened up the possibility that the silicon-germanium chip was a reachable goal.

Such external sources of knowledge could come from outside of the company as a whole, such as a university, professional conference, and outside personal contacts. It can also come from another division or department within a company. The knowledge transfer could come from one person to another, from a researcher remembering an outside experience or publication, and from a formal presentation or a chance remark made by an office mate working in another discipline in a different department.

Table 14.10 highlights that critical piece of information that was needed in this early research phase by many of our advanced materials innovations. For each of these projects, the table explains what pivotal knowledge was needed, what information was obtained to reduce the intellectual risk at this point, where it came from, and the ultimate result of that information on the project, whether "Go" or "NG." We indicate when there was no intellectual risk—that is, no major external insight or information was needed in order to move the project ahead—by writing "NN" (not needed) in the columns. For those projects that never got past the initiation stage (phase I), we simply write "NA" in the columns indicating that they have been terminated before reaching phase II.

In looking over the table, we note a couple of interesting situations. In the case of Nucor steel, for example, no risks were involved in the research phase at all because the company licensed its plants from the Germans (thin slab casting) and Australians (thin strip casting). When we write "Go" in this case, we are simply saying that Nucor by its policy of buying technology avoids having to deal with the costs and uncertainties that research entails. The case of the polycarbonates at DuPont is a very different matter. By designating "NG" for both phases I and II, we mean that these two stages in this instance are so closely related that they cannot be disentangled:

TABLE 14.10 Aboveground Advanced Materials Innovation Phase II: Early Research—Intellectual Risks

Advanced Materials Innovation	Company	Informational Risk	Critical Information Obtained to Eliminate Risk	Source	Result "Go/No Go"	Additional Comments
Thin slab steel	Nucor	None	NN	NN	Go	No risks involved here since Nucor did not have to undertake research: it purchased a prototype plant
Thin slab steel	"Big steel"	NA	NA	NA	NA	—
Ultrathin steel + microalloys	Nucor	None	NN	NN	NN	Again, no risks involved since Nucor did not have to undertake research: it purchased a prototype plant
High-pressure chemicals: first generation	DuPont	High-pressure catalysts needed	New types of metal-based heterogeneous catalysts	Hired chemists from government laboratory (Fixed Nitrogen Research Laboratory)	Go	The Fixed Nitrogen Research Laboratory was the government repository for advanced catalytic research
Nylon	DuPont	Need to find ways to grow long molecules in the laboratory	Molecular still apparatus	Scientific conference	Go	Without the molecular still, Carothers' research would have foundered because he would not have been able to make sufficiently long molecules to reach nylon

(continued overleaf)

TABLE 14.10 (*Continued*)

Advanced Materials Innovation	Company	Informational Risk	Critical Information Obtained to Eliminate Risk	Source	Result "Go/No Go"	Additional Comments
Orlon	DuPont	None	NN	NN	Go	For Orlon, no real need for external information. Consequently, no risks involved by not having access to needed external information
Dacron	DuPont	None	NN	NN	Go	For Dacron, also no real need for external information. Consequently, no risks involved by not having access to needed external information
Lycra	DuPont	Needed a new type of process to make it + better understanding of how to structure an "elastic" molecule	German information related to interfacial polymerization + Orchem insights on block copolymers	Postwar allied reports on German chemicals + direct visits to fibers from Orchem experts	Go	In this case, external information came from two sources: the Germans and another department within DuPont
Kevlar	DuPont	None	NN	NN	Go	No intellectual risks involved
Polycarbonates	DuPont	Fibers Department needed "external" information from Polychemicals Department that would link interfacial process and synthetic plastics	Not obtained	Did not obtain	NG	In this case, the failure of one department from getting vital information from another within a company hurt DuPont's chances for leading another polymer revolution

High octane (catalytic) fuels: fixed bed	Sun Oil	Information on catalysts and their ability to crack oil + how to regenerate spent catalyst	Special clay catalysts + cyclic reactor system	Agreement with Eugene Houdry to bring in and develop his knowledge on cracking catalysts	Go	Sun Oil did not have chemical capability inside company. Company needed Houdry and his chemical research
High octane (catalytic) fuels: fluid bed	Sun Oil	How to move solid particles from lower reactors to upper reactors for a continuous system	Required outside chemical engineering research on laws of interactions between solid powders and air currents	Did not obtain	NG	Houdry dominated Sun Oil's cracking program, and he was more inventor-chemist than a chemical engineer. He did not have the capability to do the basic research necessary to come up with the best way to go for moving bed systems
High octane (catalytic) fuels: fixed bed	Jersey Standard	NA	NA	NA	NA	NA
High octane (catalytic) fuels: fluid bed	Jersey Standard (Exxon)	Needed a totally different way to move catalyst powders	Empirical laws on interactions between solid powders and air	Lewis' research program at MIT	Go	Jersey Standard did not have in-house chemical engineering research capability on the basic level needed to get this vital information. Needed Lewis and MIT

(continued overleaf)

TABLE 14.10 (*Continued*)

Advanced Materials Innovation	Company	Informational Risk	Critical Information Obtained to Eliminate Risk	Source	Result "Go/No Go"	Additional Comments
Advanced polyolefins and metallocenes	DuPont	DuPont needed to adapt its own earlier work on fluid movement of solids to the polyethylene problem	This vital information did transfer over to—or was not seriously considered by engineers at—polychemicals for two reasons: isolation of polychemicals from other departments + too strong influence of high-pressure ammonia management on polychemicals	Did not obtain	NG	Here is a similar situation as with the polycarbonates: lack of communication between departments + inability to break free of a past core competency
Transistor: point contact	Bell Labs	Needed insight into why Shockley's field-effect transistor didn't work	Bardeen develops theory of trapped electrons on semiconductor surface creating a barrier to incoming electric fields	Bardeen recalled the earlier (pre-Bell) work he did on his doctoral dissertation	Go	Here is a similar situation with GE and the polycarbonates: critical information coming from past work done by inventor/researcher in graduate school

Device	Company	Need	Technology	Source	Decision	Comments
Diode: four layered	Shockley Semiconductor	Thermal diffusion + crystal growing techniques	Thermal diffusion + crystal growing techniques	Knowledge brought from experience at Bell Labs	Go	Since Shockley Semiconductor was a start-up, much of the information and technology it needed were transferred by the people themselves who started the company
Transistor: three-layered junction (bipolar)	Shockley Semiconductor	Thermal diffusion + crystal growing techniques	Thermal diffusion + crystal growing techniques	Knowledge brought from experience at Bell Labs	Go	Same as above
Transistor: silicon junction (bipolar)	Fairchild Semiconductor	Need a way to make identical, high-quality silicon junction transistors + need a way to protect each transistor from environmental contaminants	Invention of the "step-and-repeat" camera and its application to photolithographic technique + invention of the planar process	Information and technology on "step-and-repeat" cameras obtained directly by Noyce by him going to Kodak himself + Jean Hoerni, through his contacts at Bell Labs, borrows and adapts Frosch and Derick's silicon dioxide layer technique they developed a few years earlier at Bell Labs	Go	In this case, we have two important pieces of external information transferred directly (Noyce goes to Kodak) and indirectly (Hoerni uses his Bell network)

(continued overleaf)

TABLE 14.10 (*Continued*)

Advanced Materials Innovation	Company	Informational Risk	Critical Information Obtained to Eliminate Risk	Source	Result "Go/No Go"	Additional Comments
Metal-on-oxide (MOS) transistors	Bell Labs	Had to find a way to eliminate the surface barrier to incoming electric fields (so that electric field can penetrate into silicon)	Create silicon dioxide layer to eliminate the surface barrier to incoming electric fields	Borrow and adapt Bell's Frosch and Derick's silicon dioxide layer technique. Learn about Bell's because of fluid communication between research projects and departments at Bell	Go	Here we see a second application of Frosch's work at Bell (the first one being to make high-quality silicon transistors). In this case, the work was done within the same organization but for a different project within a different department
Integrated circuit: bipolar	Fairchild Semiconductor	None	NN	Planar process + photolithography	Go	The basic components of what were to become Noyce's integrated circuit—planar process and photolithography—already existed in-house (these components, as noted, were themselves based on outside knowledge)

Integrated circuit: MOS	Fairchild Semiconductor	Needed information on how to make stable, workable MOS devices		Brought in a new man, Frank Wanlass, with extensive experience in MOS technology from his work in other companies	Go	In this case, Wanlass is the source of outside knowledge into TI. Provided fresh perspective
Integrated circuit: bipolar	Texas Instruments	Needed solution to micromodule (or "tyranny of numbers") problem	The silicon integrated circuit	Hired a new man, Jack Kilby—who had been working at a firm called Centralab—to put fresh eyes on problem	Go	In this case, Kilby is the source of outside knowledge into TI. Provided fresh perspective
Integrated circuit: MOS	Texas Instruments	Needed approach to integrated circuits that would make them more easily manufacturable	MOS techniques	Through their information network in Silicon Valley	Go	The "MOS" group at TI similar to the "traitorous eight" at Shockley Semiconductor in that they bucked the mainstream thinking of their organization. TI allowed them to proceed but only up to a point
Memory chip: MOS	Mostek	Needed a better memory chip than could be made from a bipolar design	Produce chips using MOS process	Company started by a group of engineers who had worked on MOS integrated circuits at Texas Instruments + licensed ion implantation technology	Go	This small group of MOS engineers allowed to continue with their work because, with this special knowledge they acquired from outside Mostek, they had chance to produce competitive chips for TI

(continued overleaf)

TABLE 14.10 *(Continued)*

Advanced Materials Innovation	Company	Informational Risk	Critical Information Obtained to Eliminate Risk	Source	Result "Go/No Go"	Additional Comments
Memory chip: DRAM	Intel	Needed to find out how to significantly improve yields of chips using the MOS-silicon gate process	New type of etching process using an innovative chemical rinse	Head of project (Tom Rowe) remembered the work done on etching and rinses during his years at Fairchild	Go	Here is another example of critical information obtained externally that allowed project to go forward because of increased chances of success
Memory chip: EPROM	Intel	None	NN	NN	Go	In this case, the inventor (Dov Frohman) did not rely on external information. The immediate problem to be solved ("floating gate") was internal. Also, the MOS-silicon gate process itself was sufficiently flexible to suggest the solution and resolve the issues that arose
Micro-processor	Intel	None	NN	NN	Go	In this case, the inventors (Ted Hoff and Federico Faggin) did not rely on external information. The immediate problem to be solved (create "computer on a chip" for client's new electronic calculator) depended on the MOS-silicon gate process and was sufficiently flexible to suggest the solution and resolve

Micro-processor	Mostek	Needed a process that would have allowed it to design a more complex logic chip than they had ever done before	Knowledge and technique specific to silicon gate technology	Did not obtain	NG	Without this knowledge, Mostek was limited to certain memory chips. Mostek's chance for entering a new phase in its company had evaporated
Semiconductor laser (850 nm)	Bell Labs	Needed to find a way to make a semiconductor laser with greater luminescence for much longer periods of time	Bell researcher learned of the importance of matching the lattices of the different materials making up the laser + adaptation of RCA's earlier liquid-phase epitaxial process	Information on the importance of lattice matching obtained from presentation by an IBM researcher at an academic conference + tapping professional network and literature to discover IBM's liquid-phase epitaxial process	Go	As the case with IBM's silicon-germanium's chip, an example of vital information obtained through the presentation of current research by another company at a national conference
Semiconductor laser (1.3 μm)	Bell Labs	None	NN	NN	Go	Bell Labs could begin to create laboratory test models of these longer-wavelength lasers using the knowledge and techniques it had in hand in making devices using vapor-phase techniques

(continued overleaf)

TABLE 14.10 (*Continued*)

Advanced Materials Innovation	Company	Informational Risk	Critical Information Obtained to Eliminate Risk	Source	Result "Go/No Go"	Additional Comments
Semiconductor laser (1.3 μm)	MCI	Need a laser that would work in the most advanced optical fiber network so that signals can travel very far along the fiber without distortion	Adopt a semiconductor 1.3-μm wavelength laser	Purchased/licensed from the Japanese	Go	As in the case of Nucor, this is an example of a company purchasing (or licensing) a foreign technology
Silicon–germanium chip	IBM	Needed way to clean surface without use of high heats	The use of hydrofluoric acid baths	Obtained from discovery of publication in the technical literature of university research conducted a few years earlier	Go	A case where recent information is obtained by scanning the technical literature
LCDs	RCA	Needed liquid crystal materials that would operate at room temperatures	Discovery of mixtures of "nematic" organic compound using three-dimensional–phase diagrammatic analysis	Hired organic chemists from the outside to bring their knowledge to RCA	Go	Here is another example of obtaining external knowledge through people transfer

LCDs	ILIXCO	As a start-up, needed a new way to position itself competitively in the market	Based itself on twisted nematic LCD technology	Technology transferred from Kent State University via its inventor: James Fergason	Go	Another example of technology transfer through people transfer
LCD: electronic calculators	Sharp	Needed a new way to improve watches + homegrown type of liquid crystals + various specialized knowledge	Obtained information from RCA and what it was doing with LCDs + identified liquid crystal research being done internationally + absorbed specialized knowledge from different departments at Sharp	Direct visit to and reading about RCA labs + searched technical and patent literature + organized research effort to optimize people and information flow across Sharp's divisions	Go	Here is an example of a project needing three pieces of outside intellectual help: two from outside the company (visit to another company + patent literature) and one from other departments within the same company
LCD: watches	Seiko	Need a proprietary source of room-temperature liquid crystal material + economical and effective way to make the LCD	Develops own liquid crystal material + champions twisted nematic liquid crystal technology	Scans technical and patent literature on liquid crystal materials to help his research + licenses TN technology from the United States with the aid of his professional network	Go	An example of a project needing two important pieces of outside help: patent literature + licensing

(continued overleaf)

TABLE 14.10 *(Continued)*

Advanced Materials Innovation	Company	Informational Risk	Critical Information Obtained to Eliminate Risk	Source	Result "Go/No Go"	Additional Comments
Thin-film transistors—LCD	Westinghouse	None	NN	NN	Go	Champion of technology, Brody, carried on his research internally. He came upon his "active-matrix" idea himself early in his research, and it was promising enough to convince his superiors to keep project going
Thin-film transistors—active-matrix displays	Panelvision	To obtain investment funding, new company has to be based on a new competitive technology	Brody's new company bases itself on inventor's work he did on liquid crystal active-matrix displays while at Westinghouse	New company licenses Brody's technology	Go	Here is another case of a start-up licensing prior work of founder
Thin-film transistors—LCD: flat-screen TVs	Sharp	The old material it was using for its thin-film transistors (tellurium) did not work. Needed new materials for its large LCDs	Found out about the existence of hydrogenated amorphous silicon	Codiscoverer of the materials (Walter Spear from Scotland) gave a talk in Japan	Go	Sharp obtained its critical information through direct contact (via lecture) with a foreign inventor

Thin-film transistors— LCD: computer screens	IBM	The old material it was using for its thin-film transistors (tellurium) did not work. Needed new materials for its large LCDs	Found out about the existence of hydrogenated amorphous silicon	Through a published paper by University of Dundee scientists	Go	Unlike Sharp, IBM obtained its critical information—the same information Sharp obtained and from the same source—indirectly through publications
Nanotubes: single walled	Carbon Nanotechnologies, Inc. (CNI)	Smalley (without consciously knowing it at the time) needed to focus his attention away from semiconductors and toward carbon molecules as an interesting area of research	A researcher (astrochemist) from the United Kingdom came to Rice to use the AP2 and, by so doing, got Smalley's attention focused on carbon molecules	This important piece of external intellectual stimulation was through direct contact	Go	While Smalley didn't know beforehand that this was the information he needed, once he realized its importance, he saw that he had a good chance of discovering something important. In other words, the risks of failing to do so now seemed small, and so it was now a project worth pursuing
Nanotubes: multiwalled (Baytubes)	Bayer Advanced Materials	None	NN	NN	Go	Because Bayer was not developing a new material, it did not have to deal with the risks associated with advanced materials basic research

the fact that DuPont did not solve its intellectual risks—specifically, its Polychemicals Department (plastics) did not access vital knowledge about the interfacial process from an outside organization (DuPont's Fibers Department)—resulted in it not even recognizing the potential relevance of the polycarbonates to the department and company.

Leaving these two anomalies aside, and not counting the projects that ended during the initiation phase, we find, out of the rest of the projects, intellectual risk played a major role in 83% of the innovations during their early research period. That is to say, the vast majority of the projects fall into one of the following two situations: (i) they required an important piece of external information of some kind, obtained it in some way, and by so doing, reduced the immanent risk to the point that they could move onto the next phase or (ii) they required an important piece of external information of some kind, did not obtain it, could not therefore rid itself of the immanent risk, and as a result, could not move onto the next stage in the process. The ones that ended because of intellectual risk include (besides the polycarbonates at DuPont) the microprocessor at Mostek, the advanced polyolefins at DuPont, and the fluid bed cracking at Sun Oil. In no case did a project that faced intellectual risk and that could not reduce that risk go on to fight another day. As for those few cases that did not face any obvious intellectual risks, they of course moved on to the next phase automatically. Any knowledge that they needed to proceed to the next level they obtained internally and without any undue drama, generally through ideas that came to inventors on the spot (Fairchild and the integrated circuit); through continuous, painstaking experimentation that added knowledge incrementally (such as with DuPont's Dacron and Kevlar); or through incorporation of technology and skills that were close at hand (also Fairchild and its bipolar integrated circuit).

But for most of our projects, finding pivotal knowledge and technical inputs outside the walls of their small world was a make-or-break—a now or never—situation. In some cases, this contribution came in the form of a scientific instrument (nylon's molecular still) and in others a technique (catalytic cracking's fluidization), scientific insight (Lycra's flexible molecule), and even new perspective (buckyball's "space dust"). And these insights entered into these innovations in many ways: hiring specially selected personnel, attending professional conferences, doing (and recalling) doctoral dissertations, discussing research with outsiders, visiting plants of clients, contracting university research, and so forth. A project then cannot generally be isolated and closed off from the world. Informational networks must be formed. At the same time, if a project is to proceed past early research, it cannot simply be a repository for all sorts of knowledge that may come through, for much of it is irrelevant if not downright destructive. The time and money spent in sorting out and finding the true signal from all the static can in itself add to the perceived risks of an innovation in this still early stage.[8]

Underground projects also require important pieces of information. They too must rely on extensive informational networks that may extend beyond a department, division, or the company itself. But unlike aboveground ventures, they do not have to prove anything to superiors. They have much more freedom than do their aboveground counterparts to proceed without having actually to minimize perceived risks.

They are not being closely watched by corporate leaders and they themselves have risk-embracing personalities and are perfectly willing to move ahead without having all or even most uncertainties locked down. Their hope is that the situation will become clearer and less dangerous as they proceed and learn more about the technical and market terrain. Certainly, we can see this sort of cowboy mentality in Union Carbide's Joyce. Without it, it is fairly certain this organizational maverick would never have succeeded in bringing the Unipol process to light.

14.4.3 Phase III: Late Research—Resource Minimization Risks

The project began (phase I) by being a good little soldier and towing the corporate line and has been rewarded for this by beginning to build up momentum. In some cases, the project is being supported—and financed—from the top: corporate leaders have championed the research from the beginning. This is true, for example, of Nucor and its two Crawfordsville plants, of DuPont and some of its synthetics (nylon, Lycra, and Kevlar), and of a number of other companies we have examined. In the majority of cases however, at this point innovative researchers have become renegades in their own company. They have veered from their official assignments. Harnessing the momentum generated by starting a project approved by corporate leaders, they have, for various reasons, taken advantage of their situation and have now made a sharp left turn away from company-approved research and have begun to enter into unchartered territory. They are like teenagers who have been given access to the family car to buy groceries but instead have exploited their good fortune and are driving toward some adventure on their own. The fact that our rebellious scientists have secured critical knowledge (phase II) that has given the effort a direction and even a purpose attracts some additional converts to the cause. This has added mass and a certain additional impetus to the work. But we should not deceive ourselves that the project has gained the confidence of corporate leaders who are watching very closely. Their immediate response is to shut the venture down before it "wastes" significant money and man-hours. What is required now from the champion is reducing the size of the project so that it consumes as few resources as possible. Senior executives who know about the project can comfortably allow this very uncertain research to continue feeling safe in the belief it will not cause much of a dent in the department's (or company's) budget. They may also not want to unnecessarily antagonize a valued scientist, engineer, or manager who intensely desires to pursue this track. As long as it does ask much of the company, why not let those who support it continue tinkering with their idea? At the same time, corporate leaders are keeping a close eye on what these would-be pioneers are doing: the latter must begin to at least show some progress to prove to their bosses they have some sense they know what they are doing. Corporate leadership cannot allow crackpot efforts or dead-end ventures sully the reputation of the company, something that could discourage the existing workforce, hinder the hiring of top-notch researchers, and play havoc with the loyalty of customers should word of such failed efforts extend beyond company walls. For example, even after he convinced IBM to allow him to continue with his work despite its no longer having immediate relevance to the company, by severely cutting down on his research team, Bernie Meyerson knew very well he had to begin obtaining interesting data soon or else his project would be terminated permanently.

Statistically, the percentage of all ongoing aboveground projects that require a resource minimization strategy is close to 50% (not counting Nucor and MCI, who do not have research programs). While in a number of cases no resource minimization strategy was required, this was simply because top managers actively supported the project and made resources readily available to the venture. In all cases, project leaders considered scavenging very important in one way (it was necessary because of lack of resources) or the other (it was not necessary because resources were available but would have been so if resources had not been so easily secured).

Our three underground champions also scavenge for resources but all they have to do is proceed "on cat's feet" so that they do not rock the boat, as it were, and attract the attention of their corporate bosses. The abovegrounders have a greater responsibility: they must not only be able to work under de minimis conditions but also show immediate, short-term progress while doing so. Their degree of freedom is far less than their underground counterparts. Table 14.11 highlights resource minimization strategies during this phase of innovation.

14.4.4   Phase IV: Early Development—Prototyping Risks

The development phase as a whole encompasses the task milestone of spanning for both above- and underground innovations. From early through late development, the project crosses over from the laboratory environment into the commercial space where technology and markets meet. As has been true for earlier phases and task milestones, project leaders working underground have a different agenda from that of their aboveground colleagues when it comes to development. The secret project is little concerned with being able to build a complex prototype as a stepping stone to full commercialization. In fact, to the extent that such a structure costs money and could focus the corporate spotlight on the still-clandestine activities, underground champions would just as soon forego this step. Much more important for them is to put into action a strategic forcing strategy to begin creating market interest in their technology and to be able to respond to customer demands for product modifications as soon and with as little fanfare as possible. In doing this, they look for new market arenas and new customers. Whereas a prototype is meant to design in production efficiencies to be expanded during scale-up, underground personnel are willing to take any process that works—even a ragtag contraption—and get it quickly to market by having it turn out products cheaply for customers to test out. In the case of the GE's polycarbonates, we have seen how Pechukus and his team put together a dirt cheap, makeshift, and thoroughly noncompetitive process (melt polymerization) that made sufficiently acceptable polymer that could be used to reel in new customers. Years later over at Union Carbide, Joyce had the same goal: keep your head low and silently get markets growing any way possible. His tool for this was the fluidization process. The technology was so giving and pliable that he could very cheaply get it to produce different varieties of polyethylene for many customers and quickly respond

TABLE 14.11 Aboveground Advanced Materials Innovation Phase III: Late Research—"Resource Minimization" Risks

Advanced Materials Innovation	Company	Resource Minimization Issue	(RMS)	Result: Go/No Go
Thin slab steel	Nucor	Project supported by top management from the start + bulk of research and its risks already dealt with by German firm from whom Nucor is a licensing pilot plant	No RMS needed	Go
Thin slab steel	"Big steel" companies	NA	NA	NA
Ultrathin steel + microalloys	Nucor	Project supported by top management from the start + bulk of research and its risks already dealt with by German firm from whom Nucor is a licensing pilot plant	No RMS needed	Go
High-pressure chemicals: first generation	DuPont	Depression + ammonia overproduction forced belt tightening	Bootstrapping	Go
Nylon	DuPont	Depression + the fact that the project was "blue-sky" research + main researcher untested forced restriction in funding	Bootstrapping	Go
Orlon	DuPont	A larger than expected pilot plant needed to supply enough fiber for test marketing + unexpected technical and market problems soured top management to the project and forced the project leaders to take belt-tightening measures	Bootstrapping	Go
Dacron	DuPont	For Dacron also, a larger than expected pilot plant is needed forcing project leaders to go take belt-tightening measures	Bootstrapping: in this case, resource minimization took the form of adapting an existing facility—DuPont's Seaford nylon plant—for the pilot plant	Go

(continued overleaf)

TABLE 14.11 (*Continued*)

Advanced Materials Innovation	Company	Resource Minimization Issue	(RMS)	Result: Go/No Go
Lycra	DuPont	Management believes in Fibers Department due to its track record in nylon, Orlon, and Dacron	No RMS needed	Go
Kevlar	DuPont	Management believes in Fibers Department due to its track record in nylon, Orlon, and Dacron	No RMS needed	Go
Polycarbonates	DuPont	NA	NA	NA
High octane (catalytic) fuels: fixed bed	Sun Oil	Top management supported project from the start + Houdry had already done much of the work prior to coming to Sun	No RMS needed	Go
High octane (catalytic) fuels: fluid beds	Sun Oil	NA	NA	NA
High octane (catalytic) fuels: fixed bed	Jersey Standard (Exxon)	NA	NA	NA
High octane (catalytic) fuels: fluid beds	Jersey Standard (Exxon)	Competition for resources from other research projects	Bootstrapping: use of subcontracting Lewis and MIT chemical engineering labs	Go
Advanced polyolefins and metallocenes	DuPont	NA	NA	NA
Transistor: point contact	Bell Labs	Top management (Kelly and AT&T) supports project from the start + safety net in the form of money is available to keep project going without belt tightening	No RMS needed	Go
Diode: four layered	Shockley Semiconductor	Top management (Shockley) supports project from the start + perceived competitive advantage of Shockley's expertise and "genius"	No RMS needed	Go

Transistor: three-layered junction (bipolar)	Shockley Semiconductor	Bootstrapping: top management (Shockley) wants four-layered device; Noyce and his colleagues want three layers. They must try to work under the radar. They try but fail to do so as Shockley increasingly wants all company resources devoted to his four-layered diode	Failed RMS effort	NG
Transistor: planar silicon junction (bipolar)	Fairchild Semiconductor	Top management supported project from the start + competitive advantage of the planar process	No RMS needed	Go
MOS transistors	Bell Labs	Bootstrapping: at this point in his MOS research, inventor project director (Atalla) faced competition for resources from the epitaxial transistor, which had greater currency with top management	Failed RMS effort	NG
Integrated circuit: bipolar	Fairchild Semiconductor	Venture capitalists (who believe in Noyce and Moore) and top management support project from the start + has competitive advantage of planar process	No RMS needed	Go
Integrated circuit: MOS	Fairchild Semiconductor	MOS quickly gained high-level support (especially from Moore) + safety net of successful bipolar technology	No RMS needed	Go
Integrated circuit: bipolar	Texas Instruments	Kilby takes new approach to IC different from TI policy. Has to make IC himself on his own time (during company vacation). His approach to IC design soon dominates thinking at TI	Bootstrapping	Go
Integrated circuit: MOS	Texas Instruments	Bootstrapping: by this time, Kilby's bipolar IC controls thinking at Texas Instruments and few resources available for MOS ICs. Group supporting MOS gives up the effort to bring MOS to TI and jump ship to start own spin-off firm (Mostek)	Failed RMS effort	NG
Memory chip: MOS	Mostek	Top management supports project from the start + perceived competitive advantage of ion implantation process	No RMS needed	Go

(continued overleaf)

TABLE 14.11 (Continued)

Advanced Materials Innovation	Company	Resource Minimization Issue	(RMS)	Result: Go/ No Go
Memory chip: DRAM	Intel	No issues: top priority of management + competition with bipolar to see which would win and become Intel's major technology that provides safety net	No RMS needed	Go
Memory chip: EPROM	Intel	Not on immediate agenda of top management	Bootstrapping	Go
Microprocessor	Intel	Not on immediate agenda of top management	Bootstrapping	Go
Microprocessor	Mostek	NA	NA	NA
Semiconductor laser (850 nm)	Bell Labs	AT&T suffers financial downturn: wants to shut down project	Bootstrapping	Go
Semiconductor laser (1.3 μm)	Bell Labs	While top management supports the 850-nm laser, group at Bell continues to work on 1.3-μm technology as a parallel research effort	Bootstrapping	Go
Semiconductor laser (1.3 μm)	MCI	MCI purchasing technology and so not taking R&D risks + top management supports the 1.3-μm laser from the start	No RMS needed	Go
Silicon-germanium chip	IBM	Major change in context: IBM switches from bipolar to MOS chips for its mainframes: wants to shut down project, which is geared only to bipolar technology	Bootstrapping	Go
LCDs	RCA	Loss of support from the top (Sarnoff's influence wanes) + lack of interest in a business unit forced champion (Heilmeier) to go into survival mode	Bootstrapping	Go
LCDs	ILIXCO	Investors and top management (Fergason) support project from the start + possess competitive edge: twisted nematic patents	No RMS needed	Go

Product/Technology	Company	Description	RMS	Decision
LCD: electronic calculators	Sharp	Champion (Wada) had to convince top management LCDs still important to pursue after joint venture deal with RCA fell through	Bootstrapping	Go
LCD: watches	Seiko	Not on agenda of top management. Champion had to convince superiors that LCDs worthwhile by developing on his own new liquid crystal materials	Bootstrapping	Go
Thin-film transistors–LCD	Westinghouse	Financial difficulties for the company threatened champion's (Brody) work. Champion has to convince top management of viability of project by working under the radar and finding outside source of research money	Bootstrapping	Go
Thin-film transistors–active-matrix LCDs	Panelvision	At this point, investors and inventor (Brody) believed firmly in Brody's active-matrix patents and his ability to use them to create superior and competitive LCDs for televisions	No RMS needed	Go
Thin-film transistors–LCD: flat-screen TVs	Sharp	No issues: both top management and Japanese government (MITI) supported the project. Belief that any problems that researchers encountered could be surmounted if enough money and talent were made available	No RMS needed	Go
Thin-film transistors–LCD: computer screens	IBM	Top management did not support research on thin-film transistors. Would let champion (Webster Howard) pursue it only if he could do so on his own time and under stripped-down research conditions	Bootstrapping	Go
Nanotubes: single walled	Carbon Nanotechnologies, Inc. (CNI)	Top management (Smalley and Gower) supports project from the start + at this point, venture capitalists perceive the risks to be minimal due to reputation of Smalley (Nobel Prize winner) and Gower (high-tech entrepreneur)	No RMS needed	Go
Nanotubes: multiwalled (Baytubes)	Bayer Advanced Materials	Top management supports project from the start + scientific research risks minimal since Bayer planning to make a well-known material (multiwalled nanotubes)	No RMS needed	Go

to their suggestions for adjusted specifications. The setup he used for this successful strategic forcing was barely more than a laboratory model. It was the perfect apparatus to speedily capture market territory without sacrificing the benefits of project secrecy.

The situation is quite different within the realm of aboveground innovation. Building a successful prototype is immensely important because it alerts customers that the company is ready to scale processes up and so will be prepared to supply them with the new material that in turn allows them to plan their own production schedule. Unlike underground projects that seek out new customers as inexpensively as possible, aboveground ventures target existing clients with advanced prototype technology that is as close to a full-scale plant as possible. The following from a popular global marketing text gives us some idea of the importance and risks involved in prototyping:

> Developing a demonstrable physical product requires committing significant financial resources. Typically, prototype development, with all of its technical complexities, requires about three times as much money and time as the screening and business analysis [tasks] combined. Nevertheless, the company must determine whether it can manufacture the product at an acceptable cost level and whether it is able to develop the required technology.[9]

While corporate leaders understand the importance of creating an advanced prototype, they will not proceed if they feel it is unlikely project leaders can deliver on such an investment, for the stakes are really too high. For these reasons, it is crucial that project leaders who find themselves involved in aboveground innovation find any way possible to relieve their superiors—or investors—of the belief that prototyping will be a difficult row to hoe. The table in the following (in which "NA" designates those projects eliminated during the three previous phases) underscores the importance of this approach. It tells of the type of risk-reducing strategy used and whether it was successful as noted in the column "Degree of Perceived Risk," where "Low" (or "High") indicates success (or failure). If the ploy is successful, the perceived risk is low and the project moves ahead; if not, then the perceived risk is high and the project is more than likely terminated at this point. The impact on a firm failing to reduce the risk factor in prototyping can be devastating. Take the case of Shockley Semiconductor. The cream of the crop scientists and engineers in his firm did not believe that Shockley could ever build a working prototype of his beloved four-layered diode, and Shockley failed to convince them otherwise. As a result the "traitorous eight" defected, taking with them the company's major talent. After this point, Shockley's firm drifted into eventual oblivion. Similarly, Heilmeier left RCA when he could not convince RCA's management that his team could come up with anything of significance beyond his laboratory displays. He then left for a position in Washington, thus gutting the project. Then too, believing the risks were too great to play a lead role in the 1.3-μm semiconductor laser, Bell Labs fell behind this advanced technology and came in second place behind MCI in fiber optics communications.

As for the successes, every one of the aboveground cases that made it through the prototyping round did it through project leaders lowering the level of uncertainty perceived by a closely watching corporate management. From the table, we can discern a number of risk-reducing arguments made to keep the project going: the pilot plant has been made elsewhere and is licensed or acquired by the company, thus entailing few risks; there are technical shortcuts that can reduce time and costs; the prototype is compared or in other ways likened to past technical successes; external funds are obtained to defray cost of prototyping; the company has an infrastructure that can be trusted to quickly and efficiently build the pilot plant; and the relative risk argument, that is, it would be more dangerous not to proceed to a prototype whatever the costs than to move forward. Finally, please note that GE's polysilicone project now resurfaces as its project team is once again being closely monitored by the powers that are in the company. Up to this point, the team members worked secretly as long as they could; when they per force resurfaced, they had already reached a level in their work that convinced their superiors to keep the project going, at least into development. As shown in Table 14.12, the flexibility of their fluid method combined with the relative ignorance of top managers in chemical engineering made it easier for Reed and his researchers to make their superiors believe (somewhat optimistically) that prototyping would be an easy matter, something that was actually far from the truth. But here, as in other aboveground examples, perception trumped reality and ultimately determined whether a project would live to see another day.

14.4.5 Phase V: Late Development—Technology–Market Interaction Risks

During the late development stage, the task milestone of spanning continues to move forward in both underground and aboveground innovations. In the former case, project leaders continue to find new markets via strategic forcing using any reasonably cheap and sufficiently effective process with improvements to the technology taking place later on; in the latter, technology and markets develop at the same time: the semicommercial prototype is put into place and prepared to reach out to known customers in the hope of convincing them that the new technology holds promise. The following discussion is concerned with this latter situation.

The aboveground prototype, once in place, plays an essential role in the next stage of the innovation process, namely, to interact with the market and find out what customers think about the new material and what they need changed and then to conform to these needs. It is then at this stage that the technology meets the market forcing a dialog so that both sides clearly understand one another, what the market requires and how the technology can meet these demands. The prototype may change accordingly during this phase, in some cases substantially, and the market better defined and also alerted that an important innovation that may significantly impact the industry is on the way. The part played by the prototype in helping integrate the evolving technology to the market is of great importance and not often appreciated with the general

TABLE 14.12 Aboveground Advanced Materials Innovation Phase IV: Early Development—"Prototyping" Risks

Advanced Materials Innovation	Company	Nature of Prototype	Underlying Process	Degree of Perceived Risk	Go/No Go	Explanation
Thin slab steel	Nucor	Special mold + German pilot plant	Thin slab casting	Low	Go	Purchased prototype from German firm
Thin slab steel	"Big steel" companies	NA	NA	NA	NA	NA
Ultrathin steel	Nucor	Special mold + Australian pilot plant	Thin strip casting	Low	Go	Purchased prototype from Australian firm
High-pressure chemicals: first generation	DuPont	French (Claude ammonia process) pilot plant	High-pressure catalytic	Low	Go	Licensed prototype from French firm
Nylon	DuPont	Pilot plant	High-pressure catalytic + bulk/solution polymerization + melt spinning	Low	Go	Correct belief that Carothers' laboratory setup could be readily turned into a working pilot plant
Orlon	DuPont	Pilot plant	Bulk/solution polymerization + solution spinning	Low	Go	Recent success with nylon + champion (Charch) convinced superiors a less perfect, makeshift pilot plant could be built cheaply (such as through cannibalization of parts) and supply enough fiber for test marketing

Dacron	DuPont	Pilot plant	Bulk/solution polymerization + melt spinning	Low	Go	Top executives saw low risks involved when they found out Dacron could skip pilot plant entirely and move directly from laboratory to semicommercial phase by adapting nylon process at Seaford nylon facility. Saved DuPont much time and money and provided enough fiber for market testing
Lycra	DuPont	Pilot plant	Interfacial polymerization	Low	Go	Like nylon, Lycra's pilot plant could be built up (more or less) from laboratory setup
Kevlar	DuPont	Pilot plant	Interfacial polymerization	Low	Go	Recent success of Lycra and its application of interfacial polymerization made top executives feel comfortable with moving ahead with Kevlar pilot plant
Polycarbonates	DuPont	NA	NA	NA	NA	NA
High octane (catalytic) fuels: fixed bed	Sun Oil	Pilot cracking plant	Fixed-bed catalytic cracking	Low	Go	Eugene Houdry brings basic equipment and techniques with him from France to Sun
High octane (catalytic) fuels: fluid bed	Sun Oil	NA	NA	NA	NA	NA
High octane (catalytic) fuels: fixed bed	Jersey Standard (Exxon)	NA	NA	NA	NA	NA

(continued overleaf)

TABLE 14.12 *(Continued)*

Advanced Materials Innovation	Company	Nature of Prototype	Underlying Process	Degree of Perceived Risk	Go/No Go	Explanation
High octane (catalytic) fuels: fluid bed	Jersey Standard (Exxon)	Pilot cracking plant	Fluidized bed catalytic cracking	Low	Go	Lewis' MIT research already outlined pilot plant and showed relatively easy to use fluidization principles to scale up lab equipment to pilot facility
Polysilicones	GE	Pilot plant	Fluidized bed	Low	Go	Demonstration: easy to scale laboratory setup into pilot plant + GE management ignorant of what was really entailed
Advanced polyolefins and metallocenes	DuPont	NA	NA	NA	NA	NA
Transistor: point contact	Bell Labs	Stable and reliable transistors	Thermal diffusion (to make "p–n" junction)	Low	Go	Bell's leader Mervin Kelly has faith in Jack Morton and Western Electric to create prototype
Diode: four layered	Shockley Semiconductor	Stable and reliable diodes	Thermal diffusion (to make "p–n" junction)	High	NG	Investors lose faith in Shockley and his company for two reasons: failure of Shockley to create a viable, working prototype of his four-layered diode (or any other marketable product) + defection of the "traitorous eight"
Transistor: three-layered junction (bipolar)	Shockley Semiconductor	NA	NA	NA	NA	NA

Transistor: planar silicon junction (bipolar)	Fairchild Semiconductor	Stable and reliable transistor	Planar process	Low	Go	Planar process facilitated making of planar silicon transistor
MOS transistors	Bell	NA	NA	NA	NA	NA
Integrated circuit: bipolar	Fairchild Semiconductor	Stable and reliable ICs	Planar process	Low	Go	Fairchild had planar process (the key to the integrated circuit)
Integrated circuit: MOS	Fairchild Semiconductor	Stable and reliable ICs	MOS process	Low	Go	Top management technically oriented and open to new ideas + trust in work of champion Frank Wanlass
Integrated circuit: bipolar	Texas Instruments	Stable and reliable ICs	Thermal diffusion (to make "PNP" layers) + photolithography	Low	Go	Belief (partially correct) that Kilby's initial breakthrough was the main thing in realizing a commercial integrated circuit; once TI had this, its executives assumed (also partially correct) that it would be a relatively easy matter to construct a pilot manufacturing plant to make the chips using existing facilities that made transistors. Moreover, TI execs assumed (somewhat optimistically) that the market would grow quickly thus further aiding design and workings of both pilot and commercial plants
Integrated circuit: MOS	Texas Instruments	NA	NA	NA	NA	NA

(continued overleaf)

TABLE 14.12 (*Continued*)

Advanced Materials Innovation	Company	Nature of Prototype	Underlying Process	Degree of Perceived Risk	Go/No Go	Explanation
Memory chip: MOS	Mostek	Stable and reliable chips	Ion implantation process	Low	Go	Mostek sees its ion implantation process as giving them an edge in creating prototype
Memory chip: DRAM	Intel	Stable and reliable chips	MOS-SG process	Low	Go	Top management knew a lot about prototyping technology from experience at Fairchild + hired experts in silicon gate technology over from Fairchild to Intel
Memory chip: EPROM	Intel	Stable and reliable chips	MOS-SG process	Low	Go	Prototype chip presented to management as "fait accompli" using Intel's own MOS-SG process
Microprocessor	Intel	Stable and reliable chips	MOS-SG process	Low	Go	Prototype chip presented to management as "fait accompli" using Intel's own MOS-SG process
Microprocessor	Mostek	NA	NA	NA	NA	NA
Semiconductor laser (850 nm)	Bell Labs	Stable and reliable laser	Liquid-phase epitaxy	Low	Go	Management trusted the skills of project champion DeLoach + progress toward prototype allowed to proceed one step at a time, as long as progress seen + effective bootstrapping by DeLoach when resources cut + Bell had an alternative R&D path (millimeter waveguides) to further cushion risks

Semiconductor laser (1.3 μm)	Bell Labs	Stable and reliable laser	Vapor-phase epitaxy	High	NG	Had already spent much time and money on the 850-nm laser prototype. Bell skeptical about proceeding to another type of laser. Delayed its entrance into field
Semiconductor laser (1.3 μm)	MCI	Stable and reliable laser	Liquid-phase epitaxy	Low	Go	Did not have to create prototype. Licensed prototype from the Japanese
Silicon–germanium chip	IBM	Stable and reliable chip	Vapor-phase epitaxy: ultrahigh vacuum	Low	Go	Trust in Meyerson + use of one of IBM's core competencies (vapor-phase epitaxy) + effective and creative bootstrapping + continued progress
LCDs	RCA	Functioning display	Product oriented	High	NG	While working prototype difficult to make, there were other factors that made the prospect of successful prototyping seem even more distant and unlikely (i.e., risky): LCD technology seen as alien—and therefore risky—by RCA departments + loss of project's champion (Heilmeier)
LCDs	ILIXCO	Functioning display	Product oriented	Low	Go	Prototype based on company's superior and proprietary "twisted nematic" liquid crystal technology
LCD: electronic calculators	Sharp	Functioning display for electronic calculators	Product oriented	Low	Go	Champion of project (Wada) minimized perceived risks by taking a number of steps: possibility of placing circuits and LCD on common substrate + developed own liquid crystal materials + multidisciplinary organization to increase chances of success + funding from MITI to build prototype liquid crystal cell

(continued overleaf)

TABLE 14.12 *(Continued)*

Advanced Materials Innovation	Company	Nature of Prototype	Underlying Process	Degree of Perceived Risk	Go/No Go	Explanation
LCD: watches	Seiko	Functioning display for watches	Product oriented	Low	Go	Finding of proprietary liquid crystal materials stable at room temperatures + deciding to use twisted nematic process as that would shorten time and costs for prototype + Seiko had significant discretionary funds for research
Thin-film transistors–LCD	Westing-house	NA	NA	NA	NA	NA
Thin-film transistors–active-matrix LCDs	Panelvision	Functioning active-matrix display	Product oriented	Low	Go	Venture funding backs creation of prototype because of Brody's expertise in the technology + the progress he made toward a prototype while at Westinghouse
Thin-film transistors–LCD: flat-screen TVs	Sharp	Functioning 14-in. display for flat-screen TV	Product oriented	Low	Go	Great promise of α-Si TFT and active-matrix technology + Sharp's previous experience in LCD technology for electronic calculators and small-screen TVs + Sharp believed itself the most advanced in the LCD field and most capable of making a working prototype

Thin-film transistors–LCD: computer screens	IBM	Functioning displays for computers	Product oriented	Low	Go	Promise of the research coming out of University of Dundee (Scotland) + outside consultants hired by IBM told company it would have competitive advantage in such displays + relative risk issue: fear that the Japanese would take over global market in computer displays if company didn't get into the game quickly
Nanotubes: single walled	Carbon Nanotechnologies, Inc. (CNI)	Pilot plant	High-pressure fluidization (CNI's "HiPco" process)	Low	Go	Venture capitalists fund company and the development of prototype reactor because they perceive the risks to be small: Smalley as world-class expert in the field + early work on HiPco already undertaken by Rice University + joint arrangement made between CNI and KBR, which would reduce risks for CNI
Nanotubes: multiwalled (Baytubes)	Bayer Advanced Materials	Pilot plant	High-pressure fluidization (Bayer's "Baytube" process)	Low	Go	Support from the top and full resources of Bayer given to project + Bayer management believed nanotubes would follow the path of one of its past successes: polycarbonates, including successful and relatively quick design and building of pilot plant

belief that it exists merely to "iron out the wrinkles" of an already accepted design. But the reality is very different:

> For breakthrough products and services, establishing an effective business model frequently proves to be the most challenging task. Often, the customer segment that was originally considered to be the best for innovation fails to work, and only after trial and error is the right customer segment discovered. The core of the learning activity is designing experiments (including prototypes), learning from them, and adjusting the discovery process to the learning that accrues over time ... [10]

Indeed, a number of companies in many different industries have used prototypes to explore technology–market interactions. A classic example is Nestlé's Nespresso coffee system that took time to find the appropriate market. Other instances include Internet businesses (releasing so-called beta models of innovative products as test balloons measuring customer interest).[11]

The dangers involved during this stage are significant and hinge on two factors. One is the degree of stability of the target market. The pivotal questions here are (i) whether there is a clearly defined market toward which the prototype has been designed and, (ii) if so, does this market remain in place or does it fluctuate during the prototype–market interaction period? The greatest risks occur when there is fluctuation. When the market for the new material is fixed, in other words if it is predictable, the prototype will not change, and it then can be scaled up knowing there will be a customer at the end of the project. Of course, there are always issues that come up as the prototype is expanded into a full-scale commercial facility. But if there is one large, predictable customer, the advanced materials firm can begin to enjoy the cost advantages of economies of scale and so start to turn a profit sooner rather than later. For example, in the case of nylon, there were no surprises when it came to predicting the "killer" market for the new material. DuPont's original target customer—women's hosiery—turned out to be nylon's first important source of demand. The fact that DuPont stuck with one process and one large market was important in the success that the company enjoyed with its first synthetic fiber.

But what if interaction between prototype and market turns up some surprises and, like the case of Nespresso, the planned demand does not materialize and new markets have to be found or targeted customers demand significant changes to be made to the material and the process that makes it? This situation amplifies the risk factor significantly. New markets may (and often do) insist that fundamental changes be made before the innovation process can proceed. Delays and cost overruns now become a distinct possibility, and top management begins to get anxious that the company will miss out on the opportunity to be first mover in a market and that the time necessary to begin earning a profit from the new material will be extended beyond what had been planned. Investors too become nervous, as do employees who may desert the company to start their own firms and become competitors.

Since it is vitally important that prototypes are able to respond quickly and cheaply to changes in the market situation, the question of process flexibility once again comes into play. In addition to the ability of an advanced materials process to

TABLE 14.13   Elastic Versus Inelastic Processes: Concept and Illustrations from the Mechanical World

Type of Process	Description	Illustration
Elastic process	Responds quickly and easily to changes in the external environment	Computer-aided design
Inelastic process	Responds slowly and with difficulty to changes in the external environment	Henry Ford's model T production line

move rapidly and efficiently from the laboratory to prototype, the latter, which now embodies that process, has to move nimbly and swiftly in adjusting to the signals from a still uncertain, changeable target market. In this fluctuating situation, learning about, and responding to, customers must be done in an agile fashion or all could be lost. In the mechanical world, a classic example of an inflexible process was Henry Ford's early production line. Designed to make one standard product cheaply, it was not made to accommodate changes in that product, which the public increasingly wanted. The high costs, significant delays, and upheavals in the plant when Ford attempted to institute model changes are well documented. On the other end of the flexibility spectrum has been the development of computer-aided design technology within the manufacturing plants of certain industries—such as footwear—allowing rapid design alteration to meet vacillating demands (refer to Table 14.13).[12]

In a similar way, we identify a range of process flexibilities within advanced materials. Table 14.14 distributes these processes accordingly. The technologies in the left column are highly flexible with respect to "conversion" (from laboratory to prototype), "market response" (adjusting prototype to demands of the consumer), and "scaling" (expanding prototype to commercial plant). These processes reduce both

TABLE 14.14   The Elasticity Spectrum of Advanced Material Processes

Elastic Processes	Intermediate Elastic Processes	Inelastic Processes	Product Oriented
Low-pressure fluidization	High-pressure catalytic process (first generation) Thin strip (CASTRIP) casting	Thin slab casting	Liquid crystal displays: watches and pocket calculators
MOS-silicon gate (MOS-SG) process	Interfacial polymerization	High-pressure catalytic process (second generation)	Liquid crystal displays: flat-panel TVs and computers
Vapor-phase epitaxy/chemical vapor deposition	Planar process	Ion implantation High-pressure fluidization (HiPco process)	

perceived and real risks associated with prototyping, technology–market interactions, and scaling. The technologies on the right are the least flexible (the most rigid). They pose the greatest risks over these three stages of innovation for costs and delays are most likely to occur when attempting to transform a laboratory result to a proto-type and respond to a sifting target market or enlarge a prototype to a full-sized plant. Finally, in the far right column are the product-oriented technologies, which are the least malleable of all. For these inelastic and product-oriented technologies, the best-case scenario in terms of risk reduction is a fixed and stable target market following the nylon model.

Table 14.15 captures the various risks as perceived by top managers involving aboveground technology–market interactions. The degree of market instability for each innovation is indicated by comparing the third and fourth columns, initial and final target markets, respectively. If the two are not the same—that is, if the origi-nal markets turn out not to be the eventual "killer" market upon first introduction of the material—then the market is deemed unstable and therefore risky. In this case, the prototype has to work overtime to adjust to particular demands of the new mar-ket environment prior to scale-up. Here, the more flexible the process upon which the prototype is based, the less risk involved, that is, the prototype can quickly and cheaply adjust to the new market and get ready for scaling. If, however, the market is variable and the prototype is inflexible, the greater is the risk that costs and delays will occur—or be projected to occur—with the more likely chance that the project will be abandoned. If however an inflexible prototype is matched with a stable, unmoving tar-get, the promise of economies of scale reduces perceived risks of proceeding. (In this static case as well, having a flexible process does not hurt and can in fact be a benefit.)

In the table, I have identified low market stability (high market instability) with a high risk that has to be dealt with. As is seen, and not surprisingly, market insta-bility is more the rule than exception: for our cases, there are 4.3 times as many unstable (more risky) than stable (less risky) situations, reflecting the difficult hur-dle put up by this phase of the innovation process. Then, in the second column on the right, I summarize the type of strategy used by project leaders to minimize this perceived risk if they succeeded or failed in their strategy and the results ("Go" or "NG"). In the table, I have noted "NA" for those projects terminated during phases one through four, "NN" in the strategy's column for those cases where market is stable, and "NG" for those projects that ended their journey in this later develop-ment phase. There are actually four such projects that failed to deal with the risks involved with a shifting target market: ILIXCO and twisted nematic (TN) LCDs, Pan-elvision and thin-film transistor/active-matrix LCDs, Carbon Nanotechnologies and single-walled nanotubes (SWNTs), and Bayer Advanced Materials and multiwalled nanotubes (MWNTs). In all cases, the prototype did not readily adjust to the moving markets, a situation that piled on costs and delays and was particularly destructive to the smaller firms.

What about the projects that were able to overcome the risks—both the real risks and those perceived by management to be daunting—attending an unstable target? The best-case scenario in this situation was to have a prototype based on a flexible process so that responses to market signals could be quickly met and, in the process,

TABLE 14.15 Aboveground Advanced Materials Innovation Phase V: Late Development—Technology–Market Interaction Risks

Advanced Materials Innovation	Company	Initial Target Market	Final Target Market(s)	Market Stability	Risk	Perceived Risk Minimization Strategy	Go/No Go
Thin slab steel	Nucor	Flat, low carbon steel: automotive and appliance	Same	High	Low	NN	Go
Thin slab steel	"Big steel" companies	NA	NA	NA	NA	NA	NA
Ultrathin steel + microalloys	Nucor	Stainless steel: appliance	Nonstainless, ultrathin low carbon steel: construction	Low	High	Process flexibility via thin strip casting (CASTRIP)	Go
High-pressure chemicals: first generation	DuPont	Ammonia and nitrates for DuPont + explosives and fertilizer industries	Alcohols for automotive and industrial	Low	High	Process flexibility	Go
Nylon	DuPont	High-end women's hosiery	Same	High	Low	NN	Go
Orlon	DuPont	Textile fibers	Wool applications	Low	High	Link to "nylon model"	Go
Dacron	DuPont	Tire cord fiber	Wool applications	Low	High	Link to "nylon model"	Go
Lycra	DuPont	Cotton-wrapped rubber	Elastic fiber	Low	High	Process flexibility via interfacial polymerization	Go

(continued overleaf)

TABLE 14.15 (*Continued*)

Advanced Materials Innovation	Company	Initial Target Market	Final Target Market(s)	Market Stability	Risk	Perceived Risk Minimization Strategy	Go/No Go
Kevlar	DuPont	Tire cord fiber	Niche applications (e.g., body/vehicle armor)	Low	High	Link to "nylon model" + recent success of Lycra	Go
Polycarbonates	DuPont	NA	NA	NA	NA	NA	NA
High octane (catalytic) fuels: fixed bed	Sun Oil	Automotive fuel	Higher octane automotive fuel; aviation fuel; synthetic rubber	Low	High	Houdry's technical capabilities + reliance on Sun's shipbuilding facilities	Go
High octane (catalytic) fuels: fluid beds	Sun Oil	NA	NA	NA	NA	NA	NA
High octane (catalytic) fuels: fixed bed	Jersey Standard (Exxon)	NA	NA	NA	NA	NA	NA
High octane (catalytic) fuels: fluid beds	Jersey Standard (Exxon)	Automotive fuel	Higher octane automotive fuel; aviation fuel; synthetic rubber	Low	High	Process flexibility via fluidization	Go
Polysilicones	GE	Insulating rubber (e.g., for electrical wires)	Growing demand for different types of polysilicones	Low	High	Process flexibility via fluidization	Go
Advanced polyolefins and metallocenes	DuPont	NA	NA	NA	NA	NA	NA

Transistor: point contact	Bell Labs	Telecommunications	Consumer electronics: hearing aids and radios	Low	High	Trusted Morton and Western Electric to produce transistors for different applications	Go
Diode: four layered	Shockley Semiconductor	NA	NA	NA	NA	NA	NA
Transistor: three-layered junction (bipolar)	Shockley Semiconductor	NA	NA	NA	NA	NA	NA
Transistor: planar silicon junction (bipolar)	Fairchild Semiconductor	Computer manufacturers	Variety of applications	Low	High	Process flexibility via planar process	Go
MOS transistors	Bell Labs	NA	NA	NA	NA	NA	NA
Integrated circuit: bipolar	Fairchild Semiconductor	Unknown	Different circuits for computers	Low	High	Process flexibility via planar process	Go
Integrated circuit: MOS	Fairchild Semiconductor	Unknown	Different circuits for computers	Low	High	Technically oriented top management continues to push the project despite process difficulties	Go
Integrated circuit: bipolar	Texas Instruments	Unknown	Customized circuit applications, space and military and pocket electronic calculators	Low	High	Despite inherent inflexibilities with Kilby's concept for the integrated circuit, top management supports project based on its continued excitement over invention of the IC and confidence in its Jack Kilby's reputation and skills as an inventor who can "make things happen"	Go
Integrated circuit: MOS	Texas Instruments	NA	NA	NA	NA	NA	NA

(continued overleaf)

TABLE 14.15 (*Continued*)

Advanced Materials Innovation	Company	Initial Target Market	Final Target Market(s)	Market Stability	Risk	Perceived Risk Minimization Strategy	Go/No Go
Memory chip: MOS	Mostek	Memories for mainframe computers	Increasing demand for smaller chips with greater capacity	Low	High	Competitive advantage of ion implantation process	Go
Memory chip: DRAM	Intel	Memories for mainframe computers	Increasing demand for smaller chips with greater capacity	Low	High	Process flexibility via MOS-SG	Go
Memory chip: EPROM	Intel	Demand by computer makers for small quantities for prototyping computers still in production	Demand by computer buyers for large (mass produced) quantities	Low	High	Process flexibility via MOS-SG	Go
Microprocessor	Intel	Processor for electronic calculators	Smaller, faster, more powerful processor chips for personal computers	Low	High	Process flexibility via MOS-SG	Go
Microprocessor	Mostek	NA	NA	NA	NA	NA	NA
Semiconductor laser (850 nm)	Bell Labs	850-nm laser for fiber optics	Same	High	Low	NN	Go
Semiconductor laser (1.3 µm)	Bell Labs	NA	NA	NA	NA	NA	NA

Technology	Company	Current application	New application			Comments	Decision
Semiconductor laser (1.3 μm)	MCI	Fiber optics communications	Same	High	Low	NN	Go
Silicon–germanium chip	IBM	Improved junction (bipolar) chip to increase speed and power of next-generation mainframes	Advanced wireless chip for mobile devices	Low	High	Process flexibility via ultrahigh vacuum chemical vapor deposition (UHV CVD)	Go
LCDs	RCA	NA	NA	NA	NA	NA	NA
LCDs	ILIXCO	Small LCDs	Medium–large LCDs: TVs, computers, mobile devices	Low	High	ILIXCO dealt with a highly inflexible product-oriented technology in a highly dynamic market. This reality finally knocked it out of contention as state-of-the-art LCD company	NG
LCD: electronic calculators	Sharp	Electronic calculators	Same	High	Low	NN	Go
LCD: watches	Seiko	Electronic watches	Same	High	Low	NN	Go
Thin-film transistors–LCD	Westinghouse	NA	NA	NA	NA	NA	NA
Thin-film transistors–active-matrix LCDs	Panelvision	Small-to-medium-sized LCDs	Large-screen, flat-panel LCDs	Low	High	Like ILIXCO, Panelvision dealt with a highly inflexible product-oriented technology within a highly dynamic market. This reality finally ended Panelvision's attempt to become a major player in the LCD market	NG

(continued overleaf)

TABLE 14.15 (*Continued*)

Advanced Materials Innovation	Company	Initial Target Market	Final Target Market(s)	Market Stability	Risk	Perceived Risk Minimization Strategy	Go/No Go
Thin-film transistors–LCD: flat-screen TVs	Sharp	Smaller, black-and-white LCD TVs	Larger, color flat-panel LCD TVs	Low	High	While technology inflexible, Sharp and Japan take steps to reduce the risks: MITI takes over to provide country-wide resources to cover high costs and know-how required to supply an unstable market	Go
Thin-film transistors–LCD: computer screens	IBM	Computer screens for IBM PCs	Wider variety of LCD screens with increased functions for different types of PCs and growing family of mobile devices	Low	High	While technology inflexible, project champion at IBM takes steps to reduce the risks: joint venture with Japanese firm to spread the costs and know-how required to supply an unstable market	Go
Nanotubes: single walled	Carbon Nanotechnologies, Inc. (CNI)	Nanopolymer composites	Many small, specialized markets	Low	High	The HiPco process (fluidization + high pressures) proved too unstable and could not effectively or efficiently supply a highly dynamic SWNT market	NG
Nanotubes: multiwalled (Baytubes)	Bayer Advanced Materials	Nanopolymer composites	Many small, specialized markets	Low	High	As with CNI's technology, Bayer's own process (also based on a combination of fluidization + high pressures) proved too unstable and could not effectively or efficiently supply a highly dynamic MWNT market	NG

convince corporate executives that risks in proceeding to the next stage were minimal. We certainly see evidence in the table that many of these entries fit this description including Nucor (thin strip casting), Jersey Standard (fluidization—catalytic cracking), GE (fluidization—polysilicone), Fairchild (planar process), Intel (MOS-SG process), and IBM (vapor-phase epitaxy/chemical vapor deposition). But in the face of high market risks (unstable demand) and low process flexibility, project leaders hoping to keep their efforts funded had to convince top executives that their venture did not entail great hazards and that their technology could effectively satisfy market requirements and without posing undue dangers to the company. They resorted to many of the stratagems they had used in the previous phase of the project: linking the project to past successes (DuPont), trusted personalities (Bell Labs), proven company infrastructure (Bell, Sun Oil), competitive and proprietary patents (Mostek), and external funding sources (Sharp, IBM). Even when such an approach did not actually reduce real risks—as in the case of DuPont's Orlon and Dacron—they at least bought the project more time to show that it could satisfy the test market and, hopefully, to allow it to begin constructing a full-scale plant.

14.4.6   Phase VI: Scale-Up Phase—Scaling Risks

Both underground and aboveground advanced materials innovations continue to pursue their different agendas as they proceed to the scaling-up phase. They both of course use this stage to extend and solidify their prior accomplishments, but they do so in ways that reflect their different situations. Underground projects, still trying to stay out of the corporate eye while continuing to build their case, have to be able to show that their idea can be turned into a commercial process. It is one thing to take any sort of technology no matter how inefficient just to get into and start building a market—as was done during the development phases—and it is another to be able to demonstrate the ability to turn this backwoods contraption into a workable, large-scale unit capable of operating successfully in a highly competitive environment. Working underground means of course doing this without calling attention to the project from corporate headquarters. In a large diversified firm—especially one where the clandestine plant is going to be built geographically distant from headquarters—this is certainly possible as long as the scaling effort does not engross substantial resources in the form of materials, personnel, and cash. Latching on to a simple, elastic, and readily scaled process is the key, one in which output can be rapidly and efficiently expanded through relatively small (and inexpensive) changes in chemical and mechanical conditions—slight modifications to catalyst, small adjustments in the rate of materials flow, and so forth—rather than through the application of extremely rigorous conditions requiring large capital equipment expenditures as is the case in high-pressure synthesis. And in fact the scaling of both the polycarbonates and the Unipol system, both underground efforts, depended on cheap and easy to expand processes, specifically low-temperature interfacial polymerization (polycarbonates) and low-pressure fluidization (Unipol).

But for those aboveground initiatives, project leaders had no choice but to convince their corporate bosses one way or another that scaling up the prototype would not involve a great deal of uncertainty and that the path was clear for an easy, pain-free march toward a fully realized and commercially superior operating plant.

Project leaders proffered a number of arguments to heighten corporate's comfortability factor in scaling up the new process, most of which attempted to link the current effort to the company's accepted core competency. A particularly important and powerful approach was to show how a company's familiar, deeply rooted process technology could be employed to make radically new products through economies of scope (Intel's EPROM and microprocessor; IBM's silicon-germanium chip). Closely related strategies on the part of venture champions included incorporating into the current effort procedures and personnel closely associated with past successes (DuPont's Rayon), showing that production of the new material or device could be carried out on existing equipment (Intel's MOS-SG; IBM's silicon-germanium chip; DuPont's Dacron), reminding sheepish corporate leaders that the scaling process would require similar techniques and skills used in previously successful ventures (Nucor's thin slab and thin strip casting), and leveraging the fact that the department where the scaling effort would take place has strong ties to other trusted, successful departments (DuPont's/Ammonia Department's nylon; Bell's/Western Electric's transistor). Of course, in a few projects, such core competency-based arguments simply did not fit. Champions for these ventures had to find other ways to make top managers feel sufficiently comfortable to agree to the scale-up procedure. They did so by resorting to the other arguments used in previous phases: technical and managerial expertise of the main actors (including trusted contractors), forming joint ventures to share the costs of the project, and making sure there was a "safety net" in case of failure (such as Intel's shift register business providing a net for development and scale-up of the DRAM chip). To be noted is the fact that process elasticity was important for both underground and aboveground projects but for different reasons: for undergrounders, flexibility meant being able to proceed with scaling without being noticed by the top corporate powers; for our abovegrounders, it was a psychological tool with which to persuade an already clued-in and severely judgmental CEO that building a full-sized plant will be a fairly harmless affair.

Table 14.16 summarizes these trends for the various innovations, with NA indicating those projects already terminated in the first five phases of the innovation and the right-hand column ("Explanation") describing how, for projects allowed to continue past this phase, perceived scaling risks were minimized to the satisfaction of corporate leadership (in such cases, as is seen, the "Degree of Perceived Risk" is rated "Low").

14.4.7   Phase VII: Commercialization Phase—"Cultural-Strategic" Risks

It is in this last phase of the innovation cycle that the abovegrounders and heretofore undergrounders—who have surfaced and are now recognized by top executives—are judged by the corporate powers. Cultural issues now become the dominant force for both types of project champions.

The culture of an organization, of course, pervades the innovation process from the start. Culture certainly played a role, for example, in "Big Steel" decision not to undertake such an important innovation as thin slab casting. However, it is when an innovation reaches—or completes—the scale-up phase that cultural considerations become most intense. Now the entire organization is involved and a final decision has to be made whether to embrace, isolate, or outright reject the new technology.

TABLE 14.16   Aboveground Advanced Materials Innovation Phase VI: Scale-Up Phase—"Scaling" Risk

Advanced Materials Innovation	Company	Degree of Perceived Risk	Result: Go/No Go	Explanation
Thin slab steel	Nucor	Low	Go	Past experience scaling steel minimills + trusted contractors
Thin slab steel	"Big steel" companies	NA	NA	NA
Ultrathin steel + microalloys	Nucor	Low	Go	Past experience scaling steel minimills and the thin slab plant + trusted contractors + economies of scope strategy
High-pressure chemicals: first generation	DuPont	Low	Go	Economies of scope strategy
Nylon	DuPont	Low	Go	Close link with trusted department: Ammonia Department
Orlon	DuPont	Low	Go	Close link with "nylon model"
Dacron	DuPont	Low	Go	Close link with "nylon model"
Lycra	DuPont	Low	Go	Close link with "nylon model" + confidence in interfacial polymerization
Kevlar	DuPont	Low	Go	Confidence in interfacial polymerization + recent memory of Lycra success
Polycarbonates	DuPont	NA	NA	NA
High octane (catalytic) fuels: fixed bed	Sun Oil	Low	Go	Faith in Houdry and his process + Sun shipbuilding facilities
High octane (catalytic) fuels: fluid bed	Sun Oil	NA	NA	NA
High octane (catalytic) fuels: fixed bed	Jersey Standard (Exxon)	NA	NA	NA
High octane (catalytic) fuels: fluid bed	Jersey Standard (Exxon)	Low	Go	Lewis' research at MIT gives blueprint for scaled plant + World War II (government will pay for scale-up)

(continued overleaf)

TABLE 14.16 *(Continued)*

Advanced Materials Innovation	Company	Degree of Perceived Risk	Result: Go/No Go	Explanation
Polysilicones	GE	Low	Go	Impressive demonstration to managers of ease of scale-up of fluid process
Advanced polyolefins and metallocenes	DuPont	NA	NA	NA
Transistor: point contact	Bell Labs	Low	Go	Trust in Jack Morton + faith in Western Electric
Diode: four layered	Shockley Semicon-ductor	NA	NA	NA
Transistor: three-layered junction (bipolar)	Shockley Semicon-ductor	NA	NA	NA
Transistor: planar silicon junction (bipolar)	Fairchild Semicon-ductor	Low	Go	Planar process easily scaled up with economies of scope
Transistors: MOS	Bell Labs	NA	NA	NA
Integrated circuit: bipolar	Fairchild Semicon-ductor	Low	Go	Planar process easily scaled up with economies of scope
Integrated circuit: MOS	Fairchild Semicon-ductor	High	Go	Antagonism between R&D in Palo Alto and manufacturing in Mountain View over MOS increases risks of failure of scale-up. However, technically skilled top managers continue to see potential in MOS and so continue to support project
Integrated circuit: bipolar	Texas Instruments	Low	Go	Faith in Kilby pushes TI to design or acquire equipment for scaling up his prototype
Integrated circuit: MOS	Texas Instruments	NA	NA	NA
Memory chip: MOS	Mostek	Low	Go	Mostek puts its hope on ion implantation technology, which readily scales up + promises economies of scope benefits

TABLE 14.16   *(Continued)*

Advanced Materials Innovation	Company	Degree of Perceived Risk	Result: Go/No Go	Explanation
Memory chip: DRAM	Intel	Low	Go	Progress made in making silicon gate (SG) process work + ready scalability of the SG process + promise of economies of scope benefits + shift register safety net
Memory chip: EPROM	Intel	Low	Go	Comfort in the fact that EPROM chip based on MOS-SG process + faith in scalability of MOS-SG process + promise of economies of scope
Microprocessor	Intel	Low	Go	Comfort in the fact that EPROM chip based on MOS-SG process + faith in scalability of MOS-SG process + economies of scope promise
Microprocessor	Mostek	NA	NA	NA
Semiconductor laser (850 nm)	Bell Labs	Low	Go	Faith in DeLoach + continual progress made + faith in Western Electric
Semiconductor laser (1.3 µm)	Bell Labs	NA	NA	NA
Semiconductor laser (1.3 µm)	MCI	Low	Go	Contracted out development and production of lasers (and optical fibers)
Silicon-germanium chip	IBM	Low	Go	Faith in Meyerson + proven scalability of his ultrahigh vacuum (UHV) chemical vapor deposition (CVD) technique
LCDs	RCA	NA	NA	NA
LCDs	ILIXCO	NA	NA	NA
LCD: electronic calculators	Sharp	Low	Go	A number of factors reduced perceived risk of scale-up: secured supply of critical liquid crystal + new type of liquid crystal cell designed that was easily manufactured + funding and knowledge transfer from MITI + perceived effectiveness of interdisciplinary teams + focused on a single mass-produced product

(continued overleaf)

TABLE 14.16   (*Continued*)

Advanced Materials Innovation	Company	Degree of Perceived Risk	Result: Go/No Go	Explanation
LCD: watches	Seiko	Low	Go	In this case as well, a number of factors come into play to reduce perceived risks: clear advantages and promise of the preferred (twisted nematic) process + focus on only one product (LCD for watches) + active and clear communication between project leaders and top management
Thin-film transistors–LCD	Westing-house	NA	NA	NA
Thin-film transistors–LCD	Panelvision	NA	NA	NA
Thin-film transistors–LCD: flat-screen TVs	Sharp	Low	Go	MITI orchestrates industry-wide consortium to spread financial and technical risks + focus on one product
Thin-film transistors–LCD: computer screens	IBM	Low	Go	IBM joint ventures with Japanese firm (Sharp) to spread technical and financial risks
Nanotubes: single walled	Carbon Nanotech-nologies, Inc. (CNI)	NA	NA	NA
Nanotubes: multiwalled (Baytubes)	Bayer Advanced Materials	NA	NA	NA

In the case of isolation, a spin-off entity might be formed to specialize in the new material. In the case of aboveground projects, it might seem strange that a company would allow the innovation process to go so far and then back away from it. This can happen for a number of reasons. The management itself may have changed during the time of scale-up and does not see eye to eye with the previous administration on the strategic value of the new material and so will cancel the project. It may be that project leaders have been so successful in convincing management to proceed with the venture that it is only now, once a plant is in progress, that they realize they made a mistake in letting it proceed to this point, and so they put a stop to it (a situation akin to prosecutors doing their jobs so well they convict innocent people). Another possibility is that only now, with a commercial plant in sight, do forces within the organization understand the implications of what the new material will mean to the company, and they, believing it conflicts with their values and goals, rise up against it. When the old guard plant managers at Jersey Standard fought against new catalytic cracking technology, they came close to forcing the company to abandon its new refining units.

Cultural conflict involves risk: when a culture opposes something (or someone) different as "the other," it feels itself—its values and beliefs—threatened and avoids or rejects it.[13] A new technology on the verge of commercialization may or may not threaten the organization. A company that is internally oriented tends not to license out its technologies but prefers to work it itself. This is an important consequence of its cultural makeup. If corporate leaders find out rather late in the game that they will need to license it if it is to be profitable, they will see this as a risk—an attack on its cultural identity—and may very well jettison the technology. It is then of interest to identify what the culture of a company is at the time of scale-up and commercialization and then see if that new technology accords well with that culture. If it does, then the corporate powers will perceive it as a familiar entity and that no risk is involved with finally "allowing it to happen" within the company.

How then to characterize a company's culture at this crucial time for an advanced materials innovation? As the guru of organizational culture, Geert Hofstede, tells us, there are a number of cultural dimensions in any company (or society)—for example, individualism versus collectivism.[14] From our case histories, we can identify three powerful distinct cultural traits. One of these measures the degree of isolation of an organization, that is, whether it is "internally" or "externally" oriented. Companies that are vertically integrated and work the technologies they create (they do not license out) are internal organizations; those that establish extensive networks with outside suppliers, license out their innovations, and so forth are external. A second dimension tells us whether an organization has a "dynamic" or "static" culture. A dynamic company believes in constantly creating new materials technologies to replace the old ones. Such a culture actively cannibalizes its existing products and processes and bases its competitive advantage on its ability to excel at economies of scope.[15] A static culture focuses in on one or two major products. It does not like to cannibalize its existing products. It competes through a moderate level of product innovation but even more through pumping economies of scale. Finally, there is the third dimension that involves the perception of time: "long-term" oriented firms are willing to tackle difficult, "blue-sky" research; they spend the time and money necessary to solve long-range problems in the hope of coming up with earth-shaking new technologies. "Short-term" oriented companies want new technology quickly; they do not want to wait for the rare great-leap discoveries but are interested only in projects that produce results fast (we might even characterize these cultures as having a form of attention deficit disorder); these quick innovations may be only incremental in nature but may also be rather major, very important products or processes.[16]

Five points about these dimensions are in order. First, any company's culture contains all these (as well as other) dimensions at different times and to varying degrees. However, we identify only that (those) dimension(s) that appears to be the most prominent at the time when one of our advanced materials innovations is on the verge of entering the market. Second, as this last sentence suggests, a corporate culture may have more than one prominent characteristic, as has been indicated in the narratives. Thus, for example, Nucor at the time of thin slab casting can be described as both "static" and "internal." Intel's culture is even more protean being as it was of equal measure, "dynamic," "internal," and "short term." Third, the dimension assigned to a company is relative to the behavior of other leading companies within the same industry. A company like Nucor that may appear static when compared to other firms in, say, the semiconductor industry was (at the time of scaling up its CASTRIP process) quite dynamic when placed beside other steel producers. Fourth, there are no better or worse dimensions when it comes to breakthrough innovation, for example, a static or short-term company can turn out important new materials. The point is whether a company's culture is compatible with the requirements of a given innovation, for this decides if the firm will accept or reject its own creation. Fifth, a company's culture can—and often does—change over time. It may not be the same at the time of scale-up as it was a few years earlier.

Various forces internal as well as external can come into play that alter an organization's cultural makeup. The culture of a company emerges from the beliefs and values of its founders and the type of business with which it is involved. But this root culture—the company's cultural DNA, as it were—is not immutable. Externally, government action has played an important part in our stories. The threat of antitrust action by the Justice Department—and the military's attempt to supply its own explosives—had an enormous impact on DuPont. These threats forced the company to modify its culture from static to dynamic as embodied in its diversification strategy. We have seen similar external pressures alter the cultures of IBM, Bell Labs, and other companies. Internally, a management shake-up, the aftermath of a previous innovation experience, or the efforts of a project champion to change a culture so that the company would be more amenable to a favored technology are also powerful cultural game changers.[17]

Table 14.17 summarizes the cultural issues at play within the surviving firms around the time that each innovation is being scaled that have been described in our narratives. Columns 3–8 describe the major cultural dimensions (D = dynamic, E = external, I = internal, LT = long term, S = static, and ST = short term). An "X" is placed in the column(s) that most accurately describes a particular corporate culture. On the right-hand side, columns 9–11, the table shows whether the corporate culture at that time (as indicated in columns 3–8) was part of firm's original DNA or whether an environmental factor (En) or efforts of a champion (Ch) forced the root culture to change. In some cases, such as when an entrepreneur/champion founds a company based on that particular innovation, the table marks both the "DNA" and "Ch" columns as relevant. Then, column 12 shows whether the corporate culture accepted the innovation with "PC" (project completed). As usual, "NA" in this column says that the project terminated prior to reaching this stage. In one case, a project was ended at this stage (indicated by "NG") because of a cultural shift, specifically, when Sherman Fairchild and "Mahogany Row" moved in

TABLE 14.17 Aboveground Advanced Materials Innovation Phase VII: Commercialization Phase—"Cultural Risks"

1	2	3	4	5	6	7	8	9	10	11	12	13
Innovation	Company	D	S	E	I	LT	ST	DNA	En	Ch	Result: PC/NG	Comments
Thin slab steel	Nucor		X		X			X		X	PC	Nucor's culture at this time can be described as "internal" and "static" under Iverson. It had developed a stance that it had unique qualities and a manifest destiny to succeed. Nucor at this time was also a static company in the sense that it did not want to jeopardize existing operations with new technology; rather new technology should add to the company's capabilities without sacrificing current units. Moreover, new technology should not change too rapidly but serve a large, stable market. This "internally static" culture was imposed on Nucor by president/founder Iverson in the early years of Nucor's history. Thus, we can also say that it is part of Nucor's DNA. Thin slab steel technology met the requirements of Nucor's culture. Plant built in-house and given its own location (Crawfordsville) so as not to replace an existing plant. Most importantly, one type of steel made—low carbon flat steel—for appliance and automotive industries

(continued overleaf)

TABLE 14.17 (*Continued*)

1	2	3	4	5	6	7	8	9	10	11	12	13
Innovation	Company	D	S	E	I	LT	ST	DNA	En	Ch	Result: PC/NG	Comments
Thin slab steel	"Big steel" companies										NA	The "big steel" companies never reach this stage in the innovation process. They terminated any consideration of pursuing process at the initiation stage
Ultrathin steel + microalloys	Nucor	X		X					X		PC	By the time Nucor entered upon the thin strip process in the 1990s, the company had shifted from being internally to externally oriented culture. It had also become a more dynamic company. The economic problems with chasing a low-value thin slab market made the company anxious to be able to be more nimble in entering numerous high-value steel markets.
												Nevertheless, Nucor also continued to search outside its walls for existing prototype processes to scale up.
												Nucor's revised culture ("external dynamism") came from environmental pressures.
												Thin strip steel technology met the requirements of Nucor's modified culture: the prototype came from an outside source (Australia), and thin strip casting provided the product stream and market flexibility Nucor now demanded

High-pressure chemicals + advanced fibers: nylon, Orlon, Dacron, Lycra, Kevlar	DuPont	X		X		X	X	PC

DuPont's initial—its DNA—culture could be described as essentially static: it made one type of product (explosives) for a limited market (military, construction) and rarely brought in new technology to replace the old. By the 1920s, DuPont was becoming much more dynamic and externally oriented company: it diversified its product line through acquisition and cannibalized when necessary. By the 1930s, under the potent influence of DuPont champion Charles Stine, high-technology culture had become more intensely dynamic but now more internalized: it wanted new products it created in-house through the application of science to technology.

Moreover, despite protestations from time to time, top management was ultimately willing to replace older products (such as rayon) with the newer and more promising technologies (nylon).

This new culture of "internal dynamism" was compelled by environmental developments, particularly loss of the military as customers, threats of federal antitrust actions, and the example of internally generated R&D coming from other industries (e.g., GE and the electrical industry), as well as through the influence of a champion (Stine).

High-pressure chemicals and advanced fibers met the requirements of DuPont's culture of "internal dynamism" as embodied in its diversification strategy and were therefore commercialized by DuPont

(continued overleaf)

TABLE 14.17 (*Continued*)

1	2	3	4	5	6	7	8	9	10	11	12	13
Innovation	Company	D	S	E	I	LT	ST	DNA	En	Ch	Result: PC/NG	Comments
Polycarbonates	DuPont										NA	DuPont never reaches this stage in the polycarbonate innovation process. They terminated any consideration of pursuing the technology at the early research stage
Polycarbonates	GE		X						X		PC	The major theme in the evolution of GE's high-technology culture has been "internalism" versus "externalism" traditionally, and GE research had been internally oriented. The company viewed the output of internally generated GE research as only for increasing the competitiveness of its in-house electrical products. Until the 1940s, top management in fact did not want its R&D output to escape GE's corporate walls as it could then be incorporated into competitors' products and aid them in taking market away from GE. But a decade before the polycarbonate project, the champion of the polysilicone project (Jeffries) forced a change of culture at GE. GE now saw itself as a player in the chemical industry. Its new venture group (the Chemical Department) was to look for any promising technology—whether in-house or outside its walls—that the company could use to achieve a competitive advantage in the external market for a particular chemical product. In other words, advanced synthetics were to become a profit center in and of themselves.

GE's "externalized" attitude toward the polycarbonates directly benefitted from the earlier efforts of the polysilicone champion (Jeffries).

Just like the polysilicones, the polycarbonate process required a mass market in order to be competitive. It also is a synthetic with a broad range of market applications and so had the demand to justify mass production. Accordingly, the polycarbonates met the requirements of GE's new "externalist" culture and were therefore commercialized by GE

		X	PC
High octane (catalytic) fuels: fixed bed	Sun Oil	X	

Traditionally, Sun Oil had had a rather static culture in the sense that it nurtured tried-and-true thermal refining processes innovated by others and merely improved them to increase efficiencies and product quality. When he became president, J. Howard Pew changed the culture. He was open to radically new processes that would replace the old, as long as the breakthrough innovations provided the company with a competitive advantage. Under Pew, then, Sun's culture shifted from static (noncannibalistic) to dynamic (cannibalistic). As far as Pew was concerned, it didn't much matter if the source of the new process was internal or external.

(continued overleaf)

TABLE 14.17 (*Continued*)

1	2	3	4	5	6	7	8	9	10	11	12	13
Innovation	Company	D	S	E	I	LT	ST	DNA	En	Ch	Result: PC/NG	Comments
												Sun's new culture came from Pew, and we can say it originated with one of the company's important champions. However, because J. Howard Pew did not impose this culture at the beginning of the company's history, we cannot say it is part of Sun's DNA.
												Houdry's fixed-bed process amply met the requirements of Sun's new dynamic culture. (Please note that the coming of Houdry and his process to Sun would further alter Sun's culture so that it became much more internalized as well as dynamic. This new internally dynamic culture meant that only Houdry and his approach to refining held sway at Sun. While this impacted Sun's approach to innovation later on, it did not have much influence on Sun in 1936, when Houdry first approached Pew about supporting his research.)
High octane (catalytic) fuels: fluid bed	Sun Oil										NA	DuPont never reaches this stage in the fluid cracking process. They terminated any consideration of pursuing the technology at the early research stage
High octane (catalytic) fuels: fixed bed	Jersey Standard (Exxon)										NA	Exxon never reaches this stage in the fixed-bed cracking process. They terminated any consideration of pursuing the technology at the initiation stage

Technology	Company				Description
High octane (catalytic) fuels: fluid bed	Jersey Standard (Exxon)	X	X	X PC	The coming of Frank Howard as head of Jersey's R&D organization pushed Jersey into a more externally oriented dynamism. Howard wanted Jersey to be more aggressive in developing new and advanced refining technology that would replace the old, existing operations. Moreover, he actively searched outside for these superior processes. Thus, he engineered the patent exchange agreement with Germany's IG Farben and looked to MIT to undertake fundamental research that would lead to fluid cracking. Howard's new culture clashed sharply with the traditional internally static culture—its cultural DNA—that was defended by the old guard plant managers. Frank Howard was the champion who forced this cultural shift onto Jersey and its R&D. Fluid catalytic cracking amply satisfied the requirements of Howard's new cultural regime. (A less radical process that did not incorporate the latest outside research and knowledge that would have been rejected by Howard and Jersey.)
Polysilicones	GE		X	X PC	Creative efforts by champion (Jeffries) causes GE research to shift from internal to external culture (refer to the polycarbonates above). GE corporate embraces polysilicones, which requires external markets to realize the new material's full benefits via fluidization
Advanced polyolefins and metallocenes	DuPont			NA	DuPont never reaches this stage in achieving Unipol process. They terminated any consideration of pursuing the technology in the early research stage

(continued overleaf)

TABLE 14.17 (*Continued*)

1	2	3	4	5	6	7	8	9	10	11	12	13
Innovation	Company	D	S	E	I	LT	ST	DNA	En	Ch	Result: PC/NG	Comments
Advanced polyolefins and metallocenes	Union Carbide	X		X						X	PC	Union Carbide's culture underwent a radical shift as the Unipol process scaled up. To be able to accept the Unipol process commercially, Carbide culture had to shift its thinking along two cultural dimensions: it had to embrace both (i) new dynamism (the new process would have to replace the old, existing high-pressure units) globally and (ii) new external orientation regarding licensing policy. A middle-level champion for the process (Joyce) was the key influence convincing top management to adopt the new "external dynamism" in order to accommodate the new technologies of the 1980s, such as Unipol. Unipol of course satisfied the requirements of the new culture that it itself had been instrumental in creating. It was therefore warmly accepted by Carbide and rapidly commercialized
Transistor: point contact	Bell Labs		X	X				X	X		PC	A root cultural attribute of AT&T/Bell Labs was that it was technologically static. Despite its impressive scientific discoveries, the organization at heart did not like rapid technical advance. This is because AT&T, Bell's parent, could not afford to cut into and revamp its telephone network very often. They thought in terms of a well-run, stable communications system.

Another major cultural trend that came later (1940s) was a shift from an internal to external mentality. While the "not invented here" syndrome continued in force at Bell, AT&T increasingly wanted Bell to invent for the benefit of outside companies. The reason was that the growing threat of federal antitrust action against the telephone monopoly forced it to show that it was willing to share the fruits of its R&D with other companies. By doing so, AT&T hoped to use Bell to quash government lawsuits. AT&T shared its technology in two ways: nonexclusive licenses at low rates to any interested company and selling the technology to the government—especially the military.

The source of Bell's "externally static" culture then came from both its DNA (the "static" part) and, as with DuPont, from environmental forces, that is, external pressures exerted by the threat of government action (the "external" component).

AT&T embraced the transistor because it conformed closely to Bell's externally static culture. It did not threaten the current network as it would be years before the device would work well enough or be cheap enough for such use, but, at the same time, it offered an immediate way to reach outward to other companies and research organizations. It was a well-publicized, major innovation that greatly intrigued many electronic companies who were only too happy to become licensees. The transistor also was of great interest to the military, especially for use in advanced missile systems. Both Bell (working with Western Electric) and its licensees could provide this technology

(continued overleaf)

TABLE 14.17 (*Continued*)

1 Innovation	2 Company	3 D	4 S	5 E	6 I	7 LT	8 ST	9 DNA	10 En	11 Ch	12 Result: PC/NG	13 Comments
Diode: four layered	Shockley Semiconductor										NA	Shockley Semiconductor never reaches this stage in the innovation process for the four-layered diode. The company terminated any consideration of pursuing the technology by the early development stage
Transistor: three-layered junction (bipolar)	Shockley Semiconductor										NA	Shockley Semiconductor never reaches this stage in the innovation process for the three-layered bipolar (junction) transistor. The company terminated any consideration of pursuing the technology by the later research stage
Transistor: planar silicon junction (bipolar)	Fairchild Semiconductor	X					X	X		X	PC	The founders of Fairchild (Noyce and Moore) begin to create a "short-term dynamic" culture that looked to apply a robust and elastic process in creating new and more powerful semiconductor products on a regular basis (every 18 months or so). In this case, the planar process is the central technology. By itself, the planar process was flexible up to a point. This culture was imposed on the company by its founders/champions. It is thus a part of company's DNA. Fairchild accepted the planar silicon transistor, even though it cannibalized its existing mesa transistor. This technology fit very nicely into Fairchild's "short-term dynamic" culture

MOS transistor	Bell Labs				NA	Bell Labs never reaches this stage in the innovation process for the MOS transistor. The company terminated any consideration of pursuing the technology by the later research stage
Integrated circuit: bipolar	Fairchild Semiconductor	X	X	X	X PC	The founders of Fairchild (Noyce and Moore) expand and solidify a "short-term dynamic" culture that looks to apply a powerful and elastic process (again, the planar process) in creating new and more robust semiconductor products on a regular basis (every 18 months or so). This culture was imposed on the company by its founders/champions. It is thus a part of company's DNA.

Fairchild Semiconductor embraced Noyce's invention because it fit perfectly into the culture that Noyce himself helped create.

It is clear that if Noyce had invented the integrated circuit along the lines Jack Kilby came up with, Fairchild would have rejected it as too bulky and "messy" and not capable of keeping up with the short-term dynamic requirements of Moore's law.

TI only learned how to turn Kilby's "frog" of an IC into a prince by adopting many aspects of Fairchild's process (e.g., planarization, photolithography)

(continued overleaf)

TABLE 14.17 (*Continued*)

1	2	3	4	5	6	7	8	9	10	11	12	13
Innovation	Company	D	S	E	I	LT	ST	DNA	En	Ch	Result: PC/NG	Comments
Integrated circuit: MOS	Fairchild		X				X		X	X	NG	Fairchild Semiconductor's culture changed suddenly when Sherman Fairchild brought in "Mahogany Row" personnel to take charge. This move was, in part, a result of the difficulties Noyce and Moore were having with MOS circuits.

This action hanged the company's culture from one of "short-term dynamism" to one that can be described as "short-term static," which emphasizes the rapid improvements to existing technology.

This cultural revolution was in part imposed by one person (Sherman Fairchild) and in part was the result of external conditions (Fairchild Semiconductor's declining position within an increasingly competitive industry) that caused Sherman Fairchild to act.

Noyce and Moore, when in charge, were willing to keep this project going despite its many problems because they—especially Moore—understood that if successful, MOS would provide the company a process that was superior to bipolar technology in terms of being able to create new devices on a regular basis. It would, in other words, be the perfect tool to reflect the company's culture of "short-term dynamism."

However, Mahogany Row imposed a new culture on the company, which we call short-term static. This new culture rejected further development and scale-up of MOS technology. |

Integrated circuit: bipolar	Texas Instruments	X	X	X	PC

This cultural shift was brought about by Sherman Fairchild and his team from Motorola in response to environmental pressures.

This new culture rejected pursuing MOS technology

From its founding (as Geophysical Service Inc.)—as an oil services company—Texas Instruments R&D can be described as having an "internally static" culture.

With respect to its "static component," it has been a company that likes to service single, large-volume markets using relatively static technologies. Moreover, it has always been an intensely internal company: it (generally) developed its own components and built equipment and products from in-house. The final products were then sold to outside customers.

(Thus, TI R&D echoed the internal culture of the R&D of places like GE.)

TI's "internally static" culture was an integral part of TI research's DNA.

TI immediately embraced Kilby's invention because it clearly fit very nicely into the corporate culture. It clearly solved the tyranny of number problem, which made it a perfect circuit for new, improved, and more easily produced consumer products, such as electronic calculators. Top management soon understood that it would become an important component in products and technologies that TI could make to sell to the military and, even more importantly, mass produce for consumers. Thus, TI initially made computers out of Kilby's IC and sold them to the military for the missile program.

(continued overleaf)

TABLE 14.17 *(Continued)*

1	2	3	4	5	6	7	8	9	10	11	12	13
Innovation	Company	D	S	E	I	LT	ST	DNA	En	Ch	Result: PC/NG	Comments
Integrated circuit: MOS	Texas Instruments										NA	Most importantly for TI, it went for the mass market by incorporating ICs as the basic components of TI's new line of electronic calculators
												Texas Instruments never reaches this stage in the innovation process for the MOS transistor. The company terminated any consideration of pursuing the technology as a first mover in the field by the later research stage. (TI would eventually make MOS devices after Intel led the field in this area.)
Memory chip: MOS	Mostek	X			X			X		X	PC	Mostek founded on strength of its ion implantation technology to make its version of MOS circuits and components.
												While a powerful process, it did not give the company the flexibility and range that the MOS-SG gave Intel. While Mostek used its ion implantation advantage to initially outcompete Intel in memories, its process simply did not have the elasticity needed, and Mostek could only come out with a narrow range of memory chips. Essentially, Mostek R&D was from the start an "internally static" culture: it relied on its own in-house process for competitive advantage, and, while the company did sell its chips to outside equipment makers, it relied on making its own electronic equipment and products for the mass market: telephone tone and pulse dialers, touchtone decoders, watch circuits, and electronic calculators—from its MOS semiconductor circuits. These products did not demand rapid change in circuit components.

Mostek's "internally static" culture is created at the beginning by its founders and thus is also part of its DNA. Memory chips conformed closely to Mostek's culture, which was formed from its total reliance on ion implantation as a competitive weapon. Mostek could make memory chips—and excel at them—using its homegrown ion implantation technology, and they could be used to make in-house products without the need for rapid change to new generation memory ICs for electronic products for the mass market

Intel's founders (Noyce and Moore) brought with them the "short-term dynamic" culture they had nurtured at Fairchild, a culture that could be satisfied with Intel's development of the MOS-SG process. The early addition of Andy Grove and his fierce competitiveness imposed an added cultural component to Intel: an intense internalism. More than any one, Grove had no desire to release the secrets of their success—especially with the silicon gate process—to competitors. He disliked having to license/second source and eventually stopped doing it. Intel now was interested only in those R&D projects that used its MOS-SG process. Accordingly, from the beginning Intel had what might be called a culture of "internal short-term dynamism."

Noyce, Moore, and Grove imprinted the culture of "internal short-term dynamism" on Intel almost from the start and all championed going after the DRAM.

The DRAM chip fit very nicely into Intel's culture: it was created in-house using Intel's MOS-SG process. It could be made smaller and denser every 18 months using this process

(continued overleaf)

Memory chip: DRAM						
Intel	X	X	X	X	X	PC

TABLE 14.17 (*Continued*)

1	2	3	4	5	6	7	8	9	10	11	12	13
Innovation	Company	D	S	E	I	LT	ST	DNA	En	Ch	Result: PC/NG	Comments
Memory chip: EPROM	Intel	X		X	X		X	X		X	PC	Intel's top management—especially Gordon Moore—believed in the future of the EPROM right after he saw its demonstration by Dov Frohman. The EPROM fit very closely to Intel's internal short-term dynamism: it was internally created, it was rooted in Intel's silicon gate core competency, and it was a radical leap forward in Intel's memory technology
Microprocessor	Intel	X		X		X	X		X	PC			Intel's top management had much more difficulty accepting the viability of the microprocessor as a potentially important Intel product for three reasons: (i) it was not a memory product (and Intel saw itself as primarily a memory company); (ii) it was instead a logic product, which meant that it required the talents of circuit designers (Intel did not see itself with a core competency in circuit design); and (iii) it appeared useful only in a single application (electronic calculators). Intel began to support commercialization of the microprocessor once it understood that (i) it came from the same MOS-SG process that was at the real heart of Intel's core competency and (ii) the microprocessor had numerous potential applications above and beyond calculators and that these applications would grow as new and more powerful microprocessors came along, following closely the dictated path of Moore's law.

Technology	Company					Description
Microprocessor	Mostek				NA	The two champions of the microprocessor—Ted Hoff and Federico Faggin—were responsible for actively convincing Intel's top management that their invention fit perfectly within the company's internal short-term dynamic culture: it was created within the company using the MOS-SG process, which would continue to turn out new generations of the device every 18 months. Mostek never reaches this stage in the innovation process for the microprocessor. The company terminated any consideration of pursuing the technology by the early research stage. (Mostek lacked the basic technology—i.e., the MOS-SG process—to be able to invent the microprocessor.)
Semiconductor laser (850 nm)	Bell Labs	X	X X	X	PC	The semiconductor laser (850 nm) was embraced by Bell; it well suited AT&T/Bell's "externally static" culture (see "Transistor: Point Contact"). Firstly, it did not threaten AT&T's existing telephone network but was going to be used in a totally new, parallel network, one made of optical fibers rather than copper coaxial cable. Secondly, finding a stable, long-lasting, working laser took Bell a long time. Bell had every reason to believe it would be many years before a new improved laser would come along, which is what it wanted since it did not want to have to revamp its costly fiber optics network any time soon. Thirdly, Bell believed constructing such a network—one external to its traditional copper system—would allow it to keep up with future demand and, as a result, keep the federal government satisfied that AT&T should retain its communications monopoly.
Semiconductor laser (1.3 μm)	Bell Labs				NA	Bell Labs never reaches this stage in the innovation process for the 1.3-μm laser. The company stopped trying to be the first mover in the field by the early development stage. (Bell would never be more than an also-ran in this technology.)

(continued overleaf)

TABLE 14.17 (*Continued*)

1	2	3	4	5	6	7	8	9	10	11	12	13
Innovation	Company	D	S	E	I	LT	ST	DNA	En	Ch	Result: PC/NG	Comments
Semiconductor laser (1.3 μm)	MCI			X				X		X	PC	MCI had an externalist culture. It aggressively set out to identify and license from outside innovators the most advanced optical fiber technology and commercialize it. This externalist culture was imposed on the start-up by founder (Gower) and so is part of company's DNA. The 1.3-μm laser was perfectly suited to Gower and MCI. It was an outside technology researched and partially developed by other firms. It also was superior to AT&T's 850-nm laser and so would give MCI the competitive edge
Silicon-germanium chip	IBM	X		X						X	PC	While not part of its initial DNA (as imposed on the company by its founder Thomas Watson, Sr.), IBM began to evolve into a company that was both dynamic and externally oriented. This process started with the presidency of Thomas Watson, Jr. (who succeeded his father) and accelerated with IBM's association with Microsoft (software) and Intel (microprocessor).
												It was reinvigorated by the presidency of Louis Gerstner, who reopened the idea of IBM departments working together and the formation of external alliance and partnerships.
												Champion (Meyerson) understood the possibilities of IBM's external dynamism and fully exploited it to advance his project.
												Meyerson's silicon-germanium chip project fit into this culture in two ways: (i) it was a device that promised rapid progress in its power and range through new generation chips and (ii) its application meant reaching out to strategic partners in the mobile communications field

Technology	Company			
LCDs	RCA		NA	RCA never reaches this stage in the innovation process for LCDs. The company terminated any consideration of pursuing the technology by the early development stage
LCDs	ILIXCO		NA	ILIXCO does not reach this stage in the innovation process for LCDs. The company terminated any consideration of pursuing the technology as a first mover by the later development stage
LCD: electronic calculators	Sharp	X X	X PC	The culture at Sharp—as well as other Japanese firms—combines an external orientation (departments work together to create a product) with a static mentality (focus on a single product for one large market).

This "externally static" culture is imbedded in the DNA of Sharp (as well as other Japanese companies). The project champion (Wada) also played a role in convincing top management that his pet project fit very well into Sharp's culture.

The LCD pocket calculator fit very well into—and was readily accepted by—Sharp's culture. It required the input of many of Sharp's departments. Due to Sharp's external culture, all the departments contributed to the technology and there was no conflict between departments over the technology when completed that could have derailed its further commercialization.

Moreover, the static nature of Sharp's culture eagerly accepted into the fold a new technology with a single growing market and one that values detailed precision

(continued overleaf)

TABLE 14.17 (*Continued*)

1	2	3	4	5	6	7	8	9	10	11	12	13
Innovation	Company	D	S	E	I	LT	ST	DNA	En	Ch	Result: PC/NG	Comments
LCD: watches	Seiko		X	X				X		X	PC	The culture at Seiko—as well as other Japanese firms—combines an external orientation (departments work together to create a product) with a static mentality (focus on a single product for one large market). This "externally static" culture is imbedded in the DNA of Seiko. Again, we cannot forget the role of the champion (Yamazaki) in convincing top management that his pet project fit very well into Sharp's culture. The LCD watch fit very well into—and was readily accepted by—Seiko's culture and what we said about Sharp and its pocket calculator applies to Seiko and its LCD watch. The project required the input of many of Seiko's departments. Due to Sharp's external culture, all the departments contributed to the technology, and there was no conflict between departments over the technology when completed that could have derailed its further commercialization. Moreover, the static nature of Seiko's culture, the fact that the LCD watch did not "creatively destroy" other products and that there was one single market for the technology, meant that it conformed very closely to Seiko's deeply rooted culture

Technology	Company					Description
Thin-film transistors–LCD	Westinghouse				NA	Westinghouse never reaches this stage in the innovation process for thin-film transistors and large LCDs. The company terminated any consideration of pursuing the technology by the later research stage
Thin-film transistors–active-matrix LCDs	Panelvision				NA	Panelvision never reaches this stage in the innovation process for thin-film transistors–large active-matrix LCDs. The company terminated any consideration of pursuing the technology by the later development stage
Thin-film transistors–LCD: flat-screen TVs	Sharp	X	X	X	PC	This technology conformed with Sharp's "externally static" culture (see "LCD Electronic Calculator"). It required participation both from different departments within Sharp and from different companies within Japan. Also, like Seiko's watch and Sharp's LCD pocket calculator, and in line with the company's static culture, the LCD flat-screen TV served one large consumer market (60-in. LCD flat-panel televisions) and did not cannibalize Sharp's existing products as the company did not make picture tube TVs. In this case, it was a government agency—MITI—that served as champion for the project and coordinated the national effort in developing the new technology
Thin-film transistors–LCD: computer screens	IBM	X	X	X	PC	While not part of its initial DNA (as imposed on the company by its founder Thomas Watson, Sr.), IBM began to evolve into a company that was both dynamic and externally oriented. This process started with the presidency of Thomas Watson, Jr. (who succeeded his father) and accelerated with IBM's association with Microsoft (software), Intel (microprocessor), and Meyerson's silicon-germanium chip. Thus both external environment and champion created new culture.

(continued overleaf)

TABLE 14.17 (*Continued*)

1	2	3	4	5	6	7	8	9	10	11	12	13
Innovation	Company	D	S	E	I	LT	ST	DNA	En	Ch	Result: PC/NG	Comments
												IBM's champion (Howard) exploited IBM's relatively new "externally dynamic" culture in order to advance his computer LCD technology.
												Howard's thin-film transistor–computer LCD technology conformed closely to IBM's culture in terms of relying on a joint partner to make LCDs for its computers (the external component) and on the need to rapidly develop new and more powerful computer screens to keep up with advanced computer chips and wireless devices (the dynamic component)
Nanotubes: single walled	Carbon Nanotechnologies, Inc. (CNI)										NA	Carbon Nanotechnologies never reaches this stage in the innovation process for single-walled nanotubes (SWNTs).
												The company terminated any consideration of pursuing the technology by the later development stage
Nanotubes: multiwalled (Baytubes)	Bayer Advanced Materials										NA	Bayer Advanced Materials never reaches this stage in the innovation process for multiwalled nanotubes (MWNTs).
												The company terminated any consideration of pursuing the technology by the later development stage

on Noyce and Moore and essentially gutted the MOS project at Fairchild Semicon-ductor. Finally, column 13 provides commentary on the intersection of culture and technology for that project. It explains why the company had the culture it did at that point (columns 3–8), the mechanism of how that culture came to be (columns 9–11), and the degree of compatibility between that culture and the project in question and thus the reason for acceptance or rejection by the organization. For example, in the case of Nucor and thin slab casting in the 1980s, the technology fit very nicely into the company's then dominant "internally static" culture. By the 1990s, environmen-tal (competitive) pressures forced a cultural shift on the company that compelled it to become "externally dynamic" and the CASTRIP process closely conformed to this new cultural regime. Column 13 in Table 14.17 provides a narrative, taken from the case studies, describing the cultural context greeting each innovation and implica-tions for final acceptance or rejection of the advanced materials technology by the company involved.

In reviewing this data, we find that project acceptance at this last stage of the innovation process is contingent upon top management perceiving minimal cultural risks. The dissonant case of Fairchild and MOS technology shows what happens when culture and technology clash (i.e., Fairchild declined and Intel was born). We should also be aware that, in some of the cases successful harmonization between a corporate culture and its own innovation only came about through the conscious, active, and strenuous efforts of a project champion to convince corporate leaders to alter their deeply rooted culture and strategic thinking in order to accommodate a new and promising—possibly game-changing—advanced materials technology. This occurred at GE (silicones and polycarbonates) and at Union Carbide (the Unipol pro-cess). This "cultural forcing" is typical of projects that have spent a considerable part of their time underground. Leaders of these renegade ventures ignore a company's official structural context and create their own out of sight of the established power structure. By the time they come to the surface, they (hopefully) have a product the company not only cannot ignore but has to bow down to by reorienting its existing cultural and strategic makeup. In contrast, projects that have been aboveground all along, although they also can turn out breakthrough materials, tend not to require its advocates to culturally or even strategically change a firm. Rather, their job is to show their bosses that the new product or process conforms very closely to an organi-zation's existing—and sometimes long-standing—cultural personality. The statistics reflect this. Out of all the aboveground projects we have examined, three-quarters (75%) of them were accepted by top management because they were consistent with the firm's cultural DNA. (This doesn't mean this acceptance was automatic as project leaders often had to take management by the hand and show them what was not imme-diately obvious and that their innovation conforms very closely within the company's cultural climate, such as with Intel and both the EPROM and microprocessor.)

14.5 THE STRUCTURAL CONTEXT AND ADVANCED MATERIALS INNOVATION

As our case studies show, underground and aboveground projects within companies achieve their goals in two very different ways. The former does so by ignoring the firm's official and severely conservative structural context and collection of rules and criteria through which an organization selects projects that conforms to its existing ways of doing things.[18] But aboveground projects do not have that luxury. They have to proceed with senior managers looking over their shoulders, watching everything they do and with the power of terminating the work at any point. The job of these wary managers is to support and uphold the company's structural context and to weed out those outlier ventures that do not conform to and even threaten the status quo. How then can a radically different project survive and grow in such an environment? How are the most important new materials to be created when so many firms are organized in such a way as to wipe them out before they can see the light of day? As we have seen time and again in our case histories, advocates of new and potentially radical technology who cannot escape underground cannot ignore the fact that they will be judged by a fundamentally conservative power structure. They must face this reality head on, learn the rules of the game intimately, and actively plan on how to negotiate the various obstacles that the existing organization will throw their way. Specifically, and paradoxically, they must find any way possible to make top management believe their venture is merely incremental in nature. They do this by convincing management that at any point in time along the road to innovation, the risks to be taken are negligible. In some cases, this will in fact be true; in others it will not. Whatever the case, the important point is that management believes these risks to be trivial. Persuasion, hard data, clever manipulation, and even outright deception are the basic techniques by which this is accomplished. Finding important information that gives direction and sense of inevitability to a project (IBM's silicon-germanium chip), deceiving managers who are not familiar with a technology that it will be easy to scale up (GE's silicones), teaching senior executives that a new product is actually rooted in the company's core process technology (Intel's microprocessor), and making an increasingly troubled corporate leadership more confident in a project by linking it to technology and personnel closely associated with past successes (DuPont's synthetic fibers) are typical risk management methods that champions of breakthrough technology employ throughout the innovation cycle to keep their bosses from pulling the plug on their ventures.

So, successful champions of aboveground radical technology work within a firm's fundamentally conservative universe but do so by carving out their own special path within it. It is this path we have been calling the "gauntlet of (minimal) risks." Figure 14.2 depicts this scenario. Within the firm, the two types of projects enter into the company's structural context. One type of project is called "induced" because it comes from the top of the organization and is therefore incremental in nature. These sorts of improvements to existing technology easily get through the trials of a watchful firm, which is set up to support just such risk-free efforts. The second type of project is called "autonomous."[19] This is a new and radical idea coming from bench scientists and (possibly) their immediate bosses in R&D. If this project tries to enter the official (or legitimate) strategic context path and faces its strict requirements for passage, it will die a certain death for it could never get through its stringent selection criteria. What is required is that leaders of aboveground autonomous projects find ways to neutralize the "risk alarms" set off by the ultraconservative strategic context. For example, if corporate leaders want to stop a project because they begin to realize it is creating a material or device very different from anything the company had dealt with in the past, the project

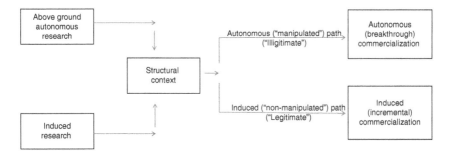

Figure 14.2 Aboveground innovation and the structural context

champion can point out that material or device is simply a by-product of a familiar and accepted process technology. If they want to call a halt because deadlines are being missed, you bring in trusted engineers and managers who have spearheaded past successes for the company to take over the project. If they fear that there will not be enough customers to make the new material profitable, you argue that the process is so flexible that one plant can create many products for many markets and that profitability will come from economies of scope. This alternative route by which project leaders directly address and "argue around" the official rules in order to get their ventures through a highly cautious selection process is what we call the "illegitimate" or "manipulated" path; the result of this path is breakthrough (or "autonomous") innovation.

Of course, not all aboveground autonomous ventures face the same level of difficulty in getting their projects through that "illegitimate" path. When top managers are themselves technically savvy, innovative thinking and linked into what is going on in their R&D department, they tend to be more patient of new ideas and the time and effort they take. We certainly saw that at Fairchild with MOS technology: despite the problems that process was having, Gordon Moore understood its potential and was willing to keep it going. Moore and Noyce also supported the EPROM and microprocessor soon after they were invented. On the other hand, when those in power are distant from the innovation process in general and from what is taking place in its own R&D backyard, they are less forgiving, as evidenced by the gutting of the MOS program at Fairchild with the invasion of the Mahogany Row crew. Generally, four factors help determine the chances that taking an illegitimate path will succeed: (i) flexibility of the technical process underlying the radical product, (ii) technical competence of top management, (iii) the degree to which top management perceives the technical process underlying the radical new product as a core and distinctive competency, and (iv) the degree of "flatness" or "integration" of the organization. Figures 14.3 and 14.4 represent, respectively, an "easy" path model (Nucor and Fairchild under Noyce and Moore) and a "difficult" path model (DuPont, Fairchild under Mahogany Row, Mostek). In these charts, a small distance between the induced and autonomous path represents the former model; a larger distance, the latter. The closer the lines, the more the autonomous, breakthrough project looks like an induced effort as far as risk perception is concerned, which is what is desired to facilitate pushing through a radical technology. The greater the distance between the paths, the more risky a radical effort appears to management. In this case, a far

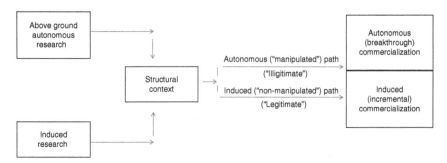

Figure 14.3   The easy path model of Nucor and Fairchild (Noyce/Moore regime)

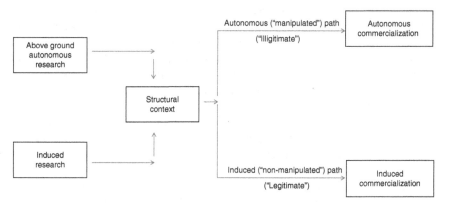

Figure 14.4   The difficult path model of DuPont and Fairchild ("Mahogany Row" regime)

greater effort—and more creativity—is needed to convince the corporate powers that the venture is not risky. This of course is not desired. Deception is more likely to be used in such cases simply because a more desperate situation requires desperate measures. This is exactly what happened with rayon, for instance. However, if the lines are too far apart, the difficulty in "working the system" becomes too great, and the only way the project will be able to survive is if its champions find a way to go underground and start building their case out of sight of senior executives. Finally, it is not necessarily true that the greater the distance of the path lines, the more radical the innovation. As we have noted, some of the great advanced materials innovation—such as microprocessor—came about by fairly easily manipulating the "legitimate" system, in this case, convincing corporate leaders that memories and microprocessor shared the same process (i.e., MOS-SG). Here the "autonomous" and "induced" path lines are very close indeed.

14.6   INVENTORS AND CHAMPIONS

One of the distinctions often made in innovation of the 19th and early 20th centuries, on the one hand, and the later 20th and 21st centuries, on the other, is the lessening importance of the individual genius and the growing role of large teams of specialists working together in corporate R&D departments.[20] Yet our narratives have shown that, even within big companies with large R&D organizations, individuals matter a great deal, as a recent book on the radical innovation process maintains:

> We were surprised by the lack of corporate attention to the critical roles played by creative technologists, entrepreneurial managers and visionary champions in the success of radical innovation projects.[21]

The following two sections review the important roles of these technologists and managers as champions in our advanced materials cases. We first look at the parts played by these leaders in easing a project through the gauntlet of risks that have to be faced and conquered over the course of the innovation process. We then turn to what exactly makes for a successful champion of advanced materials innovation.

14.6.1 Inventors, Champions, and the Gauntlet of Risks

Given the existence of the gauntlet of risks during aboveground innovation, why do some projects reach completion while others do not? Of the successes, why do some make it to the finish line faster and with less difficulty than others? Human action and particularly the role played by creative, clever, and energetic individuals really decide how quickly, efficiently, and effectively a particular project will pass through its various risk stages. Radical innovation unquestionably requires many hands. It cannot make its way through a company without the guidance, help, and approval of numerous people with different backgrounds and specialties spanning all levels of an organization. Still, some people are more involved than others. These are the central characters who conceive and initiate a project and then work to push it through a company, often in the face of indifference and even outright hostility from those who like things just the way they are. In our case histories, we can identify three types of central characters: the inventor, the champion, and the utility player. The *inventor* of the new material or process generally toils in the laboratory and by accident or design (or a little bit of both) discovers something important. Wallace Carothers (nylon), Ray Houtz (Orlon), Stephanie Kwolek (Kevlar), Daniel Fox (polycarbonates), Eugene Houdry (fixed-bed cracking), and Richard Smalley (buckyballs) are all examples of inventors. *Champions*, who may or may not be the inventors themselves, are all major supporters of a project. They are the ones that negotiate the various risks that come up over the course of a project. They see to it that, at any point in time, their superiors perceive that immediate risks facing the effort can be or have been minimized and that the project should be given the green light to move forward to the next phase where they then must once again come up with new risk minimization strategies. Finally, the *utility players* are neither inventors nor champions. They are given an assignment to do on a project and they then set on to do it. However, their importance should not be overlooked. Creative utility players can provide very important leverage to champions in their risk management strategies. Such was the case with Jersey Standard and fluid cracking when the champion (Eger Murphree) could tap the company's three most seasoned chemical engineers—Donald Campbell, Homer Martin, and Charles Tyson—to prove that the development and scaling of the process was in good hands. These utility players are really the right-hand persons of a successful champion.

Table 14.18 maps out the various roles played by inventors, champions, and utility players over a large sample of our advanced materials innovations. Most of these projects are the successful ones, that is, the ones that reached project completion. The entries are grouped according to innovation and company. Each major participant is identified as either inventor (I), champion (C), or utility player (U). In those cases where the inventor also acted as champion, they are identified as (I/C). For each participant, an "X" is placed in the space corresponding to those phases in which he or she had the most important impact. That is to say, the risk management strategies they employed were essential in the project moving successfully past that particular stage in the innovation process. The table shows how wide a span each inventor, champion, and utility player had an influence on the project and tells us where in the process "hand-offs" had to be made from one person to another to keep the project moving forward. The table also includes a column devoted to "context." This refers to the building of an organizational entity within a company that directly led to and supported project initiation. By creating such a contextual space, the inventor or champion makes it easier for an innovative effort to take root within a company. For example, DuPont's Charles Stine created a Central Research Department; the result of which led directly to nylon (and other synthetic fibers). Ken Iverson prepared a totally new facility in Crawfordsville that resulted in Nucor's two innovative plants, Bell Labs' Mervin Kelly's actions led to a semiconductor group resulting in the transistor, and Zay Jeffries was behind GE's venture group that nurtured the polycarbonates and silicones. All of these visionaries built a climate on which these major innovations could grow.

From this mapping, we can discern four general trends. First, establishing a preinitial context for the creation of a specific type of innovation—which helps to reduce the risks of a new idea getting started—appears to be important. Over 63% of our successful projects started out with the creation of such a contextual platform. Second, many important inventors seemed to have little will or capacity to play an active role in the innovation process beyond the early development stage. Approximately half of them demonstrated a fairly impressive ability to champion their inventions within a company up to a point. A number of them who were not champions still performed work that went beyond the first or second phase of the project. However, only a few of them managed to advance their cause beyond the prototype phase. Out of all the inventors listed, only 37% of them extended their efforts to this early commercial stage. Those who did—men like Eugene Houdry, Jack Kilby, Warren Lewis, Tomio Wada, T. Peter Brody, and Federico Faggin—tended to have engineering backgrounds (with the pure scientists Daniel Fox and Bernie Meyerson being the exceptions).

Third, those deemed to be project champions—whether inventors or not—influenced a fairly broad swath of activity linked to the innovation process. They spanned the technical and the business worlds and so were effective bridges between

TABLE 14.18 Inventors and Champions and Their Role in the Phases of Innovation

Champion	Company	Role	Innovation	Context	Initiation	Intellectual	Resource Minimization	Prototyping	Technology–Market Interaction	Scaling	Cultural
Ken Iverson	Nucor	C	Thin slab steel	X	X			X	X	X	X
Ken Iverson	Nucor	C	Ultrathin steel + microalloys	X	X			X	X	X	X
Roger Williams	DuPont	C	High-pressure chemicals: first generation				X	X	X	X	X
Charles Stine	DuPont	C	Nylon	X	X						
Wallace Carothers	DuPont	I	Nylon			X	X				
Elmer Bolton	DuPont	C	Nylon					X	X	X	
Ray Houtz	DuPont	I	Orlon		X						
Hale Church	DuPont	C	Orlon				X	X	X	X	
Edgar Spanagel	DuPont	I	Dacron		X						
Emmette Izard	DuPont	C	Dacron		X						
Hale Church	DuPont	C	Dacron		X		X	X	X	X	
Emerson Wittbecker	DuPont	I	Lycra		X						
Joseph Shivers	DuPont	I	Lycra		X		X				
Hale Church	DuPont	C	Lycra					X	X	X	

(continued overleaf)

TABLE 14.18 (*Continued*)

Champion	Company	Role	Innovation	Context	Initiation	Intellectual	Resource Minimization	Prototyping	Technology–Market Interaction	Scaling	Cultural
Stephanie Kwolek	DuPont	I	Kevlar		X						
Hale Charch	DuPont	C	Kevlar					X	X	X	
Zay Jeffries	GE	C	Polysilicones and polycarbonates	X							X
Daniel Fox	GE	I/C	Polycarbonates			X	X	X			
Alphonse Pechukus	GE	C	Polycarbonates						X	X	X
J. Howard Pew	Sun Oil		High octane (catalytic) fuels: fixed bed	X							X
Eugene Houdry	Sun Oil	I/C	High octane (catalytic) fuels: fixed bed			X	X	X	X	X	
Frank A. Howard	Jersey Standard (Exxon)	C	High octane (catalytic) fuels: fluid bed	X							X
Warren K. Lewis	Jersey Standard (Exxon)	I	High octane (catalytic) fuels: fluid bed			X	X	X			

Name	Company		Technology							
Eger Murphree	Jersey Standard (Exxon)	C	High octane (catalytic) fuels: fluid bed	X	X	X				
Donald Campbell, Homer Martin, and Charles Tyson	Jersey Standard (Exxon)	U	High octane (catalytic) fuels: fluid bed		X	X				
Eugene Rochow	GE	I/C	Polysilicones		X	X				
Charles E. Reed	GE	C	Polysilicones		X	X	X	X	X	
Zay Jeffries	GE	C	Polysilicones	X				X	X	X
Frederick Karol	Union Carbide	I	Advanced polyolefins and metallocenes		X	X	X			
William H. Joyce	Union Carbide	C	Advanced polyolefins and metallocenes	X		X	X	X		X
Mervin Kelly	Bell Labs	C	Transistor: point contact	X	X					X
John Bardeen Walter Brattain	Bell Labs	I	Transistor: point contact		X	X				
William Shockley	Bell Labs	C	Transistor: point contact		X	X				
Jack Morton	Bell Labs	U	Transistor: point contact			X	X	X		

(continued overleaf)

TABLE 14.18 (*Continued*)

Champion	Company	Role	Innovation	Context	Initiation	Intellectual	Resource Minimization	Prototyping	Technology–Market Interaction	Scaling	Cultural
Robert Noyce	Fairchild Semiconductor	I/C	Integrated circuit: bipolar	X	X	X	X	X	X	X	X
Jean Hoerni	Fairchild Semiconductor	I	Integrated circuit: bipolar		X						
Patrick Haggerty	Texas Instruments	C	Integrated circuit: bipolar	X						X	
Jack Kilby	Texas Instruments	I/C	Integrated circuit: bipolar		X	X	X	X			X
Willis Adcock	Texas Instruments	C	Integrated circuit: bipolar	X				X			
Richard Alberts	Texas Instruments	C	Integrated circuit: bipolar						X	X	
Robert Noyce	Intel	C	Memory chip: DRAM	X	X					X	X
Gordon Moore	Intel	I/C	Memory chip: DRAM	X	X					X	X
Andy Grove	Intel	C	Memory chip: DRAM								X
Les Vadasz	Intel	I	Memory chip: DRAM			X					

Name	Company	Code	Product	1	2	3	4	5	6	7	8
Tom Rowe	Intel	U	Memory chip: DRAM			X	X	X			
Gene Flath	Intel	C	Memory chip: DRAM						X		
Dov Frohman	Intel	I/C	Memory chip: EPROM			X	X	X			X
Robert Noyce/Gordon Moore/Andy Grove	Intel	C	Memory chip: EPROM	X						X	
Joe Friedrich	Intel	U	Memory chip: EPROM						X		
Robert Noyce/Gordon Moore/Andy Grove	Intel		Micro-processor	X							
Ted Hoff	Intel	I/C	Micro-processor			X	X	X			
Federico Faggin	Intel	I/C	Micro-processor			X	X	X	X		X
John Pierce	Bell Labs	C	Semiconductor laser (850 nm)		X	X					
Mort Panish and Izuo Hayashi	Bell Labs	I	Semiconductor laser (850 nm)			X					

(continued overleaf)

TABLE 14.18 (*Continued*)

Champion	Company	Role	Innovation	Context	Initiation	Intellectual	Resource Minimization	Prototyping	Technology–Market Interaction	Scaling	Cultural
Barney DeLoach	Bell Labs	C	Semiconductor laser (850 nm)				X	X	X	X	X
Bill McGowan	MCI	C	Semiconductor laser (1.3 µm)	X					X	X	X
Louis Gerstner	IBM	C	Silicon-germanium chip							X	X
Bernie Meyerson	IBM	I/C	Silicon-germanium chip		X	X	X	X	X	X	X
David Sarnoff	RCA	C	LCD: TVs	X						X	
Richard Williams	RCA	I	LCD: general		X						
George Heilmeier	RCA	I/C	LCD: general			X	X				
Tomio Wada	Sharp	I/C	LCD: Electronic Calculators		X	X	X	X	X	X	
Yoshio Yamazaki	Seiko	C	LCD: Watches		X	X	X	X	X	X	

William Coates	Westing-house	C	Thin-film transistors–LCD							X
T. Peter Brody	Westing-house	I/C	Thin-film transistors–LCD				X	X	X	
Isamu Washizuka	Sharp	C	Thin-film transistors–LCD: flat-screen TVs			X	X	X	X	
MITI	Sharp	C	Thin-film transistors–LCD: flat-screen TVs		X	X				
Webster Howard	IBM/Sharp	C	Thin-film transistors–LCD: computer screens	X	X	X	X	X	X	
Richard Smalley	Carbon Nanotech-nologies, Inc. (CNI)	I/C	Single-walled nanotubes (SWNTs)	X			X	X	X	X
Bob Gower	Carbon Nanotech-nologies, Inc. (CNI)	U	Single-walled nanotubes (SWNTs)				X			

what we may call the *laboratory environment*, which extended from initiation to making the prototype, and the *commercial environment*, which ranged from later development through scale-up and commercialization. This is an interesting result since it is often believed that a wide chasm splits these two worlds and that those that create innovation must hand over their work to other, more practically minded engineers and businessmen who know better how to shape and bend these technologies to the realities of the market and to the highly prescribed and structured requirements of the mainstream organization. Thus, a distinction is often made between the *product* (or *technology*) champion and the *organizational* champion.[22] The juncture between early and late development—between prototyping and technology–market interaction; between the end of the laboratory environment and the beginning of the corporate environment—then should be the critical point that we should see a significant number of handovers. Yet, in fact, in the successful advanced materials projects we have examined, individual champions often fill that gap—that is, they bridge both roles—and make such transfers unnecessary. In fact, out of all the projects listed in the table, over half (52%) were guided by a champion who spanned both the laboratory and the commercial environments.

The fourth and final point is that when hand-offs did occur, they often did so much earlier in the project and totally within the laboratory environment. We observe that the total number of hand-off events over all phases of the innovation cycle in all the projects included in the table came to 22. Of these, only a little over a quarter (27%) involved transitions between the laboratory stages (ending with "prototyping") and the corporate environment (beginning with "technology–market interactions"). However, we find that nearly 60% of all transfers that took place across all the projects that went to completion occurred somewhere between "initiation" and "prototyping," both of which fall within the world of the laboratory. Effective championing—which could have filled this gap and reduced the number of transitions—didn't seem to take hold during this first third of the innovation cycle. This is not surprising. Many scientists see their interests stop at proof of concept. Creating practical devices are not their concern. The transfer of their knowledge and insight into the next phase is often delayed and even lost thus threatening the success of future work on the technology. As we have seen, for example, scientists at Bell Labs refusing to release important laboratory results to DeLoach and his team of engineers dangerously delayed Bell's work on the semiconductor laser. Clearly, with the difficulty in finding champions that can minimize these momentum-killing hand-offs within the laboratory environment, these earlier stages seem to be the most fragile and where an aboveground project is most likely to come to an end. Referring back to our gauntlet of risk table displaying the progress of all our projects, a full 74% of the ventures failed in these initial phases. The ability of a firm to smoothly and quickly transfer information and skills appears to be the most critical during these first four steps of the innovation process. The failed and otherwise troubled projects—Fairchild's MOS and Bell's semiconductor laser—could not manage to do it; the successes—such Intel's MOS-SG technology and IBM's silicon-germanium chip—found ways to effectively negotiate these early critical transitions.

14.7   THE DIFFERENT TYPES OF ADVANCED MATERIALS CHAMPIONS

If we take our innovations as evidence, champions play an absolutely vital role in the creation of advanced materials. They act in ways to minimize perceived threat over a number of phases making up the gauntlet of risks. The presence of an effective champion means that the project can avoid the dangerous stops and starts—the delays and rising costs—that result from the numerous handovers that take place as the venture proceeds from one stage to the next. They bring continuity and momentum to projects. If they do not seem to be of much help in the early portions of the innovation cycle, they certainly are very evident a little further down the line, as the effort moves from laboratory to commercial environment. They build that important bridge across that valley of death that separates the two regions. Without them, many of our most successful advanced materials innovations would have tumbled into that abyss.

Yet, not all champions are the same. Some are more effective than others in getting technology into the world. It cannot be denied that William Shockley was a great scientist and an active champion for the transistor. It is fair to say that he is the father of the digital age. But in terms of guiding his invention through the innovation process, he was a very poor champion indeed. Compare his performance with that of another physicist, Bernie Meyerson, who is a prime example of how potent one person can be in cosseting a breakthrough innovation from first conception to commercial placement.

Table 14.19 gauges the competency of most of the champions we came across in our case histories over a number of personal and professional qualities. These characteristics, selected because they consistently played important roles in our stories on advanced materials, are measured from columns 5 to 10. Columns 5 (S) and 6 (EN) tell us how deeply champions were imbedded in science or engineering according to their educational background and working experiences prior to entering upon the project leading to their work on the innovation at hand. The question here is whether scientists or engineers make the more effective champions or if this is not a particularly important difference. Columns 7 (CL) and 8 (PL) consider champions' loyalty to the company and to the project, respectively, with the former measuring how willing they are to leave a company for another more enticing opportunity and the latter their inclination to take another position within the organization in the middle of the project. Technology ventures falter when effective leaders take flight for new opportunities in midproject whether that new position awaits within a different company or another department within the same firm. Column 9 (CG) gives a sense of the degree to which a champion is a technical specialist vis-à-vis corporate generalist. The relevant question here is whether a person with broad experience and interests trumps the brilliant specialist on the impact they have in nurturing a new materials technology. To evaluate this trait, we look into the individual's educational and work history as well as apparent interests and aptitude. Finally, column 9 (U) signifies whether the individual is more of a "unifier" or "divider." In the former case, the champion reaches

TABLE 14.19 Characteristics of Successful Versus Less Successful Champions

Champion	Role	Innovation	Company	S	EN	CL	PL	CG	U	Result
Ken Iverson	C	Thin slab steel	Nucor	2	3	5	5	5	5	A
Ken Iversona	C	Ultrathin steel + microalloys	Nucor	2	3	5	5	5	5	A
Roger Williams	C	High-pressure chemicals: first generation	DuPont	4	2	5	5	5	5	A
Charles Stine	C	Nylon	DuPont	4	2	5	2.5	5	5	Ch
Wallace Carothers	I	Nylon	DuPont	5	1	2	2.5	1	2.5	Ch
Elmer Bolton	C	Nylon	DuPont	2	4	5	5	5	5	A
Ray Houtz	I	Orlon	DuPont	5	1	4	4	1	4	Ch
Hale Charch	C	Orlon	DuPont	3	4	5	5	5	5	A
Edgar Spanagel	I	Dacron	DuPont	5	1	4	4	1	4	S
Emmette Izard	C	Dacron	DuPont	5	1	4	4	1	4	Ch
Hale Charch	C	Dacron	DuPont	3	4	5	5	5	5	A
Joseph Shivers	I	Lycra	DuPont	5	2	5	2	2	2.5	Ch
Hale Charch	C	Lycra	DuPont	3	4	5	5	5	5	A
Stephanie Kwolek	I	Kevlar	DuPont	5	1	5	2	1	2.5	Ch
Hale Charch	C	Kevlar	DuPont	3	4	5	5	5	5	A
Daniel Fox	I	Polycarbonates	GE	5	2	5	3	2	3	Ch
Alphonse Pechukus	C	Polycarbonates	GE	2	3	5	5	5	5	A
J. Howard Pew	C	High octane (catalytic) fuels: fixed bed	Sun Oil	1	1	5	5	5	5	A
Eugene Houdry	I/C	High octane (catalytic) fuels: fixed bed	Sun Oil	1	5	2.5	5	4	5	A
Frank A. Howard	C	High octane (catalytic) fuels: fluid bed	Jersey Standard (Exxon)	1	3	5	5	5	5	A
Warren K. Lewis	I	High octane (catalytic) fuels: fluid bed	Jersey Standard (Exxon)	2	5	2.5	5	1	4	T
Eger Murphree	C	High octane (catalytic) fuels: fluid bed	Jersey Standard (Exxon)	2	5	5	3	3	5	A
Zay Jeffries	C	Polysilicones	GE	1	3	5	5	5	5	A
Eugene Rochow	I/C	Polysilicones	GE	5	1	3	2.5	1	2.5	Ch
Charles E. Reed	I/C	Polysilicones	GE	1	5	5	3	3	5	T
Frederick Karol	I	Advanced polyolefins and metallocenes	Union Carbide	5	1	5	1	5	2.5	Ch
William H. Joyce	C	Advanced polyolefins and metallocenes	Union Carbide	1	4	5	5	5	5	A
William Shockley	I/C	Transistor: junction (bipolar)	Bell Labs/ Shockley Semiconductor	5	2	2.5	4	2	1	T

Name		Innovation	Company							
John Bardeen	I	Transistor: point contact	Bell Labs	5	1	2.5	3	2	3	Ch
Jack Kilby	I/C	Integrated circuit: bipolar	Texas Instruments	2	5	4.5	5	4	1	Ch
Robert Noyce	I/C	Integrated circuit: bipolar	Intel	4	4	5	5	5	5	A
Frank Wanlass	I/C	Integrated circuits: MOS	Fairchild	5	4	1	2	2	1	Ch
Lee Boysel	I/C	Integrated circuits: MOS	Fairchild	2	5	2	5	5	1	S
Gordon Moore	C	DRAM: MOS-SG	Fairchild/Intel	5	2	5	5	5	5	A
Andy Grove	C	DRAM: MOS-SG	Intel	4	5	5	5	5	5	A
Les Vadasz	I	DRAM: MOS-SG	Intel	2	5	5	5	5	5	A
Dov Frohman	I/C	EPROM	Intel	2	5	5	5	5	5	A
Ted Hoff	I/C	Microprocessor	Intel	3	5	5	5	2.5	2.5	Ch
Federico Faggin	I/C	Microprocessor	Fairchild/Intel	4	4	3	5	4	2.5	A
Mort Panish	I	Semiconductor laser	Bell Labs	5	1	5	2	1	1	Ch
Barney DeLoach	C	Semiconductor laser	Bell Labs	2	5	5	5	3	5	A
Bill McGowan	C	Semiconductor laser	MCI	1	2.5	5	5	5	5	A
Bernie Meyerson	I/C	Silicon-germanium chip	IBM	5	2	5	5	4	5	A
David Sarnoff	C	LCD: TVs	RCA	1	2.5	5	2.5	5	5	S
Richard Williams	I	LCD: general	RCA	5	1	5	2	1	1	S
George Heilmeier	I/C	LCD: general	RCA	5	3	2.5	2.5	4	5	Ch
Tomio Wada	C	LCD: electronic calculators	Sharp	2	5	5	5	4	5	A
Yoshio Yamazaki	C	LCD: watches	Seiko	2	5	5	5	4	5	A
William Coates	C	Thin-film transistors–LCD	Westinghouse	2.5	2.5	5.0	3.0	5.0	5.0	S
T. Peter Brody	I/C	Thin-film transistors–LCD	Westinghouse	5	4	2.5	5	2.5	4	Ch
Isamu Washizuka	C	Thin-film transistors–LCD: flat-screen TVs	Sharp	2	5	5	5	4	5	A
Webster Howard	C	Thin-film transistors–LCD: computer screens	IBM/Sharp	5	2	5	2	3	5	A
Richard Smalley	I/C	Single-walled nanotubes (SWNTs)	Carbon Nanotechnologies, Inc.	5	1	2.5	5	4	5	T

A, adult; C, child; CG, career generalist; CL, corporate loyalist; CX, context; EN, engineering guru; PL, project loyalist; S, stillborn; SG, scientific guru; T, teen; U, project unifier; V, vision.

[a] While Ken Iverson was no longer president of Nucor when the company pursued the CASTRIP process for ultrathin steel, his influence continued to pervade the company's approach to innovation.

throughout an organization and brings in information, people, and ideas from wherever they are found. In contrast, the divider is concerned with the possibility that bad information or less talented people can delay or even scuttle the project and therefore carefully selects what information and people to filter into (and out of) the project. Such an individual will tend to favor those who agree with his concepts and approach and freeze out those who don't. The behavior of our champions during their careers, as well as during the venture in question, serves as the basis for evaluating this characteristic. Based on an assessment of the qualitative evidence, champions are given a rating of between 1 (weak) and 5 (strong) for each attribute. When a trait is neutral (neither obviously strong nor weak), it gets a 2.5.

The right-hand side of the table (column 11) shows the level of innovation achieved by the champion while actively working on the project. There are four possible categories: stillborn (S), child (Ch), teen (T), and adult (A). The "stillborn" category refers to great ideas that never get past the experimental stage; "child" to inventions that reach the point of actual devices or processes on the laboratory scale; "teen" to projects that go a step further, that is, advanced prototype; and "adult" to ventures that reach commercial stage. Clearly, investors in a company would be looking to put money in a project guided by the type of champion that can produce "adults."

What qualities then make for a good champion of advanced materials innovation? In determining this, we dictate that a strong quality will be measured by results that are either 4 or 5, while ratings of 1 or 2 will be considered a weak showing. With this in mind we can say that, first, engineers perform far better than pure scientists as champions. An impressive 86% of our champions who were instrumental in guiding an innovation to adulthood received the two top scores in engineering, and further, these were higher than their scientific ratings. In contrast, only 24% of the engineers in our group were associated with either stillborn projects or children. Scientists like Shockley could be as brilliant as you please, but they are not likely to be able to carry their inventions very far. They certainly may initiate a project with a groundbreaking material, device, or process, but it is left to others to give it commercial life. Engineers have more of a sense of what works and what doesn't in the real world. Many of them also understand how laboratory models can be scaled up. Engineers typically have enough science and mathematics under their belts to be able to work comfortably with research scientists especially during device and prototyping stage. They also take business and management courses, often as part of their engineering requirements, and so can span that crucial divide between research and early development, on the one hand, and later development, scale-up, and commercialization, on the other.

In addition to being more engineering than science oriented, effective champions have a high degree of loyalty to both a company and the project. We find that 89% of our champions who commercialized their technologies got high grades for both. Only 24% of deeply loyal individuals had to live with the fact that they could not push their projects beyond the stillborn or child stages. Overall then successful champions possess personalities that like to stay where they are for long periods of time. Despite repeated disappointments in pursuing their dream, they stick with it. They stayed put and fought a good fight for their cause and were not enticed to jump from one company or one department to another. This result testifies to the uniqueness of these personalities as champions. It was a rare thing that another champion could pick up the mantle and be as effective in moving the project along. Feeling of possessing a new materials innovation tended to be concentrated in that one person who delved into the research and then carried it forward into the corporate setting. Followers who tried to fill the vacuum once the leader left—either because they didn't have the technical ability, skill in risk management, or simply "fire in the belly"—generally fell far short of the mark. RCA's Heilmeier was certainly a forceful and skillful champion for liquid crystal displays. But faced with resistance from management and an enticing Washington job, he left the project, which soon thereafter went into free fall. In stark contrast, IBM's Bernie Meyerson, a scientist with practical business and engineering experience, stood by his invention year after year until he succeeded. The results couldn't be any different: RCA never got the liquid crystal display, while IBM now dominates the increasingly important silicon-germanium chip.

When it comes to the contest between who is the more valuable champion, the deeply knowledgeable specialist or well-informed generalist, the latter easily takes the prize. Nearly 90% of our entries that nurtured their innovations to adulthood scored high as generalists, while only 29% who oversaw projects to full commercialization were specialists, either as scientists, engineers, or such business-related experts as financial analysts. The powerful champions were those that were either professional engineers with interest and training in business (Dov Frohman) or business entrepreneurs with an engineering background or at least exceptional instincts about how technology works and what innovations are the most promising ones to go after.

Finally, every one of our successful projects was steered by a champion who was a unifier. Certainly, an argument can be made for the possible benefits of leaders who form a tight group of zealots to the cause and isolate them from the less creative, more conservative parts of an organization. Yet, it does appear that integration has a greater positive impact over segregation. Approximately 70% of all of the champions listed who could not get past "child" stage of innovation could be characterized as isolationists. The problem with isolationists or dividers is that they cause an imbalance in the innovation stream: they tend to favor either the laboratory environment (initiation, research, and early development) or the corporate world (later development, manufacturing/scale-up, and commercialization) but are bedeviled with attempts to unite the two worlds, assuming they even try. Lee Boysel, who formed a tightly knit group around him at Fairchild's Mountain View site, illustrates the latter type of bias. Unifiers, on the other hand, avoid polarization in high-technology ventures altogether thus assuring a salubrious balance where both sides—the laboratory and the commercial—get equal time and consideration, a sine qua non for the successful outcome.

In considering how these results might be used by investors in deciding where to risk their money, it is well to reflect on a famous instance of how a venture capital placement went terribly wrong. A brilliant and well-known scientist wants to start his own business. He is rightly considered an exceptional thinker with the ability to conceptualize cutting-edge devices. He has carried out world-class science for government agencies and prestigious research organizations and has worked for a number of years for a large US corporate research laboratory. He has written the classic text in his field. He has enormous confidence in his abilities to think through problems. He is a charismatic individual capable of attracting other top-rank scientists and engineers into his orbit and persuading them to follow his lead in his area of expertise. He does not suffer those whom he considers fools easily, for he believes they will slow down his work and prove costly to the research effort. He enlists as colleagues only the top scientific and engineering talent with advanced degrees from the best schools in the country. His mind is very active and his professional interests vary widely. He sees no point in sticking with a project when significant difficulties arise and it appears to be heading toward a dead end. At the same time, he is willing to divide his time equally between a number of interesting and promising ventures that could prove to be multiple revenue sources for the new company. He strongly believes that his reputation, confidence, intelligence, and personality make him a superb manager of people and will serve him well as president of his new company.

These were the characteristics that attracted the businessman Arnold Beckman to supporting this exceptional individual in his proposed venture.[23] However things did not turn out so well. We now know that these very same traits—a brilliant scientist but with no actual training, experience, or aptitude in business nor even much of an appreciation for engineering work; an isolationist who only trusted a small, focused clique of devotees centered around his genius and who marginalized those with different ideas or goals; and an unfocused person who shifted from one project to another depending on difficulties encountered on the efforts in progress and the enticing possibilities of ventures not yet started—are also serious red flags that signify trouble ahead for any investors. The person we have been talking about, William Shockley, was, in fact, a disaster as a champion, manager, and entrepreneur, and he eventually ran his new company Shockley Semiconductor into the ground and ended up where most people like him end up, in academia. As for Beckman, refusing to continue to throw good money after bad, he sold Shockley Semiconductor to another company less than 4 years after the Nobel Prize winner took up the reins of leadership.

14.8 FINAL THOUGHTS AND IMPLICATIONS

Reviewing our journey into advanced materials innovation, we are obliged to address certain critical questions: how does a company maximize its chances to succeed in advanced materials innovation? How can an investor know which is the best company and innovation with which to get involved? What can a government do to guide the country to be a world-class innovator in advanced materials? What do our results tell us about America's future role as a technological leader in an increasingly globalized world?

These of course are not easy questions to answer. But we do know a few things that emerge from our case histories. Supporting large companies—via subsidies, easing antitrust laws, and so forth—will not likely be much help. We have seen too often large, resource-rich firms squander opportunities that come their way. At the same time, favoring smaller companies, spin-offs, and start-ups—through research grants, stricter antitrust laws, and tax advantages—is just as likely to be fruitless, for far too many of these have also failed, even when blessed with top talent and a forward-looking management. Nor will new and advanced science give a company the edge: Nucor's great success had little to do with scientific insights, while both Shockley and Smalley experienced significant difficulties in the commercial world despite their undisputed scientific accomplishments. As for segregating cutting-edge research from everyday corporate life by resorting to incubators, technology hubs, and corporate-sanctioned spin-offs, we have seen little evidence that these played a major part in the success or failure in our stories.

What our case histories do tell us is that whether the innovation process takes place in a large or small firm, one that is rooted in the best science or in practical, empirical engineering; one that separates radical R&D from the corporate mainstream in incubator, technology hub, or spin-off; and one that operates in the United States or Japan, there is one reality that must be faced: top managers who control the purse strings have neither the will nor obligation to support a project that promises to create a breakthrough technology. Their most immediate task is to make sure their company continues to serve their existing customers to the best of their ability. Their job is in no way to risk company resources on highly uncertain ventures in the hope of future but as yet unproven benefits to the bottom line. They certainly will back small, incremental improvement to an existing technology, for this sort of innovation is more predictable, has an excellent chance of success in the short term, and does clearly work in favor of their trusted clients. But actively backing a blue-sky innovation is another matter entirely and certainly not at all within—and indeed seriously conflicts with—their job descriptions.

Does this mean that creating breakthrough technology, such as our advanced materials examples, did not entail risks? Not at all. But the risk takers are those working below a company's power structure; they are the bench scientists and engineers and often the middle managers supervising them, such as section heads within an R&D department or managers of specially created venture groups. Of course, they don't start out being revolutionaries. Like most people, they look for praise from those above them and the best way to get this approval is to work on projects that fit very nicely within the mainstream corporate strategy. So, they initially see themselves as the ones to improve an existing product or process and be glad if they are the first to do so. But then something unexpected happens. It could be pure chance or scientific or technical curiosity or a suggestion from a colleague or a fugitive thought that comes to mind in the shower. Whatever the trigger, the inventor wants to take the idea and run with it and is willing to face the very great risk that he or she will meet head-on the ire of superiors who anger when one of their workers goes off the accepted path, the shunning of colleagues who don't want to be associated with a "loser," and the real possibility—because of their challenge to the corporate line or the fact that they simply failed and wasted company resources—they will be denied promotion or worse demoted or forced to leave the company.

At this point, the would-be champion has two options: either to go underground and silently build a case for the technology, surfacing only when technical problems have been solved, initial markets secured, and organizational links between the new product and the mainstream company forged or stay aboveground, working with corporate leaders from the start through full commercialization. Each of these situations requires different types of management skills. In the first case, project managers must be able to think strategically and be able to methodically plan out a compelling argument for their technology. They must be able to work under the radar, secretly gaining the support of the underbelly of an organization without alerting the powers that check what they are up to. They must be good at building up and leveraging this growing momentum into one powerful "attack" on the hearts and minds of the organization's ruling authorities, for they have but one chance to make their case. In the second possibility, project leaders must interact with top managers on a regular basis. Here tactical, on-the-ground logistical expertise—the ability to handle each problem that comes their way, to reduce perceived riskiness from one stage to the next—is the key to success. A certain degree of clever scrappiness and quick response time to any hurdle thrown in their way describes the successful champion of aboveground innovation. If championing a new technology in the underground venue requires the commanding efforts of a brilliant strategic thinker, then successfully pushing radical innovation under the constant watch of corporate powers demands the spirited and aggressive talents of a medieval siege commander, achieving ultimate success by conquering one castle or town at a time.

As noted, the champion may have little choice in whether he or she will have the option of which type of innovation they face, for often a new project immediately grabs the attention of senior executives and must, by definition, become an aboveground effort. Despite this willingness to take on these risks for something potentially important, the new champions of this still unformed effort will not get very far unless they are very clever. They must understand under what conditions their superiors will allow a project to proceed and when they will cut it off at the legs. They have to clearly and creatively face the gauntlet of risks that promises that pursuit of radically new materials will be a perilous journey that requires the adventurer overcoming difficult challenges along the way with full understanding that failure to solve them at any point means project extinction irrespective of any prior success. They must further appreciate that being allowed to enter into a new stage of innovation depends on perception as much as reality, specifically the belief—whether actually true or not—on the part of a basically conservative corporate management. For example, Ted Hoff and Federico Faggin actually did come up with a bold new device—basically a prototype—for a microprocessor. This was an objective achievement. Yet, even though they had this revolution right in their hands, Intel's managers felt it would be too risky to proceed because they believed (incorrectly) it was not really part of their core capability. On the other side of this same coin, scaling up rayon was a truly difficult process that did not actually resemble the nylon experience. This reality in itself should have been a red flag to abandon the high-risk project. But W. Hale Charch convinced his bosses otherwise. This "buying of time" proved crucial, for eventually (and luckily) a market appeared unexpectedly that saved the project; in time, rayon became one of DuPont's most important fibers. All of this is to say that a company succeeds or not in innovating a new material to the extent it has within its walls a dedicated champion who is agile and creative enough to persuade top executives that an emerging technology fits nicely within their comfort level and that it is part and parcel of their company's core capability and as such no more than a logical extension of the world with which they are familiar and in which they already ply their talents.

14.8.1 Implications for Companies and Investors

What then does this tell us about what can be done to accelerate important advanced materials innovation? Regarding investors, what are indicators of a promising company for investors and also what should investors who get involved do in the formation and operations of new companies? And what does a government need to know and what should it do in order to secure the latest advanced materials technologies needed to maintain its country's industrial competitiveness? To address these questions, we need to realize that there are really four concerns of any R&D strategy: planning, architecture (how R&D is organized), process (how R&D is carried out), people (who carries out R&D), and portfolio (what direction R&D should take based on what projects and technologies are suitable for a particular organization).[24] Within these categories, our results indicate there are seven major points for which investors should be on the lookout:

R&D Planning: Risk Planning Companies and investors first need to know whether the project in which they are going to invest will be undertaken as an underground venture or whether it will be carried out through its various stages "in the open" under the watchful eyes of corporate management. Different types of management skills are required, as pointed out earlier, and investors need to know if the venture team can deliver the requisite capabilities. In other words, the individuals involved—inventors, champions, and pivotal utility players—must be able to negotiate each stage of the process by successfully minimizing real and perceived risk as they surface. Whether they can handle more than one stage in succession will depend on their background and skills. Beyond mapping out who will undertake which stage and when, investors will need to be sure these individuals are actually assigned where and when they need to be.

R&D Process: Transition Managing Companies and investors have to be sure hand-offs are handled efficiently and in a timely manner. Delays and increased perceived risks need to be avoided. The company should have procedures in place clearly laying out what information will be transferred and how and to whom it will be transmitted. If it doesn't, the investor should work with the company to develop them. A certain overlap time should be required so that the new person knows what his or her predecessor has been doing even before starting on their own phase of the project. Successful transition managing is particularly important during the first four stages of the innovation process.

R&D People: Champion Targeting Companies and investors should identify a particular individual who they feel would make an excellent candidate for the main project champion. Background, personality, breadth of experience, dedication to company and project, and leadership abilities all play important roles in this selection. The proper individual should be able to span the "valley of death" separating the laboratory and corporate worlds.

R&D Portfolio Decision-Making I: Technology–Market Assessing   Companies and investors should have a clear understanding of the actual process underlying the particular material and determine its degree of flexibility. A high elasticity means that it is likely to be less risky—and appears less risky—in prototyping and scaling up. In conjunction with this, the likely market needs to be assessed. If it appears that the market is not clear and very uncertain and fluid, a flexible process is key. If the initial market appears to be large and fixed, there will be greater leeway for a less supple process. However, everything being equal, flexibility trumps rigidity. Remember that the former often can generate more important materials and devices fairly cheaply thus extending the returns from the initial investment over many years, a la Intel.

R&D Portfolio Decision-Making II: Core Competency Linking   Companies and investors should verify that the innovative, breakthrough material they are thinking of supporting is being made by a process that is not only familiar to the target company but is in fact closely linked to its traditional core competency. This will optimize the chance that the company will see the new technology as a simple extension of the familiar and perceive low risk at critical junctures during the innovation process. Intel's MOS-SG is one of our examples. If, on the other hand, the innovation in question "goes against the grain" of the company's technical know-how, it will appear as a risky proposition to the firm's leaders, and they will likely bail when problems first surface.

R&D Portfolio Decision-Making III: Cultural Monitoring   Companies and investors need to closely monitor the company's cultural shifts over the course of the innovation process to make sure that the company and technology remain compatible. If they find either the corporate culture or requirements of the technology beginning to diverge, they should work with the project's most potent champion to convince top management to adjust the cultural trajectory to be more in line with the demands of the new material or device, such as what Meyerson did at IBM. If this is not done, the company may very well reject commercializing the technology.

R&D Architecture I: Organizational Structuring   Companies and investors need to be sure a firm's organizational structure helps the project move smoothly through the gauntlet of risk chain. Making this judgment can be tricky. On the one hand, the organization cannot be too centralized, or else middle-level scientists, engineers, and managers will not have enough freedom to veer from the tried and true into breakthrough research. On the other hand, there must be a connection between top management and researchers. As our case studies show, it is particularly important that there is not too big an organizational gap between the top managers and the creative scientists and engineers conducting R&D. Corporate leadership should themselves be technical people who are not too many years away from the laboratory and pilot plant and continue to remain close to the work of the R&D department. They will then be more open to the possibilities of radically new materials, especially if they derive from the company's core processes.

R&D Architecture II: Structuring R&D Activity   Companies and investors need to carefully consider whether they should segregate breakthrough R&D activity from the rest of the organization. Our results show that it is important that radical technology creation be perceived by the corporate powers as closely related to the familiar and to what has gone on before. The ghettoization of R&D makes it difficult for this strategy to be carried out, especially when the "going gets tough." Integration of a new project with the mainstream organization can be key to reducing the risks—as perceived by upper management—of moving forward with radical innovation.

R&D Architecture III: Acquisition Assessment   Companies and investors can use our results to evaluate the chances that a possible technology acquisition—for example, a large firm wanting to acquire a hot high-tech start-up—will ultimately succeed. While top management may be enticed by the cutting-edge nature of a particular material emerging from a high-profile start-up, the technology may not actually be compatible with the acquiring company with the result that risks will mount and corporate leaders abandon the project. Two important questions that have to be asked is at what stage in the innovation process will the parent company have to take up the technology and is it prepared to complete these remaining stages? If, for instance, the firm being acquired is dealing in new materials being made by a rigid process while markets are fluid, problems will likely arise at the late development stage. Also, if the corporate culture is—or appears will be at time of scale-up—a mismatch to the technology, rejection by top management may come during commercialization. In sum, corporate leaders may be dazzled by a dynamic start-up and the possibilities of its exciting new material and be anxious to be the first one on their block to make it. However, the severe requirements of the gauntlet of risks will in time bring them back to reality. How well they handle that reality determines how far they are willing to go the distance.

14.8.2   Implications for Government

Our results also point to actions governments, both local and state-wide as well as federal, can take to facilitate advanced materials creation in order to generate wealth and strengthen a region's and country's industrial competitiveness. For example, governments can provide funding to and work with universities, professional societies, and companies to create courses, conferences, and seminars on (i) developing and implementing risk minimization strategies for the various stages of the gauntlet of risks and (ii) training effective champions specifically for advanced materials innovation stressing such topics as successfully bridging the laboratory–corporate gap and conducting research and development under crisis conditions with the aim of minimizing perceived risk at critical junctures along the R&D chain. With respect to organizational structuring, governments should activate, coordinate, and (by various means) manipulate the creation of regional technology clusters composed of companies developing advanced materials using the same and highly flexible process. Doing so would generate powerful synergies within the cluster leading to accelerated growth. The model here of course is Silicon Valley, which actually evolved around the MOS-SG process.

Government should also focus on creating and supporting new technologies and systems that reduce perceived risks throughout the innovation process. For example, government–university–industry cooperation to accelerate development of 3-D printing technology could be critical since 3-D printing systems are likely to speed the time and reduce the costs of building prototypes.[25] An ongoing government project in fact is likely to reduce risks as innovation proceeds. The Materials Genome Initiative is developing an open digitally based advanced materials innovation infrastructure that, among other things, will allow scientists and engineers working across many organizations to simulate experimental and pilot and production conditions given certain parameters. This ability will save time and costs during research, prototyping, and possibly scale-up and be an important tool to project leaders who often need to proceed with their research under very tight financial conditions.

14.8.3 A Global Perspective

It is very clear that the competitive position of both developed and newly developing nations will depend a great deal on their ability to innovate in advanced materials technologies. A nation's economic growth closely shadows its productivity and both depend mightily on technological output. As emphasized in Chapter 1, we are in a time when technological change more and more coincides with advanced materials innovation; it is becoming increasingly difficult to disentangle one from the other. This identification of advancing technology with the creation of new materials is only going to become more pronounced through the present century and beyond. Countries that are creators and major exploiters of the future infrastructure, the transformed energy economy, the new systems of transportation, the revolution in biotechnology and health care, the next frontier in information and communications, and the forthcoming advances in manufacturing and supply chain management will be the ones to thrive over the next decades; those who aren't, won't, and their fate will hinge on the ability of their companies and universities and governments to quickly invent, develop, scale-up, and commercialize the new materials that build the foundations for these rapidly evolving sectors.

At first glance, it might appear that the United States remains the leader in advanced materials. After all, so many of the innovations in this area—many of them recent—seem to come out of American companies. But Americans must be very careful of becoming too complacent. Just consider the following: Nucor's great achievements came in response to Japanese minimills trouncing US steel companies; neither DuPont nor GE has come out with a major new polymer for decades; the creator of the great Unipol process is no longer in business (and the US chemical industry overall remains in serious trouble); while Silicon Valley continues to thrive, it is living off the technological fat created in the last century and serious questions exist today as to whether Moore's Law can continue to be the guiding light of the 21st-century semiconductor business; and the great hope of US advanced materials, nanomaterials, has yet to prove it is the powerhouse technology of the future. And, lest we forget, advanced materials technologies we have invented—such as liquid crystal displays—have left our shores and ended up a major driver of the Asian economy.

Indeed, when considering the great strides Asian countries have made in capturing advanced technology markets, it is often said that we in the United States take the big risks and create totally new products and processes, while other less boldly creative countries simply take and adapt what we have done, slash production costs any way they can, and sell at very low prices.[26] In this view, this strategy is less technological achievement and more the advantage of very cheap labor and the unflinching (and unapologetic) use of predatory pricing tactics. But, as our case studies show, this is far from the case. We should remember that Japanese champions of technology also had to face the same gauntlet of risks that confronted RCA and many smaller, innovative American firms. The difference is that they successfully met the trials demanded, while the Americans didn't. What the Japanese accomplished was in fact a major technological coup involving a great deal of creativity and entrepreneurial talent. Companies like Seiko and Sharp succeeded because of the efforts of their own champions who would not take no for an answer and—unlike Heilmeier at RCA—not even think of leaving the project until it was completed. They also succeeded because these champions could leverage a culture very different from that in the United States, one more collectivist than individualistic, to greatly reduce perceived risks along the innovation chain.

If then the United States is to maintain or (some would argue) regain its status as a major force for innovation in the world, a requirement if it hopes to secure robust economic growth in the present century, it needs to shed the well-worn image of itself as the only important risk-loving country that willingly takes on the most difficult and perilous projects our more cautious, uncertainty-averse global competitors will not touch. Much is made of the American frontier mentality and individualism as the source of our love of risk taking and of entrepreneurial ventures. There may be much truth to this. But we should also remember that a company—even a small one—does not necessarily have one culture. If Americans are risk takers, this adventuresome spirit will be found "downstairs," while a very different, more conservative ethos resides "upstairs" in corporate boardrooms. We should also be very aware that, if other countries cannot boast of the same frontier aggressiveness as the United States, nevertheless, our history of liquid crystal technology clearly shows that energetic, creative, willful, and talented champions can and do appear in middle management positions globally, and when they do, they can beat us over our heads with our own inventions. We are then not so different from other countries in what we are or are not willing to change and in the types of uncertainty we in fact have to face. We cannot then pretend that risks are not daunting to us and that our high-tech firms and the people who run them refuse to be stopped by them. Those who occupy the top management levels within American companies do hesitate in the face of uncertainty. It is up to those below them to take on whatever personal and professional risks they need to clear the path to progress. Whether middle managers within our organizations are bold enough to confront these risks—these obstructions to innovation—directly and, individually and collectively, are nimble enough to find the ways necessary to remove them by substantially diminishing their threat, whether real or supposed, in the eyes of those who control the purse strings is an important question. For mitigating the fears of those who are charged with maintaining the status quo is the sine qua non of a country's materials success. The ability of our "downstairs" scientists, engineers, and managers to creatively negotiate with their superiors at those critical junctures identified in our narratives in such a way as to instill confidence and mollify their doubts and anxieties about what will come next will go far in determining just how exceptional we turn out to be as a technological power as the new materials revolution advances through the 21st century.

REFERENCES

1. Burgelman, R. (June 1983), "A Process Model of Internal Corporate Venturing in the Diversified Major Firm" *Administrative Science Quarterly*, 28(2): 223–244.

2. In the case of the Unipol process, for example, William Joyce had to find ways to keep the project under wraps until he and Karol could show impressive results. In most of our other case studies however, the upper ranks of the corporate hierarchy were in on the project at—or nearly at—its beginnings.

3. Burgelman also examined underground "internal corporate venturing" in the cases of larger, diversified firms. His process model assumes this condition. See Burgelman (June 1983).

4. Davila, T. and Epstein, M. (2014), *The Innovation Paradox: Why Good Businesses Kill Breakthroughs and How They Can Change*, San Francisco, California: Berrett-Koehler, pp. x–xiv; 1–9.

5. Drucker, P. (1985), *Innovation and Entrepreneurship*, New York, New York: Harper, p. 139.

6. Ibid., pp. 148; 151.

7. Enos, J. (1962), *Petroleum Progress and Profits*, Cambridge, Massachusetts, The MIT Press.

8. Zmud, R. (1983), "The Effectiveness of External Information Channels in Facilitating Innovation Within Software Development Groups" *MIS Quarterly*, 7(2): 43–58. For a study on the importance of information flow within a larger company, see Hansen, M. (2002), "Knowledge Networks: Explaining Effective Knowledge Sharing in Multiunit Companies" *Organization Science*, 13(3): 232–248.

9. Verhage, B. (2014), *Marketing: A Global Perspective*, Hampshire, UK: Cengage, pp. 331–332.

10. Davila and Epstein (2014), p. 115.

11. Ibid., pp. 115–118.

12. For a discussion of common mechanical processes in the 19th-century machine tool industry, see Rosenberg, N. (1963), "Technological Change in the Machine Tool Industry, 1840–1910" *The Journal of Economic History*, 23(4): 414–443.

13. See, for example, Davila and Epstein (2014), Chapter 7 ("Innovative Cultures").

14. Hofstede, G., Hofstede, G., and Minkov, M. (2010), *Cultures and Organizations: Software of the Mind*, 3rd edition, New York, New York: McGraw-Hill Education.

15. Economies of scope refer, for example, to the ability of a manufacturing plant or a particular process in quickly and economically turning out different types of products. The MOS-SG process is certainly one example of a process possessing economies of scope. Alfred Chandler similarly discusses how innovations in electric power and distribution in the late 19th and early 20th centuries " ... revolutionized the making of many metals and chemicals." See Chandler, A. (1990) *Scale and Scope: The Dynamics of Industrial Capitalism*, Cambridge, Massachusetts: The Belknap Press/Harvard University Press, pp. 62–63.

16. Hofstede, Hofstede, and Minkov (2010), pp. 235–275.

17. Davila and Epstein (2014), pp. 145–146.

18. See Burgelman, R. (1983), "A Model of the Interaction of Strategic Behavior, Corporate Context, and the Concept of Strategy" *The Academy of Management Review*, 8(1): 61–70.

19. See Burgelman, R. (2002), *Strategy is Destiny: How Strategy-Making Shapes a Company's Future*, New York, New York: The Free Press, pp. 13–15.

20. In his history of petroleum refining, John Enos stresses the growing importance of large research teams composed of narrowly focused specialists. This idea takes center stage in his chapter on fluid catalytic cracking. See Enos (1962), pp. 187–188.

21. Leifer, R. et al. (2000), *Radical Innovation: How Mature Companies Can Outsmart Upstarts*, Boston, Massachusetts: Harvard Business School Press, p. 157.

22. Burgelman (1983), "A Process Model of Internal Corporate Venturing," pp. 232–240. See also Leifer et al. (2000), "Radical Innovation ... ," pp. 171–174.

23. Reid, T. (2001), *The Chip: How Two Americans Invented the Microchip and Launched a Revolution*, New York, New York: Random House, pp. 89–90.

24. Pisano, G. (2012), *Creating an R&D Strategy: Working Paper*, Cambridge, Massachusetts: Harvard Business School.

25. See, for example, Robarts, S. (April 15, 2014), "3D Printing Drastically Reduces Development Costs of Blood Recycling Machine." http://www.gizmag.com/brightwave-hemosep-autotransfusion-3d-printing/31650/. Accessed November 30, 2015.

26. This was certainly the argument made by the US semiconductor industry in Silicon Valley in the 1980s in convincing the government to impose high tariffs against Japanese chips. In fact, as even Robert Noyce finally acknowledged, they were simply making better chips more cheaply. See Reid (2001), pp. 233–248.

INDEX

Advanced Materials Innovation: Managing Global Technology in the 21st century, First Edition.
Sanford L. Moskowitz.
© 2016 John Wiley & Sons, Inc. Published 2016 by John Wiley & Sons, Inc.